vapor-liquid equilibria
using UNIFAC
a group-contribution method

vapor-liquid equilibria using UNIFAC
a group-contribution method

AAGE FREDENSLUND
Instituttet for Kemiteknik, The Technical University of Denmark,
2800 Lyngby, Denmark

JÜRGEN GMEHLING
Abteilung Chemietechnik, Lehrstuhl Technische Chemie B,
University of Dortmund, 46 Dortmund 50, Germany

and

PETER RASMUSSEN
Instituttet for Kemiteknik, The Technical University of Denmark,
2800 Lyngby, Denmark

ELSEVIER SCIENTIFIC PUBLISHING COMPANY
Amsterdam — Oxford — New York   1977

ELSEVIER SCIENTIFIC PUBLISHING COMPANY
335 Jan van Galenstraat
P.O. Box 211, 1000 AE Amsterdam, The Netherlands

Distributors for the United States and Canada:

ELSEVIER/NORTH-HOLLAND INC.
52, Vanderbilt Avenue
New York, N.Y. 10017

First edition 1977
Second impression 1979

Library of Congress Cataloging in Publication Data

```
Fredenslund, Aage, 1941-
   Vapor-liquid equilibria using UNIFAC.

   Bibliography: p.
   1.  Vapor-liquid equilibrium--Data processing.
2.  UNIFAC (Computer program)  I.  Rasmussen, Peter,
1935-      joint author.  II.  Gmehling, Jürgen,
1946-      joint author.  III.  Title.
TP156.E65F73         660.2'9'63         77-12218
ISBN 0-444-41621-8
```

ISBN 0-444-41621-8

© Elsevier Scientific Publishing Company, 1977
All rights reserved. No part of this publication may be reproduced, stored in a retrieval system or transmitted in any form or by any means, electronic, mechanical, photocopying, recording or otherwise, without the prior written permission of the publisher, Elsevier Scientific Publishing Company, P.O. Box 330, 1000 AH Amsterdam, The Netherlands

Printed in The Netherlands

PREFACE

by the authors

The UNIFAC method is based on and at the same time the result of contributions by different groups. The authors wish to express their sincere thanks to Professor J.M. Prausnitz for his constant encouragement and helpful criticism in the further development of UNIFAC, which originated from his laboratory. We thank Professor H.C. Van Ness for carefully reading the manuscript and for constructive comments; Professor J.P. O'Connell for supplying useful information on the Hayden-O'Connell correlation; M.L. Michelsen for his collaboration in developing the distillation programs; and Professor U. Onken for his support of the project.

The following have helped us in different ways to obtain the results shown in this book: L. Grausø, P. Hansen, V.N. Hansen, S. Bay Jørgensen, L. Kierkegaard, L. Kunzner, E. Pedersen, W. Posywio, R. Rodermund, W. Schulz, and K. Thomsen. We are most grateful for their contributions. We thank the firm Uhde GmbH for their economic support, which has enabled all three of us to meet in Lyngby, and the computercenter at the University of Dortmund for the use of their facilities. Finally, we thank Mrs. H. Wolff for carefully typing the manuscript.

Lyngby and Dortmund

March 1977

PREFACE

by

J.M. Prausnitz
Professor of Chemical Engineering
University of California, Berkeley

The most common operation in chemical industry is the separation of liquid mixtures through distillation. Efficient design of distillation equipment requires quantitative understanding of vapor-liquid equilibria in multicomponent mixtures as expressed through vapor-phase fugacity coefficients and liquid-phase activity coefficients. At low or modest pressures, fugacity coefficients can be estimated easily but, except for very simple mixtures, experimental mixture data are required to calculate activity coefficients.

Since the variety of liquid mixtures in chemical technology is extremely large, it is not likely that necessary experimental data will always be available for every mixture of interest. It is therefore desirable to correlate limited experimental data in such a way as to enable the chemical engineer to predict with confidence the activity coefficients in those mixtures where little or no experimental data are available. For non-electrolyte mixtures, a procedure for establishing such a correlation is provided by the concept of group contributions. In this concept, each molecule is considered as the sum of the functional groups (e.g., methylene, nitro, keto, amino, carboxyl) which constitute that molecule; the thermodynamic properties of a solution are then correlated, not in terms of the molecules, but rather in terms of the functional groups which comprise the mixture. The large advantage of this concept is that whereas the number of different molecules in chemical technology is very large, these molecules can be constituted from a much smaller number of functional groups.

A particularly promising method for calculating activity coefficients from group contributions is provided by UNIFAC, as described in this monograph. This method is a consequence of a model for activity coefficients called UNIQUAC (Universal

Quasi-Chemical) which, in turn, represents a reasonable attempt to generalize to molecules of different size and shape Guggenheim's quasi-chemical theory of strongly nonideal liquid mixtures. The word UNIFAC stands for UNIQUAC Functional-group Activity Coefficients.

The UNIQUAC model was proposed by Denis Abrams in Berkeley in 1974 and shortly thereafter Aage Fredenslund came to Berkeley to work with Russell Jones on the group-contribution model which UNIQUAC had suggested. An early version of UNIFAC was published in 1975 but it soon became clear that a very substantial, systematic effort for data reduction was required to realize the full potential of the UNIFAC method. This effort was patiently carried out by Aage Fredenslund and Peter Rasmussen at Lyngby, Denmark, and by Jürgen Gmehling at Dortmund, Germany. As a result of their diligence and devotion, the UNIFAC method has now been significantly enlarged beyond its original scope.

As new data and additional experience become available, the utility and reliability of UNIFAC will increase. However, the present achievement of Fredenslund, Gmehling, and Rasmussen already covers a wide range of application in chemical industry. It is therefore a pleasure to welcome this monograph to the chemical design engineers and to congratulate the authors on having presented an important service toward successful design of separation equipment in chemical technology.

J.M. Prausnitz

February 1977

TABLE OF CONTENTS

	Page
PREFACES..	V
TABLE OF CONTENTS	IX
CHAPTER 1 INTRODUCTION	1
1.1 The Phase Equilibrium Problem	1
1.2 Origin and Range of Applicability of the UNIFAC Method	3
1.3 Organization of the Book	5
CHAPTER 2 VAPOR PHASE NONIDEALITY	6
2.1 Fugacity Coefficients from the Virial Equation	7
2.2 Second Virial Coefficients	8
2.3 Fugacity Coefficients from the Chemical Theory	10
2.4 Computer Program for Fugacity Coefficients	19
CHAPTER 3 LIQUID PHASE NONIDEALITY	21
3.1 Calculation of Reference Fugacities	21
3.2 Activity Coefficients	22
CHAPTER 4 THE UNIFAC GROUP-CONTRIBUTION METHOD	27
4.1 The UNIFAC Model	27
4.2 Group Interaction Parameters	36
4.3 Sample Predictions	38
4.4 Temperature Dependence of the UNIFAC Parameters	53
4.5 Alternative UNIFAC Parameters for Alcohols	59
4.6 Additional UNIFAC Parameters	63
CHAPTER 5 DETERMINATION OF UNIFAC PARAMETERS	65
5.1 The Dortmund Data Bank	65
5.2 Thermodynamic Consistency	68
5.3 Calculation of Activity Coefficients from Mutual Solubility Data	74
5.4 Estimation of UNIFAC Parameters	79

Page

CHAPTER 6	PREDICTION OF VAPOR-LIQUID EQUILIBRIA IN BINARY SYSTEMS	86
6.1	Sample Results from the Data Correlation	87
6.2	Sample Predictions	88
6.3	Predictions for Systems Containing Ethanol	88
CHAPTER 7	PREDICTION OF VAPOR-LIQUID EQUILIBRIA IN MULTICOMPONENT SYSTEMS	134
CHAPTER 8	PREDICTION OF PHASE-SPLITTING AND EXCESS ENTHALPY	150
8.1	Phase-Splitting	151
8.2	Excess Enthalpy	153
CHAPTER 9	APPLICATION OF UNIFAC TO DISTILLATION COLUMN DESIGN	165
9.1	Multicomponent Distillation Calculations by Linearization	165
9.2	The Column Calculation Procedure	174
9.3	Examples of Distillation Calculations (Simplified Version)	181
9.4	Examples of Distillation Calculations (Full Version)	190
APPENDIX 1	CALCULATION OF FUGACITY AND ACTIVITY COEFFICIENTS. THE CONSISTENCY TEST	198
	SVIR	199
	PHIB	202
	YENWO	204
	GAUSL	205
	MLMEN	207
	BIJ	208
	Thermodynamic Consistency Test	214
APPENDIX 2	UNIFAC PROGRAMS	227
	UNIFA	231
	SYSTM	232
	GRES	236
	GREF	237
	GCOMB	237
	HCON	238

	Page
Calculation of Activity Coefficients	239
Prediction of Liquid-Liquid Equilibrium Compositions	243
UNIQUAC Parameters from UNIFAC	250
APPENDIX 3 PARAMETER ESTIMATION PROGRAM	256
APPENDIX 4 DISTILLATION PROGRAMS	275
Full Version	285
Simplified Version	327
APPENDIX 5 List of Phase Equilibrium Data used in the Determination of UNIFAC Parameters	347
NOMENCLATURE	378

CHAPTER 1

INTRODUCTION

A large part of chemical engineering design is concerned with separation processes. Many of these are diffusional operations of the phase-contacting type; distillation, absorption, and extraction are probably the most common. For rational design of such separation processes, we require quantitative information on the phase equilibria in the binary or multicomponent mixture under consideration.

The purpose of this book is to describe a method for predicting such quantitative information, the UNIFAC group-contribution method. In the course of the book we describe by extensive examples the quality of predictions. We give references to tested vapor-liquid equilibrium data on which the model is based, and finally examples of how UNIFAC may be used in practical engineering design calculations are listed. The computer programs associated with UNIFAC (testing phase equilibrium data, estimation of parameters, prediction of phase equilibria, and distillation calculations) are given in appendix. We hope the reader will find it easy to use and extend the UNIFAC method.

1.1 THE PHASE EQUILIBRIUM PROBLEM

Phase equilibrium thermodynamics serves as the mathematical framework for providing quantitative information on phase equilibria. Figure 1.1 illustrates schematically the problem (see [1]).

We are given a mixture with M components and two phases, and we suppose that the two phases have reached an equilibrium state. Equilibrium thermodynamics helps answer the following question: If we know the mole fractions in phase I $(x_1, x_2 .. x_M)$

and the temperature T, what are the mole fractions in phase II $(y_1, y_2 .. y_M)$ and the pressure P?

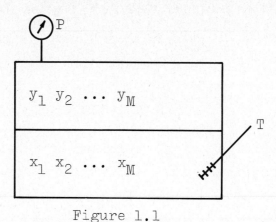

Figure 1.1

AN EQUILIBRIUM STAGE

For rational design of diffusional separation processes, this question <u>must</u> be answered, often with high degree of accuracy. Most diffusional separation processes contain many equilibrium stages in series. Therefore, and because the answer to the above question most often must be provided within an iterative procedure, the method of arriving at the answer must be relatively fast and simple.

To solve problems illustrated in Figure 1.1 we use the condition for phase equilibrium (see [1], p. 21):

$$f_i^{I} = f_i^{II} \qquad i = 1, 2, .. M \tag{1.1}$$

where f_i is the fugacity of component i in phase I or II. There are M such relationships, where M is the total number of components in the mixture.

If phase I is vapor and II is liquid, Equation (1.1) is rewritten in terms of the vapor phase fugacity coefficient φ_i, the liquid phase activity coefficient γ_i, and the liquid phase reference fugacity f_i^o:

$$\varphi_i y_i P = \gamma_i x_i f_i^o \qquad i = 1, 2, .. M \tag{1.2}$$

If both phases are liquid (mole fractions given by x_i^{I} and x_i^{II}), Equation (1.1) becomes

$$x_i^{I} \gamma_i^{I} = x_i^{II} \gamma_i^{II} \qquad i = 1, 2, \ldots M \qquad (1.3)$$

At pressures up to a few atmospheres, the fugacity coefficients and reference fugacities are readily calculated using the virial equation in conjunction with second virial coefficients obtained from experimental information or generalized correlations (see Chapter 2 of this book). Indeed, at these low pressures φ_i is often nearly unity and f_i^o is often nearly the pure-component vapor pressure at the same temperature as the mixture. To answer the question stated in connection with the phase equilibrium problem we are then left with the problem of establishing a relation for the activity coefficients as functions of composition x_i and temperature T. The UNIFAC group-contribution method is a such relation.

1.2 ORIGIN AND RANGE OF APPLICABILITY OF THE UNIFAC METHOD

The UNIFAC method for calculation of activity coefficients [2] is based on the group-contribution concept, which has been successful for estimating a variety of pure-component properties such as liquid densities, heat capacities, and critical constants. The basic idea is that whereas there are thousands of chemical compounds of interest in chemical technology, the number of functional groups which constitute these compounds is much smaller. Therefore, if we assume that a physical property of a fluid is the sum of contributions made by the molecules' functional groups, we obtain a possible technique for correlating the properties of a very large number of fluids in terms of a much smaller number of parameters which characterize the contributions of individual groups. In addition, we obtain a technique for predicting phase equilibria in systems for which no data exist. It is for example within the framework of the group-contribution approach possible to predict the equilibrium compositions in all ketone-alkane mixtures from data on, say, acetone-cyclohexane.

Any group-contribution method is approximate because the contribution of a given group in one molecule is not necessarily the same as that in another molecule. The fundamental assumption of a group-contribution method is additivity; the contribution made by one group is assumed to be independent of that made by another group. This assumption is valid only when the influence of any one group in a molecule is not affected by the nature of other groups within that molecule.

For example, we would not expect the contribution of a C=O (carbonyl) group in a ketone (say, acetone) to be the same as that of a C=O group in an organic acid (say, acetic acid). On the other hand, experience suggests that the contribution of a C=O group in acetone is close to (although not identical with) the contribution of a C=O group in 2-butanone.

Accuracy of correlation improves with increasing distinction of groups; in considering, for example, aliphatic alcohols, in a first approximation no distinction is made between the position (primary or secondary) of a hydroxyl group, but in a second approximation such distinction might be desirable. In the limit, as more and more distinctions are made, we recover the ultimate group, namely, the molecule itself. In that event, the advantage of the group-contribution method is lost. For practical utility, a compromise must be attained. The number of distinct groups must remain small but not so small as to neglect significant effects of molecular structure on physical properties.

Extension of the group-contribution idea to mixtures is extremely attractive; since the number of pure fluids in chemical technology is already very large, the number of different mixtures is still larger, by many orders of magnitude. A large number of multicomponent liquid mixtures of interest in the chemical industry can be constituted from perhaps fifty functional groups.

In its present state of development, the UNIFAC method may be used to predict the phase equilibria in non-electrolyte mixtures for temperatures in the range of 300-425 K and pressures up to a few atmospheres. All components must be condensable. Precisely which groups are covered is given in Chapter 4. In addition to its predictive capabilities, the UNIFAC method today correlates more than 70% of the published vapor-liquid equilibrium data at pressures up to a few atmospheres.

The aim of the UNIFAC method is to provide an accurate and reliable method for predicting vapor-liquid equilibria. Therefore the present table of group-contribution parameters is where possible based on vapor-liquid equilibrium data only. The equations in the UNIFAC method may also be used to predict liquid-liquid equilibria and excess enthalpies. However, in the present development of the UNIFAC method, accurate prediction of these has been of second priority.

1.3 ORGANIZATION OF THE BOOK

This book is intended for the process design engineer who needs reliable predictions of phase equilibria for designing distillation columns and other separation processes. The aim of this book is to describe the UNIFAC group-contribution method for estimating activity coefficients and to provide a manual for the design engineer so that he may use it with confidence. This book does not contain detailed theoretical considerations - for this we refer the reader to [1].

Chapters 2 and 3 describe calculation of vapor and liquid phase nonidealities in terms of, respectively, fugacity and activity coefficients. The UNIFAC method is stated in Chapter 4, and Chapter 5 deals with the determination of UNIFAC parameters, i.e. the parameter estimation procedure used in establishing the parameters needed for the model. Prediction of binary and multicomponent vapor-liquid equilibria is shown for a large number of systems in Chapters 6 and 7, and the potential UNIFAC has for predicting liquid-liquid equilibria is illustrated in Chapter 8. Application of UNIFAC to practical distillation design problems is shown in Chapter 9.

Finally, the computer programs used in connection with the UNIFAC method are listed in Appendix.

REFERENCES

1. J.M. Prausnitz, Molecular Thermodynamics of Fluid-Phase Equilibria, Prentice-Hall, Englewood Cliffs, N.J., 1969.
2. Aage Fredenslund, R.L. Jones and J.M. Prausnitz, AIChE Journal 21(1975)1086.

CHAPTER 2

VAPOR PHASE NONIDEALITY

The condition for phase equilibrium was in Chapter 1 stated as:

$$f_i^V = f_i^L \qquad i = 1,2 \ldots M \tag{2.1}$$

where f_i is the fugacity of component i in the vapor (V) or liquid (L) phase.

For a mixture where the vapor phase is an ideal gas, one may write:

$$f_i^V = y_i P \tag{2.2}$$

where y_i is the mole fraction and P is the total pressure.

Nonideality in the vapor phase is characterized by the fugacity coefficient φ_i defined in (2.3):

$$f_i^V = \varphi_i y_i P \tag{2.3}$$

Except for mixtures containing strongly associating substances as organic acids, the value of φ_i will be close to unity at the low pressures considered in this book. Nevertheless we have wanted to be sure that all vapor phase nonidealities are taken into account in the calculations for the vapor phase and not carried over to the fugacities for the liquid phase. We have therefore calculated f_i^V by means of (2.3) in all our data reductions for parameter estimation in the UNIFAC model.

In this chapter we will discuss methods for calculating φ_i.

2.1 FUGACITY COEFFICIENTS FROM THE VIRIAL EQUATION

The fugacity coefficient can be calculated by means of Equation (2.4). It is derived through a series of basic thermodynamic manipulations (see [1], p. 30):

$$\ln \varphi_i = \frac{1}{RT} \int_0^P (\bar{v}_i - \frac{RT}{P}) dP \qquad (2.4)$$

P and T are the pressure and temperature for the system, R is the universal gas constant, and $\bar{v}_i$ is the partial molar volume of component i in the mixture.

The problem of finding φ_i has now been changed to the problem of finding $\bar{v}_i$ from an equation of state. One such equation is the virial equation, which gives the compressibility factor of the mixture, z, as a function of composition, temperature, and pressure or volume.

At pressures up to a few atmospheres the volume explicit virial equation may be truncated after the second term:

$$z = \frac{Pv}{RT} = 1 + \frac{BP}{RT} \qquad i \text{ and } j = 1,2 .. M \qquad (2.5)$$

v is the molar volume of the mixture and B is the second virial coefficient accounting for two molecules interacting.

In a mixture with M components we have

$$B = \overset{i}{\Sigma} \overset{j}{\Sigma} y_i y_j B_{ij} \qquad i \text{ and } j = 1,2 .. M \qquad (2.6)$$

The B_{ij}'s represent interactions between molecules i and j. If $i = j$ one has the virial coefficient B_{ii} for a pure component.

Combination of equations (2.4), (2.5) and (2.6) gives:

$$\ln \varphi_i = \frac{P}{RT} (2 \overset{j}{\Sigma} y_j B_{ij} - B) \qquad j = 1,2 .. M \qquad (2.7)$$

It should be noted here that one very often finds a different expression for φ_i based on the virial equation comprising only pair interactions [1]:

$$\ln \varphi_i = \frac{2}{v} \sum_j^j y_j B_{ij} - \ln z \qquad j = 1,2 \ldots M \qquad (2.8)$$

This expression is derived from the virial equation in the pressure explicit form:

$$z = 1 + \frac{B}{v} \qquad (2.9)$$

For low pressures there is not much difference between the values of φ_i calculated from (2.7) and (2.8), as can be seen from the following example.

Example 2.1

The experimental value of B at 368.25 K for pure Butane is [2]: -444.2 cm^3/mol. This value is based on (2.9).

For pure component i we have from (2.7): $\ln \varphi_i = \frac{B_{ii} P}{RT}$

and from (2.8): $\ln \varphi_i = \frac{2 B_{ii}}{v} - \ln z$

Based on these equations we get the following values of φ_i.

P in atm.	φ_i	
	Eq.(2.7)	Eq.(2.8)
1	0.9854	0.9853
10	0.8632	0.8514

Up to 10 atm. the difference between the calculated values is less than 1.5% even though we have used (2.7) with a value of B based on (2.9) and not on (2.5).

2.2 SECOND VIRIAL COEFFICIENTS

It is very seldom possible in the literature to find desired experimental values of virial coefficients. It is therefore necessary to predict the virial coefficients from readily available data. We have during most of our work with the estimation of UNIFAC parameters used a method developed by Hayden and O'Connell [3].

The method is based on the assumption that the various kinds of intermolecular forces contribute to the second virial coefficient in distinct ways. The total virial coefficient can thus be taken as the sum of several contributions.

$$B_{total} = B_{free} + B_{metastable} + B_{bound} + B_{chem} \tag{2.10}$$

B_{free} represents the molecular volumes; the contribution $B_{metastable} + B_{bound}$ results from the potential energy from more or less strongly bound pairs of molecules and B_{chem} results from associating substances.

The association may be so strong that the molecular pairs resemble new constituents in the mixture, and this forms the basis for the chemical theory mentioned below.

Appendix 1 contains a program, SVIR, for calculation of second virial coefficients in binary and multicomponent mixtures. The comments in the program explain in detail the necessary input information. The program SVIR returns to the calling program values of B_{free} and B_{total} for the pure components and for the cross interactions between the components.

In order to use the method of Hayden and O'Connell one has to know some basic parameters for each component. Tables 2.1 and 2.2 comprise parameters for a number of components and mixtures.

For substances not mentioned in the tables, the parameters may be determined as follows:

<u>PC and TC.</u> Critical pressures and temperatures of the components can often be found in [4] or [5]. With no experimental values, a group-contribution method may be used as explained in [4].

<u>RD.</u> Mean radius of gyration. RD may be calculated from the parachor, P', by the following equation [3]:

$$RD = -0.2764 + 0.2697 \sqrt{P' - 48.95} \tag{2.11}$$

Values of the parachor may be obtained from a group contribution method as described in [4].

<u>DMU:</u> Dipole moments may often be obtained from [6].

<u>ETA:</u> Association and solvation parameters. These are the most difficult parameters to estimate since they must be determined empirically.

<u>Association parameters for pure substances</u>:
For hydrocarbons, for Chloroform and other halogenated alkanes, for ethers and for sulfides the association factors are set equal to zero. For components not mentioned in Table 2.1, values of ETA may be estimated by conclusions of analogy from the other components.

<u>Solvation parameters for interaction between components in a mixture</u>:
In [3] it is stated that the solvation parameters, η_{ij}, are set equal to zero unless the components of a mixture are in the same group of substances or a special solvation contribution could be justified and empirically determined.

It should be pointed out that many of the values in Table 2.2 are guessed on the basis of experience from only one system, and hence the values should be reevaluated when possible.

2.3 FUGACITY COEFFICIENTS FROM THE CHEMICAL THEORY

The virial equation truncated after the second term cannot be used for mixtures containing carboxylic acids because of very strong association effects. The socalled chemical theory is applicable for such mixtures.

In this theory it is postulated that one or more of the components in a mixture associate so strongly that the association can be compared to a chemical reaction and that such reactions are in equilibrium.

For an associating component A in a multicomponent mixture we may write:

$$2A = AA \qquad (2.12)$$

with the equilibrium relation:

Table 2.1   PURE COMPONENT PARAMETERS FOR CALCULATION OF SECOND VIRIAL COEFFICIENTS

COMPONENT NAME	PC in atm	TC in K	ZC	RD in Å	DMU in Debye	ETA
Alkanes						
Butane	37.5	425.2	0.274	2.889	0.00	0.00
2-Methylpropane	36.0	408.1	0.283	2.896	0.13	0.00
Pentane	33.3	469.5	0.269	3.385	0.00	0.00
2-Methylbutane	32.9	460.4	0.268	3.313	0.13	0.00
2,2-Dimethylpropane	31.6	433.7	0.269	3.153	0.00	0.00
Hexane	29.9	507.3	0.264	3.812	0.00	0.00
2-Methylpentane	30.0	496.5	0.270	3.809	0.13	0.00
3-Methylpentane	30.8	504.7	0.273	3.680	0.13	0.00
Heptane	27.0	540.3	0.259	4.267	0.00	0.00
Octane	24.6	568.6	0.256	4.380	0.00	0.00
Cyclohexane	40.2	553.4	0.271	3.561	0.00	0.00
Alkenes						
1-Butene	39.7	419.6	0.277	2.746	0.34	0.00
2-Methylpropene	39.5	417.9	0.270	2.828	0.50	0.00
1-Pentene	39.9	464.8	0.266	3.196	0.50	0.00
2-Methyl-1-Butene	34.5	464.2	0.268	3.224	0.60	0.00
2-Methyl-2-Butene	34.0	470.0	0.266	3.230	0.60	0.00
1-Hexene	31.1	504.0	0.261	3.647	0.50	0.00
1- and 2-Heptene	28.1	542.0	0.261	4.097	0.50	0.00

Table 2.1 cont.

COMPONENT NAME	PC	TC	ZC	RD	DMU	ETA
Aromatics						
Benzene	48.6	562.1	0.274	3.004	0.00	0.00
Toluene	41.6	592.0	0.271	3.443	0.36	0.00
o-Xylene	35.7	631.6	0.254	3.789	0.62	0.00
m-Xylene	34.7	616.8	0.258	3.897	0.40	0.00
p-Xylene	33.9	618.8	0.252	3.796	0.00	0.00
Ethylbenzene	36.9	617.1	0.272	3.821	0.58	0.00
Alcohols						
Methanol	78.5	513.2	0.222	1.536	1.66	1.63
Ethanol	63.0	516.3	0.248	2.250	1.69	1.40
1-Propanol	51.0	536.7	0.252	2.736	1.68	1.40
2-Propanol	47.0	508.4	0.248	2.726	1.66	1.32
1-Butanol	43.6	562.9	0.259	3.225	1.66	2.20
2-Butanol	41.4	535.9	0.253	3.182	1.66	1.75
2-Methyl-1-Propanol	42.4	547.7	0.257	3.140	1.64	1.90
Phenol	60.5	694.2	0.244	3.550	1.41	0.32
Ethers						
Dimethyl ether	52.6	400.1	0.285	2.127	1.30	0.00
Ethylmethyl ether	43.4	437.8	0.267	2.641	1.22	0.00
Diethyl ether	35.6	466.0	0.255	3.140	1.16	0.00
Ethylpropyl ether	32.1	500.6	0.265	3.547	1.06	0.00

Table 2.1 cont.

COMPONENT NAME	PC	TC	ZC	RD	DMU	ETA
(Ethers)						
1,4-Dioxane	51.4	588.0	0.253	3.110	0.00	0.00
Ketones						
Acetone	47.0	509.1	0.237	2.740	2.88	0.90
2-Butanone	41.0	535.0	0.249	3.139	2.70	0.90
2-Pentanone	38.4	564.0	0.250	3.500	3.10	0.60
3-Pentanone	36.9	561.0	0.269	3.482	2.88	1.50
Aldehydes						
Acetaldehyde	54.7	461.0	0.257	2.021	2.70	0.58
Acids						
Formic acid	63.0	574.0	0.157	1.800	1.52	4.5
Acetic acid	57.1	594.8	0.200	2.595	1.74	4.5
Propanoic acid	53.0	612.7	0.242	3.050	1.75	4.5
Butanoic acid	52.0	628.0	0.293	3.551	1.75	4.5
Esters						
Methyl formate	59.2	487.2	0.255	2.363	1.77	0.35
Ethyl formate	46.8	508.5	0.257	2.870	1.93	0.15
Propyl formate	40.1	538.0	0.259	3.419	1.89	0.23
Methyl acetate	46.3	506.9	0.254	2.862	1.72	0.85
Ethyl acetate	37.8	523.3	0.252	3.348	1.78	0.53

Table 2.1 cont.

COMPONENT NAME	PC	TC	ZC	RD	DMU	ETA
(Esters)						
Methyl propanoate	39.5	530.6	0.255	3.304	1.69	0.80
Amines						
Methylamine	73.1	430.1	0.279	1.662	1.31	0.13
Dimethylamine	52.4	437.8	0.280	2.264	1.03	0.17
Ethylamine	55.5	456.4	0.274	2.309	1.22	0.30
Trimethylamine	40.2	433.3	0.287	2.736	0.612	0.06
Diethylamine	36.6	496.7	0.270	3.161	0.92	0.21
Triethylamine	30.0	535.4	0.269	3.930	0.66	0.10
Aniline	52.3	698.8	0.250	3.393	1.49	0.20
Chlorides						
Dichloromethane	60.0	510.0	0.277	2.432	1.54	0.00
Chloroform	54.0	536.6	0.294	3.178	1.02	0.00
Tetrachloromethane	45.0	556.4	0.272	3.458	0.00	0.00
Chloroethane	52.0	460.4	0.274	2.281	2.05	0.00
1-Chloropropane	45.2	503.2	0.278	2.792	2.06	0.00
1,2-Dichloroethane	53.0	560.9	0.259	2.851	0.00	0.00
Chlorobenzene	44.6	632.4	0.265	3.568	1.75	0.00
Miscellaneous						
Water	218.3	647.3	0.230	0.615	1.83	1.70

Table 2.1 cont.

COMPONENT NAME	PC	TC	ZC	RD	DMU	ETA
(Miscellaneous)						
Carbon disulfide	78.0	552.0	0.293	1.424	0.00	0.34
Acetonitrile	47.7	547.9	0.184	1.821	3.94	1.65
Nitromethane	62.3	588.0	0.223	2.306	3.44	1.66
Furan	52.5	487.0	0.286	2.508	0.661	0.00
Tetrahydrofuran	51.2	541.0	0.258	2.600	1.63	0.10

Most of the parameters are taken from an extensive table in the supplementary material to [3]. The table gives comparisons of experimental second virial coefficients and calculated second virial coefficients with several correlations. The table contains information on many more substances than given here.

Most of the values of ZC are taken from [4]. Some are calculated from [5].

PC : Critical pressure
TC : Critical temperature
ZC : Critical compressibility factor
RD : Mean radius of gyration
DMU: Dipolemoment
ETA: Association parameter

Table 2.2

SOLVATION PARAMETERS, η_{ij}, FOR CALCULATION OF CROSS VIRIAL COEFFICIENTS

MIXTURES CONTAINING	η_{ij}
Nonpolar components	0.0
Hydrocarbons	0.0
Hydrocarbons and nonpolar components (except fluorcarbons)	0.0
Aliphatic hydrocarbons and a polar component	0.0

ESTIMATED SOLVATION PARAMETERS BETWEEN GROUPS OF SUBSTANCES

		1	2	3	4	5	6
1	Aromatic hydrocarbons	0	0	0	0.5	0.6	0.4
2	Alcohols		1.55	1.55	1.0	1.3	2.5
3	Water				1.0	1.3	2.5
4	Ketones				0.9	1.1	1.8
5	Esters					0.53	2.0
6	Organic acids						4.5

Table 2.2 (continued)

Further Solvation Parameters used by Hayden and O'Connell:

Benzene – Chloroform	0.12
Benzene – Nitromethane	1.00
Chloromethane – Acetone	0.20
Dichloromethane – Acetone	0.92
Chloroform – Diethylether	0.95
Chloroform – Acetone	1.26
Chloroform – Methyl formate	1.20
Chloroform – Methyl acetate	1.65
Chloroform – Ethyl acetate	1.65
Chloroform – Propyl formate	1.20
Chloroform – Diethylamine	1.57
Acetone – Nitromethane	1.63
Acetaldehyde – Acetonitrile	2.5

$$K = \frac{Z_{AA}}{Z_A^2} \cdot \frac{1}{P} \cdot \frac{\varphi_{AA}}{\varphi_A^2} \tag{2.13}$$

where K is the equilibrium constant, Z_{AA} and Z_A are the "true" mole fractions of constituents AA and A, P is the pressure and φ the fugacity coefficient.

According to Nothnagel et al. [7] it is reasonable to write

$$\frac{\varphi_{AA}}{\varphi_A^2} = \exp - \frac{bP}{RT} \tag{2.14}$$

where b is the "excluded" volume owing to the finite size of the molecules. The value of b is thus equivalent to the value of B_{free} in (2.10). In Equation (2.14) it is assumed that $b = b_{AA} = b_A$.

Equation (2.13) may be written:

$$K_t = K \cdot P \cdot \exp \frac{bP}{RT} = \frac{Z_{AA}}{Z_A^2} \tag{2.15}$$

By means of material balances for the different constituents and equation (2.15) it is now possible to calculate the true mole fractions Z_A, Z_{AA} etc.

For the associating component A in a mixture, Marek and Standart [8] found

$$Z_A = \frac{\sqrt{1 + 4K_t y_A (2-y_A)} - 1}{2K_t (2-y_A)} \tag{2.16}$$

and for a nonassociating component B in the mixture:

$$Z_B = y_B \frac{1 + 4K_t(2-y_A) - \sqrt{1 + 4K_t y_A(2-y_A)}}{2K_t(2-y_A)^2} \tag{2.17}$$

In Equations (2.16) and (2.17) the values of y_A and y_B are the measurable mole fractions of A and B.

Values for K_t can be calculated from (2.15) as a function of P and T by means of experimental values of K and values of b determined as B_{free} in (2.10). Table 2.3 gives some values for K (in mm Hg^{-1}) as a function of the temperature (in K).

Table 2.3

Dimerization constant: $-\log K = A - B/T$

Acid	A	B	Ref.
Formic	10.743	3083	[9]
Acetic	10.4205	3166	[10]
Propanoic	10.843	3316	[9]
Butanoic	10.136	3079	[11]*

* Based on table 1 in [11]

In case no experimental values exist for K, Hayden and O'Connell [3] suggest that K can be calculated from:

$$-K = \frac{B_{bound} + B_{metastable} + B_{chem}}{RT} \qquad (2.18)$$

Knowing the "true" values of the mole fractions the fugacity coefficients of each component can be calculated from:

$$\varphi_i = \frac{Z_i}{y_i} \exp \frac{B_{i,free} \cdot P}{RT} \qquad (2.19)$$

2.4 COMPUTER PROGRAM FOR FUGACITY COEFFICIENTS

Appendix 1 gives a program, PHIB, for calculation of φ_i.

For mixtures not containing organic acids, φ_i is determined from (2.7) with values of the virial coefficients predicted by the method of Hayden and O'Connell, i.e. from program SVIR.

For mixtures containing organic acids, φ_i is determined from (2.19) with Z_i's calculated from (2.16) and (2.17). Experimental values of K are used if possible. Otherwise K is determined from (2.18).

In practice one might just as well have used the method by Hayden and O'Connell for predicting all values of K, since the method is highly reliable. Example 2.2 shows the fugacity coefficients for pure acetic acid and for a mixture containing acetic acid. There is very little difference between the fugacity coefficients based on the experimental and those based on the predicted values of K.

Example 2.2

> Fugacity coefficients, φ_i, for pure Acetic acid at 351.15 K and 0.2528 atm:
>
> φ_i = 0.3033, based on the experimental K
>
> φ_i = 0.3133, based on the predicted K

> Mixture of 2-Butanone (1) and Acetic acid (2) at 351.15 K and 0.3422 atm. For y_1 = 0.3429:
>
> φ_1 = 1.3486, φ_2 = 0.2818, based on experimental K
>
> φ_1 = 1.3440, φ_2 = 0.2912, based on predicted K

REFERENCES

1. J.M. Prausnitz, Molecular Thermodynamics of Fluid-Phase Equilibria, Prentice-Hall, 1969.
2. A.E. Jones and W.B. Kay, AIChE Journal, 13(1967) 720.
3. J.G. Hayden and J.P. O'Connell, Ind.Eng.Chem.Proc.Des.Dev., 14(1975)209.
4. R.C. Reid and T.K. Sherwood, The Properties of Gases and Liquids, McGraw-Hill, 2nd Ed., 1966.
5. Landolt-Börnstein, Zweiter Band, 1. Teil, Springer Verlag, 1971.
6. C.P. Smyth, Dielectric Behaviour and Structure, McGraw-Hill, 1955.
7. K.-H. Nothnagel, D.S. Abrams and J.M. Prausnitz, Ind.Eng. Chem.Proc.Des.Dev., 12(1973)25.
8. J. Marek and G. Standart, Coll.Czech.Chem.Comm., 19(1954) 1074.
9. I. Tetsuo and F. Yoshida, J.Chem.Eng. Data, 8(1963)315.
10. J. Marek, Coll.Czech.Chem.Comm., 21(1956)269.
11. R.E. Lundin, F.E. Harris and L.K. Nash, J.Am.Chem.Soc., 74(1952)743.

CHAPTER 3

LIQUID PHASE NONIDEALITY

The fugacity of component i in the liquid solution is related to the mole fraction x_i as follows:

$$f_i^L = \gamma_i x_i f_i^O \qquad i = 1, 2 \ldots M \tag{3.1}$$

where f_i^O is the fugacity of i at an arbitrarily chosen standard state. Throughout this book, the standard state is defined as pure component i at the system temperature and pressure. A method for calculating f_i^O is given in Section 3.1. We also, but very briefly, discuss a class of models based on the "local composition concept" for calculating γ_i as functions of x_i. These models contain parameters which must be calculated using experimental phase equilibrium data.

3.1 CALCULATION OF REFERENCE FUGACITIES

The pure-component reference fugacity of i at temperature T and pressure P is related to the fugacity at the vapor pressure of i at temperature T by the following equation [1]:

$$f_i^O = f_i^S \exp \int_{P_i^S}^{P} \frac{v_i \, dP}{RT} = f_i^S \cdot POY \tag{3.2}$$

where, for pure liquid i, P_i^S is the saturation pressure, f_i^S the fugacity at saturation, and v_i the molar liquid volume, all at temperature T.

By definition of the fugacity coefficient:

$$f_i^S = \varphi_i^S P_i^S \tag{3.3}$$

where the fugacity coefficient of component i at saturation, φ_i^S, is calculated from vapor phase P-V-T properties. We use second virial coefficients to calculate φ_i^S as shown in Section 2.1.

The liquid molar volume, v_i, is obtained using the correlation by Yen and Woods [2] (see subroutine YENWO in Appendix 1). This method reproduces the literature data for liquid densities within 2%, which is satisfactory for calculating the Poynting correction factor, POY. Because of the low pressures considered in this book we may neglect the pressure dependence of the volume and write

$$POY = \exp \frac{(P-P_i^S)v_i}{RT} \tag{3.4}$$

3.2 ACTIVITY COEFFICIENTS

We now turn to the calculation of activity coefficients, γ_i. It is far beyond the scope of the book to give a detailed account of all the models available for this purpose - for this we refer the reader to [1]. In the following we shall use three models based on the local composition concept, which derives from the following consideration:

At the molecular level, the average concentration of j-molecules around a central i-molecule will in general not be the same as the average concentration of j-molecules around a central j-molecule. The local compositions in the neighbourhood of i- and j-molecules depend - in a way specified by the model - on the relative size of the i-i, i-j, and j-j molecular energetic interactions. Three local-composition models are stated below:

<u>Wilson's Equation:</u> [3]

$$\ln \gamma_k = 1 - \ln \sum_j^j x_j \Lambda_{kj} - \sum^i \frac{x_i \Lambda_{ik}}{\sum_j^j x_j \Lambda_{ij}} \tag{3.5}$$

i and j = 1,2 .. M

$$\Lambda_{ij} \equiv \frac{v_j}{v_i} \exp\left[-\frac{\lambda_{ij} - \lambda_{ii}}{RT}\right]$$

$$\lambda_{ji} = \lambda_{ij} \quad ; \quad \Lambda_{ji} \neq \Lambda_{ij}$$

Equation (3.5) gives the activity coefficient of component k in a mixture with M components. The parameters $(\lambda_{ij} - \lambda_{ii})$ can be obtained from binary data. For each possible binary pair in the multicomponent solution, two parameters are needed. Wilson's equation is in general found to represent vapor-liquid equilibrium data very well - it can not in this form be used to represent liquid-liquid equilibria.

The NRTL Equation: (Non-Random, Two-Liquid)[4]

$$\ln \gamma_i = \frac{\sum\limits_{\ell}^{j} \tau_{ji} G_{ji} x_j}{\sum\limits_{\ell} G_{\ell i} x_\ell} + \sum^{j} \frac{x_j G_{ij}}{\sum\limits_{\ell} G_{\ell i} x_\ell} \left[\tau_{ij} - \frac{\sum\limits_{r}^{r} x_r \tau_{rj} G_{rj}}{\sum\limits_{\ell} G_{\ell j} x_\ell}\right] \quad (3.6)$$

i, j, ℓ and r = 1,2 .. M

$$\tau_{ji} \equiv \frac{(g_{ji} - g_{ii})}{RT} ; \quad (g_{ji} = g_{ij} \quad ; \quad \tau_{ji} \neq \tau_{ij})$$

$$G_{ji} \equiv \exp(-\alpha_{ji} \tau_{ji}); \quad (\alpha_{ji} = \alpha_{ij})$$

The parameters $(g_{ji} - g_{ii})$ and α_{ji} $(= \alpha_{ij})$ can be obtained from binary data. For each possible binary pair in a mixture, three parameters are needed. The NRTL equation may be used to represent vapor-liquid and liquid-liquid equilibria.

The UNIQUAC model: [5]

In the UNIQUAC model, the expression for the activity coefficients contains two parts: the combinatorial part, essentially due to differences in size and shape of the molecules, and a residual contribution, essentially due to energetic interactions. This is expressed as

$$\ln \gamma_k = \ln \gamma_k^C + \ln \gamma_k^R \qquad (3.7)$$
$$ \text{combinatorial} \quad \text{residual}$$

The combinatorial contribution is given by

$$\ln \gamma_k^C = \ln \frac{\Phi_k}{x_k} + \frac{z}{2} q_k \ln \frac{\theta_k}{\Phi_k} + \ell_k - \frac{\Phi_k}{x_k} \sum^j x_j \ell_j \qquad (3.8)$$

$$\ell_k = \frac{z}{2}(r_k - q_k) - (r_k - 1); \qquad z = 10$$

$$\theta_k = \frac{q_k x_k}{\sum\limits^J q_j x_j} \quad ; \quad \Phi_k = \frac{r_k x_k}{\sum\limits^J r_j x_j}$$

$$\text{volume fraction} \qquad \text{surface fraction}$$

$$j = 1, 2 \dots M$$

Pure-component constants r_k and q_k are, respectively, measures of molecular van der Waals volumes and molecular surface areas. They are obtained from van der Waals group volumes and surface areas, given by Bondi [6].

References [5] and [7] give values for r_k and q_k for some components; for many additional compounds, r_k and q_k may be calculated using the equivalent group constants given in Chapter 4.

The residual contribution is given by

$$\ln \gamma_k^R = q_k [1 - \ln(\sum^j \theta_j \tau_{jk}) - \sum^j (\theta_j \tau_{kj} / \sum^i \theta_i \tau_{ij})] \qquad (3.9)$$

$$i \text{ and } j = 1, 2 \dots M$$

$$\tau_{ji} \equiv \exp\left[-\frac{u_{ji} - u_{ii}}{RT}\right]$$

$$u_{ji} = u_{ij} \quad ; \quad \tau_{ji} \neq \tau_{ij}$$

In the distillation calculations of Chapter 9, the temperature-dependent parameter τ_{ji} is linearized as follows:

$$\tau_{ji} = A_{ji}^{(0)} + A_{ji}^{(1)} \cdot T \qquad (3.10)$$

where $A_{ji}^{(0)}$ and $A_{ji}^{(1)}$ are assumed to be independent of temperature.

The parameters $(u_{ji} - u_{ii})$ can be obtained from binary data. For each possible binary pair in the solution, two parameters are needed. The UNIQUAC equation can be used to represent both vapor-liquid and liquid-liquid equilibria.

Some useful references for obtaining binary parameters:

Wilson: [7,8,9,10]
NRTL :   [7,10,11,12]
UNIQUAC: [5,7]

The above three models have in common that

(1) multicomponent vapor-liquid equilibria may be predicted using experimental information on binary systems only.

(2) they permit extrapolation with respect to temperature, at least within a limited range.

(3) their use depends on the availability of binary parameters calculated from experimental phase equilibrium data for all possible binary combinations in a multicomponent mixture.

REFERENCES

1. J.M. Prausnitz, Molecular Thermodynamics of Fluid-Phase Equilibria, Prentice-Hall, Englewood Cliffs, New Jersey, 1969.
2. L.C. Yen and S.S. Woods, AIChE Journal, 12(1966)95.
3. G.M. Wilson, J.Am.Chem.Soc., 86(1964)127.
4. H. Renon and J.M. Prausnitz, AIChE Journal, 14(1968)135.
5. D.S. Abrams and J.M. Prausnitz, AIChE Journal, 21(1975)116.
6. A. Bondi, Physical Properties of Molecular Crystals, Liquids, and Glasses, Wiley, New York, 1968.

7. J. Gmehling and U. Onken, "Vapor-Liquid Equilibrium Data Collection", DECHEMA Chemistry Data Series; Volume 1-8, Frankfurt, Starting 1977.
8. M. Hirata, S. Ohe and K. Nagahama, Computer Aided Data Book of Vapor-Liquid Equilibria, Elsevier, Amsterdam, 1975.
9. M.J. Holmes and M. Van Winkle, Ind.Eng.Chem., 62(1970)21.
10. I. Nagata, J.Chem.Eng. Japan, 6(1973)18.
11. H. Renon, L. Asselineau, G. Cohen and C. Raimbault, Calcul sur Ordinateur des Equilibres Liquide-Vapeur et Liquide-Liquide, Technip, Paris, 1971.
12. I. Mertl, Coll.Czech.Chem.Commun., 37(1972)375.

CHAPTER 4

THE UNIFAC GROUP-CONTRIBUTION METHOD

In this chapter we give detailed information regarding the UNIFAC model and the parameters needed for its application.

4.1 THE UNIFAC MODEL

Estimation of thermodynamic properties of liquid mixtures from group contributions was first suggested by Langmuir [1]. This suggestion, however, received little attention until Derr and co-workers [2,3] used group contributions to correlate heats of mixing, followed by Wilson and Deal [4] who developed the solution-of-groups method for activity coefficients. The UNIFAC method is based on these ideas.

The basic aim of the solution-of-groups method is to utilize existing phase equilibrium data for predicting phase equilibria of systems for which no data are available. The method entails the following: suitable reduction of experimentally obtained activity coefficient data to obtain parameters characterizing interactions between pairs of structural groups in nonelectrolyte systems, and use of these parameters to predict activity coefficients for other systems which have not been studied experimentally but which contain the same functional groups. A "group" is any convenient structural unit such as $-CH_3$, $-COCH_2-$ and $-CH_2C\ell$. Following [4], the fundamental assumptions of solution-of-groups methods are:

<u>Assumption 1</u>. The logarithm of the activity coefficient is assumed to be the sum of two contributions: a combinatorial part, essentially due to differences in size and shape of the molecules in the mixture, and a residual part, essentially due to energy interactions.

For molecule i in any solution:

$$\ln \gamma_i = \underbrace{\ln \gamma_i^C}_{\text{combinatorial}} + \underbrace{\ln \gamma_i^R}_{\text{residual}} \qquad (4.1)$$

The distinction between two kinds of contributions to $\ln \gamma_i$ is necessary, since the liquid phase nonidealities caused by size and shape effects can not be associated with group energetic interactions.

<u>Assumption 2</u>. The contribution from group interactions, the residual part, is assumed to be the sum of the individual contributions of each solute group in the solution less the sum of the individual contributions in the pure-component environment. We write

$$\ln \gamma_i^R = \sum_{\substack{k \\ \text{all groups}}}^{k} \nu_k^{(i)} [\ln \Gamma_k - \ln \Gamma_k^{(i)}] \qquad (4.2)$$

$k = 1, 2 \ldots N$, where N is the number of different groups in the mixture

Γ_k is the residual activity coefficient of group k in a solution; $\Gamma_k^{(i)}$ is the residual activity coefficient of group k in a reference solution containing only molecules of type i; and $\nu_k^{(i)}$ is the number of groups of kind k in molecule i. In Equation (4.2) the term $\ln \Gamma_k^{(i)}$ is necessary to attain the normalization that activity coefficient γ_i becomes unity as $x_i \rightarrow 1$. The standard state for the group residual activity coefficient need not be defined due to cancellation of terms.

<u>Assumption 3</u>. The individual group contributions in any environment containing groups of kinds $1, 2 \ldots N$ are assumed to be only a function of group concentrations and temperature:

$$\left. \begin{array}{c} \Gamma_k \\ \Gamma_k^{(i)} \end{array} \right\} = F(X_1, X_2 \ldots X_N; T) \qquad (4.3)$$

The same function is used to represent Γ_k and $\Gamma_k^{(i)}$. The group fraction X is defined by:

$$X_k = \frac{\sum\limits_{i}^{i} \nu_k^{(i)} x_i}{\sum\limits_{i}\sum\limits_{j}^{ij} \nu_j^{(i)} x_i} \qquad (4.4)$$

$i = 1,2 \ldots M$ (number of components)
$j = 1,2 \ldots N$ (number of groups)

According to this assumption, for example the residual activity coefficients for all ketone-alkane mixtures may be calculated from the same function F. That is, the same parameters are used to represent vapor-liquid equilibria in acetone-hexane mixtures and decane-5-nonanone mixtures.

To formulate a specific group-contribution method for prediction of activity coefficients, one needs to specify:

(1)   The equation used to calculate $\ln \gamma_i^C$
(2)   The equation used to calculate Γ_k and $\Gamma_k^{(i)}$
(3)   The definition of functional groups used to "build" the molecules ("group assignments")

In Derr, Deal, and Wilson's ASOG model (Analytical Solutions Of Groups [4-7])

(1)   The combinatorial activity coefficients are calculated using the athermal Flory-Huggins equation [8].

$$\ln \gamma_i^C = \ln r_i^{FH} + (1 - r_i^{FH})$$

$$r_i^{FH} = \nu_i^{FH} \bigg/ \sum\limits_{j}^{j} \nu_j^{FH} x_j \qquad (4.5)$$

ν_i^{FH} is the number of atoms in the molecule except hydrogen-atoms.

(2) Γ_k and $\Gamma_k^{(i)}$ are represented by the Wilson equation (see Equation (3.5)), where the independent variable is the group fraction, X_k.

(3)   All groups are assigned integer numbers associated with their size ("size counts"). For example, the groups CH_2,

CH, CO, are assigned size 1.0, the ester group COO is assigned size 3.0, etc. Only water differs from this scheme — the "group" H_2O is assigned size 1.4.

Ratcliff and co-workers [9,10] follow essentially the same development. They use various equations for $\ln \gamma_i^C$, for example in [9] the Brønsted and Koefoed expression [11].

Combining the UNIQUAC model (see Section 3.2) with the solution of groups method leads to the UNIFAC method [12,13]. In the UNIFAC method:

(1) The combinatorial activity coefficients are calculated using Staverman's potential in exactly the same manner as that used by Abrams and Prausnitz in the UNIQUAC model [14]. This combinatorial part (see Equation (3.8)) takes into account contributions from differences in both molecular size and molecular shape. These are in turn obtained from the well-defined group volume and area constants R_k and Q_k, see below.

(2) The group residual activity coefficients are represented by the residual part of the UNIQUAC equation, Equation (3.9), where the independent concentration variable is the group fraction, X_k.

(3) The constants representing the group sizes and surface areas R_k and Q_k are obtained from atomic and molecular structure data, the van der Waals group volumes and surface areas V_k and A_k [15]:

$$R_k = V_k/15.17 \text{ and } Q_k = A_k/(2.5 \cdot 10^9) \tag{4.6}$$

The normalization factors 15.17 and $2.5 \cdot 10^9$ are those derived by Abrams and Prausnitz [14].

In the ASOG method, the "size counts" are rather arbitrary. This arbitrariness does not exist in the UNIFAC method — here the group sizes and surface areas are determined using well established, rational procedures. The UNIFAC method is stated by equations (4.1), (4.2), and:

Combinatorial activity coefficient for component i:

$$\ln \gamma_i^C = \ln \frac{\Phi_i}{x_i} + \frac{z}{2} q_i \ln \frac{\theta_i}{\Phi_i} + \ell_i - \frac{\Phi_i}{x_i} \sum_j^j x_j \ell_j$$

$$\ell_i = \frac{z}{2}(r_i - q_i) - (r_i - 1) \quad ; \quad z = 10 \tag{4.7}$$

$$\theta_i = \frac{q_i x_i}{\sum_j^J q_j x_j} \quad ; \quad \Phi_i = \frac{r_i x_i}{\sum_j^J r_j x_j}$$

Molecular surface area fraction   Molecular volume fraction

j = 1,2 .. M (number of components)

The van der Waals volume: $\quad r_i = \sum_k^k \nu_k^{(i)} R_k$

and van der Waals surface area: $\quad q_i = \sum_k^k \nu_k^{(i)} Q_k$ \hfill (4.8)

k = 1,2 .. N (number of groups in molecule i)

are found by summation of the corresponding group properties. Note that γ_i^C does not depend on temperature.

Typical calculated combinatorial contributions to activity coefficients are shown in Table 4.1. Note that while the combinatorial contribution is often small, it is far from negligible when the molecules differ much in size and shape.

Residual activity coefficient for group k:

$$\ln \Gamma_k = Q_k [1 - \ln(\sum_m^m \theta_m \Psi_{mk}) - \sum^m (\theta_m \Psi_{km} / \sum_n^n \theta_n \Psi_{nm})] \tag{4.9}$$

m and n = 1,2 .. N (all groups)

Equation (4.9) also holds for $\Gamma_k^{(i)}$. The equation is similar to the one used in the UNIQUAC model for calculating γ_i^R.

$$\theta_m = \frac{Q_m X_m}{\sum_n^n Q_n X_n} \quad ; \quad X_m = \frac{\sum_j^j \nu_m^{(j)} x_j}{\sum\sum_n^{jn} \nu_n^{(j)} x_j} \tag{4.10}$$

Group surface area fraction   Group fraction

j = 1,2 .. M;   n = 1,2 .. N

Table 4.1 TYPICAL COMBINATORIAL ACTIVITY COEFFICIENTS CALCULATED FROM UNIFAC

System	Van der Waals volume		Van der Waals surface area		$x_1 = 0.0$		$x_2 = 0.5$		$x_1 = 1.0$	
	r_1	r_2	q_1	q_2	γ_1^C	γ_2^C	γ_1^C	γ_2^C	γ_1^C	γ_2^C
Pentane(1)-Octane(2)	3.83	5.85	3.32	4.94	0.930	1.0	0.977	0.982	1.0	0.909
Benzene(1)-Water(2)	3.19	0.92	2.40	1.40	13.30	1.0	—*	—*	1.0	2.358
Phenol(1)-Water(2)	3.55	0.92	2.68	1.40	15.10	1.0	1.141	1.161	1.0	2.163
Toluene(1)-Ethanol(2)	3.92	2.11	2.97	1.97	1.130	1.0	1.011	1.027	1.0	1.052
Hexane(1)-Ethylamine(2)	4.50	2.27	3.86	2.08	0.778	1.0	0.962	0.942	1.0	0.848
Cyclohexane(1)-Tetrachloromethane(2)	4.05	3.39	3.24	2.91	1.024	1.0	1.005	1.005	1.0	1.020
Heptane(1)-Acetone(2)	5.17	2.57	4.40	2.34	0.769	1.0	0.960	0.940	1.0	0.843
Heptane(1)-Acetic acid(2)	5.17	2.20	4.40	2.07	0.686	1.0	0.948	0.916	1.0	0.796
2-Butanone(1)-Acetic acid(2)	3.25	2.20	2.88	2.07	0.942	1.0	0.988	0.985	1.0	0.953
Cyclohexane(1)-Methanol(2)	4.05	1.43	3.24	1.43	0.702	1.0	0.947	0.919	1.0	0.796

* immiscible

In Equation (4.9) the parameter Ψ_{nm} is given by

$$\Psi_{nm} = \exp(-a_{nm}/T) \tag{4.11}$$

Table 4.2

SOME RESIDUAL ACTIVITY COEFFICIENTS AT 353 K
CALCULATED FROM UNIFAC

System	$x_1 = 0.0$		$x_1 = 0.5$		$x_1 = 1.0$	
	γ_1	γ_2	γ_1	γ_2	γ_1	γ_2
Pentane(1)-Octane(2)	1.0	1.0	1.0	1.0	1.0	1.0
Benzene(1)-Water(2)	107.5	1.0	–	–	1.0	88.48
Phenol(1)-Water(2)	2.239	1.0	1.040	1.151	1.0	1.257
Toluene(1)-Ethanol(2)	4.880	1.0	1.643	1.461	1.0	8.293
Hexane(1)-Ethylamine(2)	3.091	1.0	1.250	1.317	1.0	2.484
Cyclohexane(1)-Tetrachloromethane(2)	1.115	1.0	1.034	1.027	1.0	1.145
Heptane(1)-Acetone(2)	6.948	1.0	1.489	1.596	1.0	5.005
Heptane(1)-Acetic acid(2)	18.40	1.0	1.647	1.934	1.0	8.486
2-Butanone(1)-Acetic acid(2)	1.045	1.0	1.072	1.001	1.0	1.471
Cyclohexane(1)-Methanol(2)	18.95	1.0	1.944	1.981	1.0	18.06

Equation (4.11) contains the group-interaction parameter, a_{nm}. It is a measure of the difference in the energy of interaction between a group n and a group m and between two groups m. Note that $a_{nm} \neq a_{mn}$ and that the group-interaction parameters are assumed independent of temperature. There are thus two group-interaction parameters for each pair of groups; no ternary (or higher) parameters are needed. The parameters must be evaluated from phase equilibrium data.

Group-interaction parameters a_{nm} and pure group constants R_k and Q_k are given in Section 4.2 for a large number of different groups.

Table 4.2 shows the calculated residual activity coefficients corresponding to the combinatorial activity coefficients of Table 4.1. In many cases, the residual contribution is much larger than the combinatorial contribution. Interesting enough, the reverse is the case for phenol-water.

Figures 4.1 and 4.2 show Γ_k as functions of X_k for two typical cases.

The combinatorial contribution to the activity coefficient [Equation (4.7)] depends only on the sizes and shapes of the molecules present. For large chain molecules, $q_i/r_i \rightarrow$ a constant value, and in that limit, Equation (4.7) reduces to an equation similar to the Flory-Huggins equation used in the ASOG method.

The residual contribution to the activity coefficient, Equations (4.9-11), depends on group areas and group interactions. When all group areas are equal, Equation (4.9) is similar to that used in the ASOG method.

From the user's point of view, UNIFAC provides three advantages:

1 - flexibility, because UNIFAC has a well founded basis for establishing group sizes and shapes

2 - simplicity, because UNIFAC parameters are nearly independent of temperature for the temperature range considered here (see Section 4.4)

3 - large range of applicability, because UNIFAC parameters are now available for a considerable number of different functional groups.

As it stands, the UNIFAC method may be applied to nonelectrolyte binary and multicomponent mixtures at conditions where the UNIQUAC model applies, i.e. removed from the critical region. Furthermore, all components must be condensable. The temperature range considered is typically 30-125 ^{o}C. Finally, the UNIFAC method does presently not apply to mixtures containing polymers. As a rough guide, it should rarely be applied to mixtures containing components with more than ten functional

Figure 4.1

GROUP RESIDUAL ACTIVITY COEFFICIENTS FOR
MIXTURES OF CH_2- AND CH_2CO- GROUPS AT 353 K

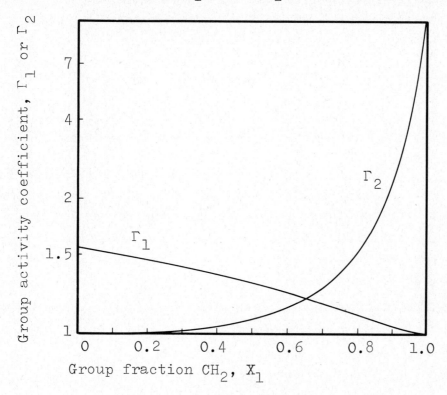

Figure 4.2

GROUP RESIDUAL ACTIVITY COEFFICIENTS FOR
MIXTURES OF CH_2- AND CH_2NH_2-GROUPS AT 353 K

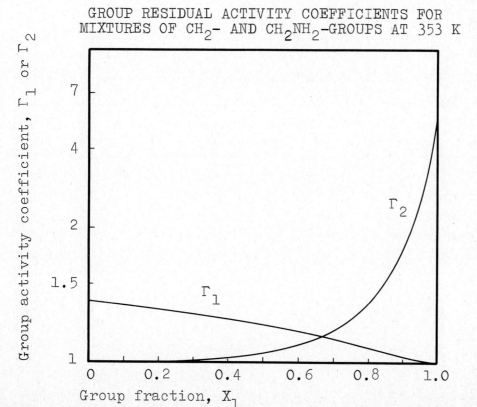

groups.

The capability of the UNIFAC method for predicting phase equilibria in systems where no data exist is illustrated in Chapters 6, 7, and 8.

As is already stated, the UNIFAC method in its present state of development correlates more than 70% of the published vapor-liquid equilibrium data.

4.2 GROUP INTERACTION PARAMETERS

All group-contribution methods are necessarily approximations because any group within a molecule is not completely independent of the other groups within that molecule. But it is precisely this independence which is the essential basis of every group-contribution method. We can allow for interdependence of groups within a molecule by our definition of what atoms constitute a group. Increasing distinction of groups, however, also increases the number of group interactions that must be characterized. Ultimately, if we carry group distinction to the limit, we recover the individual molecules. In that event, the advantage of the group-contribution method is lost. Judgment and experience must tell us how to define functional groups so as to achieve a compromise between accuracy of prediction and engineering utility. The definition of functional groups given in this section is, on the basis of experience, believed to be a reasonable such compromise.

In this section, we list the group-interaction parameters a_{nm} and the pure group constants R_k and Q_k needed in conjunction with the equations for the UNIFAC method given in the Section 4.1. Chapter 5 explains how the group-interaction parameters were calculated from experimental data and lists the literature references for the data on which they are based. The remaining chapters in the book deal with the application of the material given in Sections 4.1 and 4.2 to the prediction of phase equilibria and to separation process design.

The UNIFAC groups are given in Table 4.3. The values of R_k and Q_k are, except in few cases, calculated from [15] using Equations (4.6). While "subgroups" within the same main group have different values for R_k and Q_k (for example R_{CH_3CO} = 1.6724 and R_{CH_2CO} = 1.4457), the group-interaction parameter is the same for all "subgroups" within a "main group" (for example $a_{CH_2,CH_3CO} = a_{CH_2,CH_2CO}$). The specific choice of functional groups in Table 4.3 is based on experience gained from fitting the UNIFAC model to vapor-liquid equilibrium data. Our choice was to define as functional groups those resulting in the best fit, as long as the advantages of the group-contribution method were not seriously jeopardized. Thus methanol was made into a separate "group" and was not included in the alcohol "main group". Often the addition of a CH_2-group to the smallest possible group resulted in improved correlation (for example COOC, CH_2O, and CNH_2). For alcohols, the optimal choice was a group containing two carbon atoms. In these cases, the improvements in the correlation were significant. An example of the negative side of the compromises made in developing Table 4.3 is that the alcohol groups, CCOH, can not be used to predict the activity coefficients of mixtures containing allyl alcohol ($CH_2 = CH-CH_2OH$).

Isomer effects are to a large extent included in Table 4.3. The main alcohol group contains both primary and secondary alcohols; propylamine and isopropylamine are covered by the same main group; many other examples could be mentioned.

Table 4.4 gives the group-interaction parameters, a_{nm}. The code n.a. (not available) means that no reliable data could be found for determining the necessary parameters. Unfortunately, it does not appear possible to give a quantitative method for replacing the "n.a.'s" in Table 4.4 with group interaction parameters. The group interaction parameters for related functional groups - for example CNH_2 and CNH - do indeed follow similar trends. If one predicts activity coefficients for secondary amine - alkane systems using the CNH_2-CH_2 group interaction parameters (but R_k and Q_k for CNH), the result is reasonably good. However, this allows only for the estimation of a few of the missing group interaction parameters, and it appears that the only reliable way to eliminate the "n.a.'s" is by providing more accurate phase equilibrium data followed by data reduction as shown in Chapter 5.

It is, naturally, possible to extend Tables 4.3 and 4.4 to groups not already covered, provided experimental data are available. This needs most urgently to be done for some of the special solvents used in extractive distillation: pyridine, furfural, some glycols, and others.

An alternative to this extension can be pointed out. One may use one of the available correlations [16] for calculating extractive distillation solvent selectivity in parallel with the UNIFAC method to generate activity coefficients at infinite dilution. Then one can determine, for example, UNIQUAC parameters from these (see Section 9.2). A review of correlations available for estimating solvent selectivity is given in [17].

The number of isotherms/isobars used in the determination of each pair of group-interaction parameters and references to the data used may be ascertained from Appendix 5. If the parameters are based on limited data (see Appendix 5) they must be used with caution.

In Sections 4.4, 4.5, and 4.6 we deal with specific aspects of the UNIFAC parameters. A few examples of their use are given in Section 4.3.

4.3 SAMPLE PREDICTIONS

The use of Tables 4.3 and 4.4 is illustrated in Example 4.1. This corresponds to the detailed example shown in [12]. However, the definition of the ketone group used in the present version of UNIFAC differs from that used originally.

Example 4.1. ACTIVITY COEFFICIENT FOR ACETONE IN PENTANE

What is the activity coefficient for acetone in the mixture acetone (1) - pentane (2) system at 307 K and $x_1 = 0.047$? Pentane is made up from 2 CH_3-groups and 3 CH_2-groups. Acetone is made up from 1 CH_3-group and 1 CH_3CO-group.

Table 4.3 GROUP VOLUME AND SURFACE-AREA PARAMETERS

Main Group	Sub Group	No	R_k	Q_k	Sample Group Assignment
1 "CH_2"	CH_3	1	0.9011	0.848	Butane: 2 CH_3, 2 CH_2
	CH_2	2	0.6744	0.540	
	CH	3	0.4469	0.228	2-Methylpropane: 3 CH_3, 1 CH
	C	4	0.2195	0.000	2,2-Dimethylpropane: 4 CH_3, 1 C
2 "C=C"	CH_2=CH	5	1.3454	1.176	1-Hexene: 1 CH_3, 3 CH_2, 1 CH_2=CH
	CH=CH	6	1.1167	0.867	2-Hexene: 2 CH_3, 2 CH_2, 1 CH=CH
	CH=C	7	0.8886	0.676	2-Methyl-2-butene: 3 CH_3, 1 CH=C
	CH_2=C	8	1.1173	0.988	2-Methyl-1-butene: 2 CH_3, 1 CH_2, 1 CH_2=C
3 "ACH"	ACH	9	0.5313	0.400	Benzene: 6 ACH
	AC	10	0.3652	0.120	Styrene: 1 CH_2=CH, 5 ACH, 1 AC
4 "$ACCH_2$"	$ACCH_3$	11	1.2663	0.968	Toluene: 5 ACH, 1 $ACCH_3$
	$ACCH_2$	12	1.0396	0.660	Ethylbenzene: 1 CH_3, 5 ACH, 1 $ACCH_2$
	ACCH	13	0.8121	0.348	Cumene: 2 CH_3, 5 ACH, 1 ACCH
5 "CCOH"	CH_2CH_2OH	14	1.8788	1.664	1-Propanol: 1 CH_3, 1 CH_2CH_2OH
	$CHOHCH_3$	15	1.8780	1.660	2-Butanol: 1 CH_3, 1 CH_2, 1 $CHOHCH_3$
	$CHOHCH_2$	16	1.6513	1.352	3-Octanol: 2 CH_3, 4 CH_2, 1 $CHOHCH_2$

Table 4.3 cont.

5 "CCOH"	CH_3CH_2OH $CHCH_2OH$	17 18	2.1055 1.6513	1.972 1.352	Ethanol: 1 CH_3CH_2OH 2-Methyl-1-propanol: 2 CH_3, 1 $CHCH_2OH$
6	CH_3OH	19	1.4311	1.432	Methanol: 1 CH_3OH
7	H_2O	20	0.92	1.40	Water: 1 H_2O
8	ACOH	21	0.8952	0.680	Phenol: 5 ACH, 1 ACOH
9 "CH_2CO"	CH_3CO CH_2CO	22 23	1.6724 1.4457	1.488 1.180	Ketone group is 2nd carbon; 2-Butanone: 1 CH_3, 1 CH_2, 1 CH_3CO Ketone group is any other carbon; 3-Pentanone: 2 CH_3, 1 CH_2, 1 CH_2CO
10	CHO	24	0.9980	0.948	Acetaldehyde: 1 CH_3, 1 CHO
11 "COOC"	CH_3COO CH_2COO	25 26	1.9031 1.6764	1.728 1.420	Butyl acetate: 1 CH_3, 3 CH_2, 1 CH_3COO Butyl propanoate: 2 CH_3, 3 CH_2, 1 CH_2COO
12 "CH_2O"	CH_3O CH_2O $CH-O$ FCH_2O	27 28 29 30	1.1450 0.9183 0.6908 0.9183	1.088 0.780 0.468 1.1	Dimethyl ether: 1 CH_3, 1 CH_3O Diethyl ether: 2 CH_3, 1 CH_2, 1 CH_2O Diisopropyl ether: 4 CH_3, 1 CH, 1 CH-O Tetrahydrofuran: 3 CH_2, 1 FCH_2O

Table 4.3 cont.

13 "CNH$_2$"	CH$_3$NH$_2$ CH$_2$NH$_2$ CHNH$_2$	31 32 33	1.5959 1.3692 1.1417	1.544 1.236 0.924	Methylamine: 1 CH$_3$NH$_2$ Propylamine: 1 CH$_3$, 1 CH$_2$, 1 CH$_2$NH$_2$ Isopropylamine: 2 CH$_3$, 1 CHNH$_2$
14 "CNH"	CH$_3$NH CH$_2$NH CHNH	34 35 36	1.4337 1.2070 0.9795	1.244 0.936 0.624	Dimethylamine: 1 CH$_3$, 1 CH$_3$NH Diethylamine: 2 CH$_3$, 1 CH$_2$, 1 CH$_2$NH Diisopropylamine: 4 CH$_3$, 1 CH, 1 CHNH
15	ACNH$_2$	37	1.0600	0.816	Aniline: 5 ACH, 1 ACNH$_2$
16 "CCN"	CH$_3$CN CH$_2$CN	38 39	1.8701 1.6434	1.724 1.416	Acetonitrile: 1 CH$_3$CN Propionitrile: 1 CH$_3$, 1 CH$_2$CN
17 "COOH"	COOH HCOOH	40 41	1.3013 1.5280	1.224 1.532	Acetic acid: 1 CH$_3$, 1 COOH Formic acid: 1 HCOOH
18 "CCl"	CH$_2$Cl CHCl CCl	42 43 44	1.4654 1.2380 0.7910	1.264 0.952 0.724	1-Chlorobutane: 1 CH$_3$, 2 CH$_2$, 1 CH$_2$Cl 2-Chloropropane: 2 CH$_3$, 1 CHCl 2-Chloro-2-methylpropane: 3 CH$_3$, 1 CCl
19 "CCl$_2$"	CH$_2$Cl$_2$ CHCl$_2$ CCl$_2$	45 46 47	2.2564 2.0606 1.8016	1.988 1.684 1.448	Dichloromethane: 1 CH$_2$Cl$_2$ 1,1-Dichloroethane: 1 CH$_3$, 1 CHCl$_2$ 2,2-Dichloropropane: 2 CH$_3$, 1 CCl$_2$

Table 4.3 cont.

"CCl_3"	20	$CHCl_3$ CCl_3	48 49	2.8700 2.6401	2.410 2.184	Chloroform: 1 $CHCl_3$ 1,1,1-Trichloroethane: 1 CH_3, 1 CCl_3
	21	CCl_4	50	3.3900	2.910	Tetrachloromethane: 1 CCl_4
	22	ACCl	51	1.1562	0.844	Chlorobenzene: 5 ACH, 1 ACCl
"CNO_2"	23	CH_3NO_2 CH_2NO_2 $CHNO_2$	52 53 54	2.0086 1.7818 1.5544	1.868 1.560 1.248	Nitromethane: 1 CH_3NO_2 1-Nitropropane: 1 CH_3, 1 CH_2, 1 CH_2NO_2 2-Nitropropane: 2 CH_3, 1 $CHNO_2$
	24	$ACNO_2$	55	1.4199	1.104	Nitrobenzene: 5 ACH, 1 $ACNO_2$
	25	CS_2	56	2.057	1.65	Carbon disulfide: 1 CS_2

Table 4.4

GROUP INTERACTION PARAMETERS, a_{mn}, K

		1 CH_2	2 $C=C$	3 ACH	4 $ACCH_2$	5 CCOH
1	CH_2	0	-200.0	61.13	76.50	737.5
2	$C=C$	2520	0	340.7	4102	535.2
3	ACH	-11.12	-94.78	0	167.0	477.0
4	$ACCH_2$	-69.70	-269.7	-146.8	0	469.0
5	CCOH	-87.93	121.5	-64.13	-99.38	0
6	CH_3OH	16.51	-52.39	-50.00	-44.50	-80.78
7	H_2O	580.6	511.7	362.3	377.6	-148.5
8	ACOH	311.0	n.a.	2043	6245	-455.4
9	CH_2CO	26.76	-82.92	140.1	365.8	129.2
10	CHO	505.7	n.a.	n.a.	n.a.	n.a.
11	COOC	114.8	n.a.	85.84	-170.0	109.9
12	CH_2O	83.36	76.44	52.13	65.69	42.00
13	CNH_2	-30.48	79.40	-44.85	n.a.	-217.2
14	CNH	65.33	-41.32	-22.31	223.0	-243.3
15	$ACNH_2$	5339	n.a.	650.4	3399	-245.0
16	CCN	35.76	26.09	-22.97	-138.4	n.a.
17	COOH	315.3	349.2	62.32	180.2	-17.59
18	CCl	91.46	-24.36	4.680	122.9	368.6
19	CCl_2	34.01	-52.71	n.a.	n.a.	601.6
20	CCl_3	36.70	-185.1	288.5	33.61	491.1
21	CCl_4	-78.45	-293.7	-4.700	134.7	570.7
22	ACCl	-141.3	n.a.	-237.7	n.a.	134.1
23	CNO_2	-32.69	-49.92	10.38	-97.05	n.a.
24	$ACNO_2$	5541	n.a.	1825	n.a.	n.a.
25	CS_2	11.46	n.a.	-18.99	n.a.	442.8

n.a. = not available

Table 4.4   cont.

		6 CH_3OH	7 H_2O	8 ACOH	9 CH_2CO	10 CHO
1	CH_2	697.2	1318	2789	476.4	677.0
2	C=C	1509	599.6	n.a.	524.5	n.a.
3	ACH	637.4	903.8	1397	25.77	n.a.
4	$ACCH_2$	603.3	5695	726.3	-52.10	n.a.
5	CCOH	127.4	285.4	257.3	48.16	n.a.
6	CH_3OH	0	-181.0	n.a.	23.39	306.4
7	H_2O	289.6	0	442.0	-280.8	649.1
8	ACOH	n.a.	-540.6	0	n.a.	n.a.
9	CH_2CO	108.7	605.6	n.a.	0	-37.36
10	CHO	-340.2	-155.7	n.a.	128.0	0
11	COOC	249.6	1135	853.6	372.2	n.a.
12	CH_2O	339.7	634.2$^+$	n.a.	52.38	n.a.
13	CNH_2	-481.7	-507.1	n.a.	n.a.	n.a.
14	CNH	-500.4	-547.7$^+$	n.a.	n.a.	n.a.
15	$ACNH_2$	n.a.	-339.5	n.a.	n.a.	n.a.
16	CCN	168.8	242.8	n.a.	-275.1	n.a.
17	COOH	1020	-292.0	n.a.	-297.8	n.a.
18	CCl	529.0	698.2	n.a.	286.3	n.a.
19	CCl_2	669.9	708.7	n.a.	423.2	n.a.
20	CCl_3	649.1	826.8	n.a.	552.1	n.a.
21	CCl_4	860.1	1201	1616	372.0	n.a.
22	ACCl	n.a.	920.4	n.a.	n.a.	n.a.
23	CNO_2	252.6	614.2	n.a.	-142.6	n.a.
24	$ACNO_2$	n.a.	360.7	n.a.	n.a.	n.a.
25	CS_2	914.2	1081	n.a.	298.7	n.a.

n.a. = not available

$^+$ predictions not reliable in whole concentration range

Table 4.4   cont.

		11 COOC	12 CH$_2$O	13 CNH$_2$	14 CNH	15 ACNH$_2$
1	CH$_2$	232.1	251.5	391.5	255.7	1245
2	C=C	n.a.	289.3	396.0	273.6	n.a.
3	ACH	5.994	32.14	161.7	122.8	668.2
4	ACCH$_2$	5688	213.1	n.a.	−49.29	612.5
5	CCOH	76.20	70.00	110.8	188.3	412.0
6	CH$_3$OH	−10.72	−180.6	359.3	266.0	n.a.
7	H$_2$O	−455.4	−400.6[+]	357.5	287.0[+]	213.0
8	ACOH	−713.2	n.a.	n.a.	n.a.	n.a.
9	CH$_2$CO	−213.7	5.202	n.a.	n.a.	n.a.
10	CHO	n.a.	n.a.	n.a.	n.a.	n.a.
11	COOC	0	−235.7	n.a.	−73.50	n.a.
12	CH$_2$O	461.3	0	n.a.	141.7	n.a.
13	CNH$_2$	n.a.	n.a.	0	63.72	n.a.
14	CNH	136.0	−49.30	108.8	0	n.a.
15	ACNH$_2$	n.a.	n.a.	n.a.	n.a.	0
16	CCN	−297.3	n.a.	n.a.	n.a.	n.a.
17	COOH	−256.3	−338.5	n.a.	n.a.	n.a.
18	CCl	n.a.	225.4	n.a.	n.a.	n.a.
19	CCl$_2$	−132.9	−197.7	n.a.	n.a.	n.a.
20	CCl$_3$	176.5	−20.93	n.a.	n.a.	n.a.
21	CCl$_4$	129.5	n.a.	n.a.	91.13	1302
22	ACCl	n.a.	n.a.	203.5	−108.4	n.a.
23	CNO$_2$	n.a.	−94.49	n.a.	n.a.	n.a.
24	ACNO$_2$	n.a.	n.a.	n.a.	n.a.	5250
25	CS$_2$	233.7	79.79	n.a.	n.a.	n.a.

n.a. = not available

[+] predictions not reliable in whole concentration range

Table 4.4 cont.

		16 CCN	17 COOH	18 CCl	19 CCl_2	20 CCl_3
1	CH_2	612.0	663.5	35.93	53.76	24.9
2	C=C	370.9	730.4	99.61	337.1	4583
3	ACH	212.5	537.4	-18.81	n.a.	-231.9
4	$ACCH_2$	6096	856.5	-114.1	n.a.	-12.14
5	CCOH	n.a.	77.61	-38.23	-185.9	-170.9
6	CH_3OH	45.54	-289.5	-38.32	-102.5	-139.4
7	H_2O	112.6	225.4	325.4	370.4	353.7
8	ACOH	n.a.	n.a.	n.a.	n.a.	n.a.
9	CH_2CO	428.5	669.4	-191.7	-284.0	-354.6
10	CHO	n.a.	n.a.	n.a.	n.a.	n.a.
11	COOC	533.6	660.2	n.a.	108.9	-209.7
12	CH_2O	n.a.	664.6	301.1	137.8	-154.3
13	CNH_2	n.a.	n.a.	n.a.	n.a.	n.a.
14	CNH	n.a.	n.a.	n.a.	n.a.	n.a.
15	$ACNH_2$	n.a.	n.a.	n.a.	n.a.	n.a.
16	CCN	0	n.a.	n.a.	n.a.	-15.62
17	COOH	n.a.	0	44.42	-183.4	n.a.
18	CCl	n.a.	326.4	0	108.3	249.2
19	CCl_2	n.a.	1821	-84.53	0	0
20	CCl_3	74.04	n.a.	-157.1	0	0
21	CCl_4	492.0	689.0	11.80	17.97	51.90
22	ACCl	n.a.	n.a.	n.a.	n.a.	n.a.
23	CNO_2	n.a.	n.a.	n.a.	n.a.	n.a.
24	$ACNO_2$	n.a.	n.a.	n.a.	n.a.	n.a.
25	CS_2	n.a.	n.a.	n.a.	n.a.	-125.8

n.a. = not available

Table 4.4   cont.

		21 CCl_4	22 ACCl	23 CNO_2	24 $ACNO_2$	25 CS_2
1	CH_2	104.3	321.5	661.5	543.0	114.1
2	C=C	5831	n.a.	542.1	n.a.	n.a.
3	ACH	3.000	538.2	168.1	194.9	97.53
4	$ACCH_2$	-141.3	n.a.	3629	n.a.	n.a.
5	CCOH	-98.66	290.0	n.a.	n.a.	73.52
6	CH_3OH	-67.80	n.a.	75.14	n.a.	-31.09
7	H_2O	497.5	678.2	-19.44	399.5	887.1
8	ACOH	4894	n.a.	n.a.	n.a.	n.a.
9	CH_2CO	-39.20	n.a.	137.5	n.a.	162.3
10	CHO	n.a.	n.a.	n.a.	n.a.	n.a.
11	COOC	54.47	n.a.	n.a.	n.a.	162.7
12	CH_2O	n.a.	n.a.	95.18	n.a.	151.1
13	CNH_2	n.a.	68.81	n.a.	n.a.	n.a.
14	CNH	71.23	4350	n.a.	n.a.	n.a.
15	$ACNH_2$	8455	n.a.	n.a.	-62.73	n.a.
16	CCN	-54.86	n.a.	n.a.	n.a.	n.a.
17	COOH	212.7	n.a.	n.a.	n.a.	n.a.
18	CCl	62.42	n.a.	n.a.	n.a.	n.a.
19	CCl_2	56.33	n.a.	n.a.	n.a.	n.a.
20	CCl_3	-30.10	n.a.	n.a.	n.a.	256.5
21	CCl_4	0	475.8	490.9	534.7	132.2
22	ACCl	-255.4	0	-154.5	n.a.	n.a.
23	CNO_2	-34.68	794.4	0	n.a.	n.a.
24	$ACNO_2$	514.6	n.a.	n.a.	0	n.a.
25	CS_2	-60.71	n.a.	n.a.	n.a.	0

n.a. = not available

From Table 4.3 one gets:

Group	Number	R_k	Q_k
CH_3	1	0.9011	0.848
CH_2	2	0.6744	0.540
CH_3CO	22	1.6724	1.488

From Equations (4.8):

$$r_1 = 1 \cdot 0.9011 + 1 \cdot 1.6724 = 2.5735;$$
$$q_1 = 1 \cdot 0.848 + 1 \cdot 1.488 = 2.336$$

$$r_2 = 2 \cdot 0.9011 + 3 \cdot 0.6744 = 3.8254;$$
$$q_2 = 2 \cdot 0.848 + 3 \cdot 0.540 = 3.316$$

From Equations (4.7):

$$\Phi_1 = \frac{2.5735 \cdot 0.047}{2.5735 \cdot 0.047 + 3.8254 \cdot 0.953} = 0.03211; \quad \Phi_2 = 0.96789$$

$$\theta_1 = \frac{2.336 \cdot 0.047}{2.336 \cdot 0.047 + 3.316 \cdot 0.953} = 0.03358; \quad \theta_2 = 0.96642$$

$$\ell_1 = 5(2.5735 - 2.336) - 1.5735 = -0.3860$$
$$\ell_2 = 5(3.8254 - 3.316) - 2.8254 = -0.2784$$
$$\ln \gamma_1^C = \ln \frac{0.03211}{0.047} + 5 \cdot 2.336 \ln \frac{0.03358}{0.03211} - 0.3860$$
$$+ \frac{0.03211}{0.047}(0.047 \cdot 0.3860 + 0.953 \cdot 0.2784) = -0.0505$$

From Table 4.4:

$$a_{1,22} = a_{2,22} = 476.4 \text{ K}; \quad a_{22,1} = a_{22,2} = 26.76 \text{ K}$$
$$a_{1,2} = a_{2,1} = a_{1,1} = a_{2,2} = a_{22,22} = 0 \text{ K}$$

From Equation (4.11):

$$\Psi_{1,22} = \Psi_{2,22} = \exp\{-476.4/307\} = 0.2119$$

$$\Psi_{22,1} = \Psi_{22,2} = \exp\{-26.76/307\} = 0.9165$$

$$\Psi_{1,1} = \Psi_{2,1} = \Psi_{1,2} = \Psi_{2,2} = \Psi_{22,22} = 1$$

For pure acetone (Equation (4.10)):

$$X_1^{(1)} = X_{22}^{(1)} = 0.5 \quad \text{(Group fractions in pure acetone)}$$

$$\Theta_1^{(1)} = \frac{0.848}{0.848 + 1.488} = 0.3630; \quad \Theta_{22}^{(1)} = 0.6370$$

From Equation (4.9):

$$\ln \Gamma_1^{(1)} = 0.848 \left[1 - \ln(0.3630 \cdot 1 + 0.6370 \cdot 0.9165) \right.$$

$$- \frac{0.3630 \cdot 1}{0.3630 + 0.6370 \cdot 0.9165}$$

$$\left. - \frac{0.6370 \cdot 0.2119}{0.3630 \cdot 0.2119 + 0.6370} \right]$$

$$= 0.4089$$

Similarly, $\ln \Gamma_{22}^{(1)} = 0.1389$

For $x_1 = 0.047$ (Equation (4.10)):

$$X_1 = \frac{0.047 \cdot 1 + 0.953 \cdot 2}{0.047 \cdot 2 + 0.953 \cdot 5} = 0.4019; \quad X_2 = 0.5884; \quad X_{22} = 0.0097$$

$$\Theta_1 = 0.5065; \quad \Theta_2 = 0.4721; \quad \Theta_{22} = 0.0214$$

From Equation (4.9):

$$\ln \Gamma_{22} = 1.488 \left[1 - \ln((0.5065+0.4721) \cdot 0.2119 + 0.0214 \cdot 1) \right.$$

$$- \frac{(0.5065 + 0.4721) \cdot 0.9165}{(0.5065+0.4721) + 0.0214 \cdot 0.9165}$$

$$-\frac{0.0214 \cdot 1}{(0.5065+0.4721) \cdot 0.2119 + 0.0214}\Bigg]$$

$$= 2.2067$$

Similarly, $\ln \Gamma_1 = 0.0014$

Then $\ln \gamma_1^R = 1 \cdot (0.0014 - 0.4089) + 1 \cdot (2.2067 - 0.1389)$

$$= 1.6603$$

$\ln \gamma_1 = 1.6603 - 0.0505 = 1.6098$

or $\gamma_1 = 5.00$

This compares with an experimental value of 4.41. (T.C. Lo, H.H. Bieber and A.E. Karr, J.Chem.Eng. Data 7(1962)327).

Appendix 2 lists a computer program, which together with Subroutine UNIFA calculates activity coefficients from UNIFAC following the procedure outlined in Example 4.1. Results from calculation of activity coefficients for a large number of systems are given in Chapters 6, 7, and 8. A few examples illustrating the wide range of applicability of the UNIFAC method are given below.

Figure 4.3 shows good agreement for predicted and calculated activity coefficients for the 2,2,3-Trimethylbutane-Benzene system. This is an example of predicted vapor-liquid equilibria because no branched alkanes were included in the data base for determining the methylene-aromatic hydrocarbon group interaction parameters.

It is shown in detail in Chapter 8 that UNIFAC can predict phase-splitting. An example of this is indicated in Figure 4.4. For design of a liquid-liquid extraction cascade, one needs to know accurately the relative distribution of the solute(s) between the two liquid phases. Since the solute mole fractions are often small, the relative error in the predicted mole fractions - and hence in the solute distribution ratios - may be large. This severe problem is encountered by all methods for calculating activity coefficients, and the UNIFAC model is no exception.

Figure 4.3

ACTIVITY COEFFICIENTS FOR 2,2,3-TRIMETHYL-BUTANE(1)-BENZENE(2) AT 1 ATM.

This system was not included in the data base.

(Data by J.M. Harrison and L. Berg, Ind.Eng.Chem., 38(1946)117)

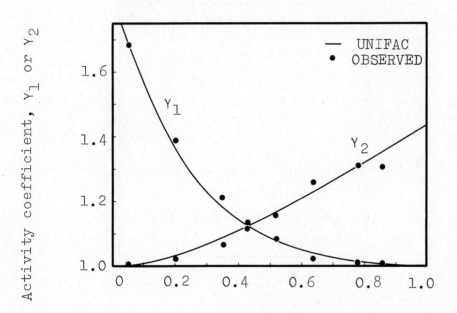

Mole fraction 2,2,3-Trimethylbutane, x_1

Figure 4.4
EXPERIMENTAL AND CALCULATED VAPOR PHASE MOLE FRACTIONS AND
PHASE-SPLIT IN THE SYSTEM HEPTANE(1)-ACETONITRILE(2)
AT 343.2 K
(Data by G. Werner and H. Schuberth,
J.Prakt.Chem., 31(1966)5)

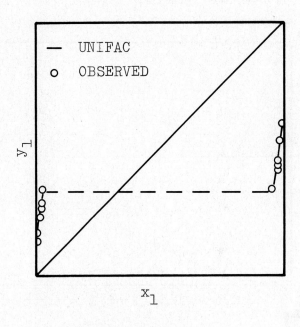

$\overline{\Delta y}$ = 0.0175

No ternary (or higher) systems were included in the data base for determining UNIFAC parameters. As shown in Chapter 7, predicted multicomponent equilibria are in good agreement with experiment. Table 4.5 shows predictions for the system Chloroform-Methanol-Ethyl acetate. This system is unusual, because it exhibits large both positive and negative deviations from Raoult's Law.

4.4 TEMPERATURE DEPENDENCE OF THE UNIFAC PARAMETERS

The various assumptions leading to the solution-of-groups concept in general and the assumptions specific to the UNIFAC method are accounted for in Section 4.1. One of the assumptions belonging to the latter category is that the group-interaction parameters a_{mn} and a_{nm} are independent of temperature. In the ASOG method, the group-interaction parameters are allowed to be temperature-dependent. For example, from Derr and Deal [5] one finds that at 60 $^{\circ}$C the ASOG OH-CH$_2$ group-interaction parameter is -1405 K, while the corresponding parameter at 80 $^{\circ}$C is -1309 K. Thomsen [18] has studied the temperature dependence of the UNIFAC group-interaction parameters, and the results of this work are summarized in this section.

An indication of the temperature dependence of the group-interaction parameters may be obtained by fitting the UNIFAC model separately to different isotherms for the same binary. This is done for 1-Propanol and 2-Propanol with Heptane, for which Van Ness et al. [19] have published highly reliable vapor-liquid equilibrium data at 30, 45, and 60 $^{\circ}$C. The resulting group-interaction parameters are shown in Table 4.6. In this and the following tables, F_{Min} is given by

$$F_{Min} = \sum^{j} \sum^{i} (\ln \gamma_i(\exp) - \ln \gamma_i(UNIFAC))^2_j \qquad (4.12)$$

where the summations are over all components (i) and data points (j) thus including all binary data sets.

Table 4.5

EXPERIMENTAL AND CALCULATED ACTIVITY COEFFICIENTS AT ONE ATMOSPHERE FOR THE SYSTEM CHLOROFORM(1)-METHANOL(2)-ETHYL ACETATE(3)

(Data from I. Nagata, J.Chem.Eng. Data, 7(1962)307)

x_1	x_2	γ_1		γ_2		γ_3	
		exp	UNIFAC	exp	UNIFAC	exp	UNIFAC
0.075	0.077	0.541	0.612	2.32	2.47	0.985	0.999
0.030	0.898	1.82	2.10	1.01	1.01	2.41	2.24
0.608	0.234	1.09	1.08	2.11	2.17	0.570	0.637
0.540	0.075	0.982	0.972	3.50	3.81	0.611	0.588
0.527	0.219	1.00	1.01	2.15	2.26	0.701	0.739

The average absolute deviation between y_i(exp) and y_i(UNIFAC) is 0.011 mole fraction

Table 4.6

UNIFAC GROUP-INTERACTION PARAMETERS AS FUNCTIONS OF TEMPERATURE
FOR 1-PROPANOL AND 2-PROPANOL WITH HEPTANE

Alcohol	Temperature K	$a_{CH_2,CCOH}$	a_{CCOH,CH_2}	Value of objective function F_{Min}
1-Propanol	303.15	813.0	−86.69	0.48
1-Propanol	318.15	787.8	−79.45	0.31
1-Propanol	333.15	753.4	−76.22	0.19
2-Propanol	303.15	764.3	−75.18	0.64
2-Propanol	318.15	731.9	−70.63	0.41
2-Propanol	333.15	695.9	−66.75	0.27

The group-interaction parameters obtained show well-defined trends with respect to varying temperature. On the surface this appears to be a point in favor of including temperature dependence in the parameters. Note, however, that the parameter changes arising from neglecting a temperature change of 30 K appears to be equal to the changes stemming from the solution-of-groups concept itself, i.e. from using the same group-interaction parameters for mixtures with 1-Propanol and 2-Propanol. If the group-contribution concept is valid, these changes are insignificant.

If the data for Heptane with 1-Propanol and 2-Propanol were combined, the difference between the two alcohols would to a great extent "wipe out" the clear temperature-trend of the parameters. If more different alcohols, for example Octanol, were included, the picture would become even more diffuse.

To further test the effect of temperature on the UNIFAC group-interaction parameters, various expressions for the parameters a_{mn} as functions of temperature, for example

$$a_{mn} = p_{mn} + s_{mn}/T$$
$$a_{nm} = p_{nm} + s_{nm}/T$$

(4.13)

were employed. All four parameters were found simultaneously by fitting the model to experimental data (see Chapter 5).

Results for various temperature-functions fitted to the Heptane-1-Propanol data in [19] are shown in Table 4.7.

Table 4.7

FIT OF TEMPERATURE-DEPENDENT GROUP-INTERACTION PARAMETERS TO HEPTANE-1-PROPANOL DATA AT 30, 45, AND 60 °C

m = CH_2 n = CCOH

Parameter expression	p_{mn}	s_{mn}	p_{nm}	s_{nm}	Value of objective function F_{Min}
a = p	785.3	0	-79.27	0	1.11
a = p + s/$T^{\frac{1}{2}}$	785.5	.007	-79.37	.111	0.99
a = p + s/T	196.0	187000.	-45.76	-10700.	0.99
a = p·T	1364.	-1.82	-106.4	.085	0.99
a = p·exp(s/T)	362.4	245.5	-54.68	118.1	0.99

The values of the objective function should be compared with the sum of the objective functions for all three isotherms in Table 4.6, i.e. 0.48 + 0.31 + 0.19 = 0.98. Thus any of the temperature-dependences tested fit the data within the overall capability of the UNIFAC model. It is striking to see, however, that using temperature-independent group-interaction parameters only increases the objective function by 10%. The same is the case when various temperature-dependent group-interaction parameters are fitted to the Heptane-2-Propanol isotherms in [19]. These results are shown in Table 4.8.

The results shown in Tables 4.7 and 4.8 indicate that in reality there is little to be gained by including temperature-dependent group-interaction parameters. Similar conclusions were reached for Ethanol-alkane mixtures and other sub-sets of data. This material is, however, too limited to form the basis for final conclusions regarding the feasibility of including temperature-dependent parameters. The real test is whether or not using temperature-dependent parameters gives rise to significant improvements when the parameters are fitted simultaneously to a large number of different mixtures. This point is illustrated below, for Ether-alkane systems in Table 4.9 and for Ketone-alkane systems in Table 4.10.

Table 4.8

FIT OF TEMPERATURE-DEPENDENT GROUP-INTERACTION PARAMETERS TO HEPTANE-2-PROPANOL DATA AT 30, 45, AND 60 $^{\circ}$C

$m = CH_2 \qquad n = CCOH$

Parameter expression	p_{mn}	s_{mn}	p_{nm}	s_{nm}	Value of objective function F_{Min}
$a = p$	731.2	0	-71.10	0	1.46
$a = p + s/T^{\frac{1}{2}}$	-260.9	17700.	-132.6	1085.	1.33
$a = p + s/T$	1.03	232000.	24.46	-30200.	1.32
$a = p + s \cdot T$	1441.	-2.23	-138.8	.213	1.32
$a = p \cdot \exp(s/T)$	295.6	287.7	-32.84	245.5	1.32

Table 4.9

FIT OF TEMPERATURE-DEPENDENT GROUP-INTERACTION PARAMETERS TO ETHER-ALKANE SYSTEMS

The data base includes the following components:
- Diethyl ether
- Dimethoxyethane
- Heptane
- Cyclohexane
- 1,4-Dioxane
- Dipropyl ether
- Bis(2-methoxyethyl) ether
- Decane
- Hexane
- Octane

$m = CH_2 \qquad n = CH_2O$

Parameter expression	p_{mn}	s_{mn}	p_{nm}	s_{nm}	F_{Min}
$a = p$	231.2	0	112.5	0	0.48
$a = p + s/T^{\frac{1}{2}}$	-703.3	17577.	-28.69	2991.	0.29
$a = p + s \cdot T^{\frac{1}{2}}$	156.3	3.979	1743.	-85.51	0.28
$a = p + s/T$	-321.8	195500.	187.0	-20200.	0.30
$a = p + s \cdot T$	39.64	.539	1136.	-2.821	0.27

Number of data points: 64

Table 4.10

FIT OF TEMPERATURE-DEPENDENT GROUP-INTERACTION PARAMETERS TO KETONE-ALKANE SYSTEMS

Acetone   2-Butanone   3-Pentanone
Pentane   Hexane   Heptane
Octane   Cyclohexane   5-Nonanone

$m = CH_2$ $n = CH_2CO$

Parameter expression	p_{mn}	s_{mn}	p_{nm}	s_{nm}	F_{Min}
$a = p$	497.9	0	11.62	0	0.61
$a = p + s \cdot T^{\frac{1}{2}}$	498.3	-.001	11.18	.003	0.61
$a = p + s \cdot T$	499.0	-.003	10.70	.002	0.61
$a = p \cdot \exp(s/T)$	497.9	-.002	11.58	.003	0.61

Number of data points: 66

For alkane-ketone systems no improvement in the objective function was obtained at all, while for alkane-ether systems the objective function was reduced by a factor of somewhat less than two. These results - and those shown earlier for alkane-alcohol systems - lead to the conclusion that the improvements gained by permitting the group-interaction parameters to vary with temperature are too small to justify the added complexity of the model. It is evident from the Dortmund Data Bank described in Chapter 5 that alkane-alcohol, alkane-ether, and alkane-ketone systems represent cases where relatively many data are available. In the majority of cases, the data would not be of sufficient quantity and/or quality to permit determination of four parameters.

With the present UNIFAC model and the data available in the literature, it appears that the temperature-independent group-interaction parameters (Table 4.4) very nearly give the best possible results. However, as more and better vapor-liquid equilibrium data become available, the parameters in Table 4.4 should be re-evaluated.

The temperature-dependence built into the UNIFAC model can possibly be improved, not by making the present parameters temperature-dependent, but perhaps by including a third parameter, by changing the mathematical expressions, or by some other means. Heat of mixing and liquid-liquid equilibrium data should prove useful in further developing the model.

4.5 ALTERNATIVE UNIFAC PARAMETERS FOR SYSTEMS WITH ALCOHOLS

Predictions of vapor-liquid equilibria using the UNIFAC method with the parameters given in Tables 4.3 and 4.4 are shown in Chapters 6 and 7. From these predictions it is evident that results for systems containing Ethanol are not of the same high standard as the results for systems containing other alcohols. An attempt has been made to improve the predictions for systems with Ethanol in the following manner:

I. The group-interaction parameters a_{mn} and a_{nm} were determined for 1) CCOH with CH_2, 2) CCOH with ACH, and 3) CCOH with H_2O. Mixtures with Ethanol were <u>not</u> included in the data base for these calculations but otherwise the data base is the same as that shown in Appendix 5. The obtained parameters are indicated by dashed lines in Table 4.11.

II. The parameters a_{mn} and a_{nm} indicated by dashed lines in Table 4.11 were now fixed, and the group constants for <u>Ethanol</u>, $R_{CH_3CH_2OH}$ and $Q_{CH_3CH_2OH}$, were calculated by fitting binary vapor-liquid equilibrium data for Ethanol with alkanes, Benzene, and Water to the UNIFAC model. The "optimal" pure-group constants for <u>Ethanol</u> were then found to be

$$R_{CH_3CH_2OH} = 1.9$$
$$Q_{CH_3CH_2OH} = 1.5$$

These values are somewhat different than those obtained according to Bondi (see Table 4.3).

III. Using $R_{CH_3CH_2OH} = 1.9$, $Q_{CH_3CH_2OH} = 1.5$, and for all other groups the R_k- and Q_k-values given in Table 4.3, new group-interaction parameters were obtained for all systems with alcohols (also Ethanol) by fitting the a_{mn} and a_{nm}'s in question to the relevant parts of the data base. This step did not significantly alter the underlined parameters of Table 4.11.

The results of the above procedure are shown as case A in Table 4.11. Case B corresponds to the results obtained in determining the equivalent parameters shown in Table 4.4. The values of the objective function are given in both cases. For alcohols with alkanes and Water case A gives overall much better correlation than case B. However, in all other cases no improvement was obtained; for alcohols with alkyl substituted aromatics ($ACCH_2$), the correlation is even considerably worsened.

Since Table 4.11 does not exhibit clear, overall improvement when the new values of $R_{CH_3CH_2OH}$ and $Q_{CH_3CH_2OH}$ are used, it was decided not to deviate from the Bondi-values of the group constants for Ethanol in the general UNIFAC parameter table. Except where noted, the results in the remainder of the book are based on the parameters given in Tables 4.3 and 4.4. However, for mixtures containing Ethanol with alkanes or Ethanol with Water, it is clearly advantageous to use the parameters given in Table 4.11.

If one maintains the values of R_k and Q_k for Ethanol given in Table 4.3, i.e.

$$R_{CH_3CH_2OH} = 2.1055$$

$$Q_{CH_3CH_2OH} = 1.972$$

the optimal group-interaction parameters for Ethanol-Water and Ethanol-alkane have been found to be

$$a_{CH_3CH_2OH, H_2O} = 15.98 \text{ K} \qquad a_{CH_3CH_2OH, CH_2} = -77.26$$

$$a_{H_2O, CH_3CH_2OH} = 121.2 \text{ K} \qquad a_{CH_2, CH_2CH_2OH} = 559.5$$

For all other interactions, the parameters of Table 4.4 must be used. The above two parameters are found to represent the vapor-liquid equilibria for systems with Ethanol extremely well. The Ethanol-Water azeotrope is, for example, predicted within

Table 4.11. ALTERNATIVE UNIFAC PARAMETERS FOR SYSTEMS WITH ALCOHOLS

$F_{min,A}$ is the objective function for the alternative parameters, $F_{min,B}$ that for the corresponding parameters from Table 4.4

For Ethanol:	$R_k = 1.9$, $Q_k = 1.5$ * (A)			$R_k = 2.1055$, $Q_k = 1.972$ (B)	$\dfrac{F_{min,A}}{F_{min,B}}$
Group m	a_{mn}, K	a_{nm}, K	$F_{min,A}$	$F_{min,B}$	
CH$_2$	755.7	−77.70	3.13	6.19	0.51
C=C	536.4	200.5	0.077	0.059	1.31
ACH	487.9	−56.85	2.28	1.96	1.16
ACCH$_2$	493.5	−14.56	3.72	1.32	2.82
H$_2$O	−191.6	294.7	1.86	7.15	0.26
ACOH	−452.6	252.7	0.10	0.10	0.99
CH$_2$CO	191.1	2.46	2.19	2.12	1.03
COOC	178.4	32.89	0.91	0.82	1.12
CH$_2$O	194.1	−51.13	1.67	1.38	1.20
CNH$_2$	−215.1	101.3	0.003	0.003	1.00
CNH	−237.0	143.7	0.19	0.14	1.41
ACNH$_2$	−246.6	414.9	0.013	0.013	1.00
COOH	−26.59	63.87	0.30	0.27	1.11

In all cases, group n = CCOH

* In other alcohols, R_k and Q_k are those shown in Table 4.3

Table 4.12. CALCULATED AND EXPERIMENTAL ACTIVITY COEFFICIENTS FOR ETHANOL(1)-WATER(2) AT 70 °C USING 3 DIFFERENT SETS OF UNIFAC PARAMETERS

	Experimental Activity coefficients		Activity coefficients predicted from UNIFAC								
			(A) $a_{CH_3CH_2OH,H_2O}$= 285.4 K a_{H_2O,CH_3CH_2OH}=-148.5 K			(B) $a_{CH_3CH_2OH,H_2O}$= 15.98 K a_{H_2O,CH_3CH_2OH}= 121.2 K			(C) $a_{CH_3CH_2OH,H_2O}$= 294.7 K a_{H_2O,CH_3CH_2OH}=-191.6 K		
x_1	γ_1(exp)	γ_2(exp)	γ_1(calc)	γ_2(calc)	%DEV	γ_1(calc)	γ_2(calc)	%DEV	γ_1(calc)	γ_2(calc)	%DEV
0.062	4.210	1.013	2.584	1.007	38	3.847	1.011	8	4.055	1.010	3
0.095	3.516	1.029	2.337	1.016	33	3.285	1.025	6	3.499	1.023	0
0.131	2.963	1.051	2.120	1.029	27	2.820	1.045	4	3.029	1.042	-3
0.194	2.317	1.102	1.835	1.057	17	2.252	1.091	2	2.437	1.086	-7
0.252	1.939	1.160	1.645	1.091	10	1.904	1.145	1	2.064	1.139	-8
0.334	1.599	1.255	1.452	1.141	1	1.580	1.236	0	1.706	1.233	-9
0.401	1.419	1.345	1.338	1.205	-5	1.405	1.323	-1	1.507	1.324	-8
0.593	1.138	1.666	1.081	1.414	-12	1.134	1.628	-2	1.185	1.674	-4
0.680	1.073	1.846	1.034	1.540	-16	1.073	1.791	-3	1.108	1.884	-1
0.793	1.024	2.104	1.028	1.747	-21	1.027	2.026	-4	1.043	2.226	4
0.810	1.019	2.143	1.003	1.784	-18	1.022	2.063	-2	1.036	2.286	5
0.943	1.001	2.413	1.002	2.135	-13	1.002	2.373	-2	1.003	2.878	16
0.947	1.001	2.418	1.000	2.148	-12	1.001	2.383	-1	1.003	2.900	17
					Avg: 17			Avg: 3			Avg: 7

$$\%DEV = 100 \left[\left(\frac{\gamma_1}{\gamma_2}\right)_{exp} - \left(\frac{\gamma_1}{\gamma_2}\right)_{calc} \right] / \left(\frac{\gamma_1}{\gamma_2}\right)_{exp}$$

(A) and (B): $R_{CH_3CH_2OH}$ and $Q_{CH_3CH_2OH}$ from Table 4.3; (C): $R_{CH_3CH_2OH}$ = 1.9; $Q_{CH_3CH_2OH}$ = 1.5

(A), (B), and (C): R_{H_2O} and Q_{H_2O} from Table 4.3

experimental uncertainty. Since no other group-interaction parameters depend on the above four parameters, they may be used directly in practical computations without any further complications. For Propanol, Butanol, etc. with Water or alkanes, however, the parameter values in Table 4.4 must be used.

Predicted activity coefficients for Ethanol-Water using different UNIFAC parameters are shown in Table 4.12. Case A corresponds to using the UNIFAC parameters of Tables 4.3 and 4.4; case B corresponds to using $R_{CH_3CH_2OH}$ and $Q_{CH_3CH_2OH}$ from Table 4.3 and the group-interaction parameters for Ethanol; case C corresponds to using the $R_{CH_3CH_2OH}$, $Q_{CH_3CH_2OH}$, and group-interaction parameters of Table 4.11. Clearly, case B gives much the best results. Similar comparisons for hydrocarbon-Ethanol systems are shown in Chapter 6.

4.6 ADDITIONAL UNIFAC PARAMETERS

Group-interaction parameters have been estimated for some mixtures with tertiary amines [18]. In the literature, there are only few data available for systems with tertiary amines. The resulting parameters are given in Table 4.13.

Table 4.13

UNIFAC PARAMETERS FOR SYSTEMS WITH TERTIARY AMINES

Group k	R_k	Q_k	Example	
CH_3N	1.1865	0.940	Trimethyl amine: $1CH_3N$, $2CH_3$	
CH_2N	0.9597	0.632	Triethyl amine: $1CH_2N$, $3CH_3$, $2CH_2$	
$Group_m$	$Group_n$	a_{mn}, K	a_{nm}, K	The interaction parameters for CH_3N are the same as those for CH_2N
CH_2N	CH_2	118.9	70.16	
CH_2N	$C=C$	106.8	126.7	
CH_2N	ACH	-3.87	-6.98	

REFERENCES

1. I. Langmuir, The Distribution and Orientation of Molecules, Third Colloid Symposium Monograph, The Chemical Catalog Company, Inc., New York, 1925.
2. O. Redlich, E.L. Derr and G. Pierotti, J.Am.Chem.Soc., 81(1959)2283.
3. E.L. Derr and M. Papadopoulos, J.Am.Chem.Soc., 81(1959) 2285.
4. G.M. Wilson and C.H. Deal, Ind.Eng.Chem., Fundamentals, 1(1962)20.
5. E.L. Derr and C.H. Deal, J.Chem.Eng.Symp.Ser. No. 32 (Instn.Chem.Engrs., London), 3:40(1969).
6. E.L. Derr and C.H. Deal, Adv. in Chemistry Series 124, Am.Chem.Soc., (1973)11.
7. K. Tochigi and K. Kopina, J.Chem.Eng. Japan, 9(1976)267.
8. P.J. Flory, J.Chem.Phys., 9(1941)660; 10(1942)51, and M.L. Huggins, J.Chem.Phys., 9(1941)440; Ann. N.Y.Acad.Sci., 431(1942)1.
9. M. Ronc and G.A. Ratcliff, Can.J.Chem.Eng., 49(1971)875.
10. M. Ronc and G.A. Ratcliff, Can.J.Chem.Eng., 53(1975)329.
11. J.N. Brønsted and J. Koefoed, Danske Videnskab.Selskab, Mat.Fys.Medd., 17(1946)1.
12. Aage Fredenslund, R.L. Jones and J.M. Prausnitz, AIChE Journal, 21(1975)1086.
13. Aage Fredenslund, J. Gmehling, M.L. Michelsen, P. Rasmussen and J.M. Prausnitz, Ind.Eng.Chem., Process Design and Development, in press (1977).
14. D.S. Abrams and J.M. Prausnitz, AIChE Journal, 21(1975) 116.
15. A. Bondi, Physical Properties of Molecular Crystals, Liquids, and Glasses, Wiley, New York, 1968.
16. G.J. Pierotti, C.H. Deal and E.L. Derr, Ind.Eng.Chem., 51(1959)95.
17. D.P. Tassios, Adv. in Chem.Ser. 115, Am.Chem.Soc.,(1972)46.
18. K. Thomsen, M.Sc. Thesis, Instituttet for Kemiteknik, The technical Univ. of Denmark, 1977. (In Danish)
19. H.C. Van Ness, C.A. Soczec, G.L. Peloquin and R.L. Machado, J.Chem.Eng. Data, 12(1967)217.

CHAPTER 5

DETERMINATION OF UNIFAC PARAMETERS

The UNIFAC model and the models described in Chapter 3 contain parameters which must be calculated using experimental phase equilibrium data.

Vapor-liquid equilibrium data for multicomponent systems do not enter into the data base for calculating UNIFAC parameters; hence in this chapter we limit our discussion to binary systems.

The number of publications on experimental vapor-liquid equilibrium for binary systems is tremendous, at least 6000. The table of UNIFAC group-interaction parameters is based on vapor-liquid equilibrium data for roughly 2500 binary isotherms or isobars. The data were chosen from the Dortmund data bank described in Section 5.1 and supplemented, where appropriate, with data from Wichterle et al. [1].

The success of the UNIFAC method rests on the availability of a large number of reliable group-interaction parameters. The values of the parameters are determined from binary vapor-liquid equilibrium data and these data must therefore be of high quality. The experimental data for binary systems used in establishing UNIFAC parameters have been consistency tested as described in Section 5.2.

For a few parameter estimations it has been necessary to rely on liquid-liquid data as explained in Section 5.3.

The procedure for estimation of the UNIFAC parameters is outlined in Section 5.4.

5.1 THE DORTMUND DATA BANK

References to data used in obtaining the UNIFAC parameters are listed in Appendix 5. For each pair of group-interactions we list the temperature range and the literature references on which the parameters are based. In all cases, the data used were what we – on the basis of the consistency test described in Section 5.2 and experience – judged "best". This does not mean, however, that all of the data referred to meet the consistency requirement. In few cases where only limited data are available, it was necessary to include data which were not found consistent.

The data referred to in Appendix 5 are included in "The Dortmund Data Bank" (DDB), see [2]. The stated literature reference numbers correspond to those used in DDB.

The Dortmund Data Bank contains at the present more than 5000 isotherms/isobars with more than 85000 sets of x-y-P-T datapoints. The data have been collected from the open literature, and presently 1500 references are included. By the end of 1977, DDB is planned to contain all of the vapor-liquid equilibrium data available in the open literature for binary and multicomponent systems for which

1) the system pressure is less than 15 atm

2) the components are Water or organic components with a normal boiling point higher than 0 $^\circ$C.

A limited amount of binary liquid-liquid equilibrium data were included in connection with the development of the UNIFAC parameter table.

The x-y-P-T data are stored on a magnetic tape together with the following additional information: number of components; number of datapoints; type of data (isothermal, isobaric ...); literature reference number; and component identification numbers. In addition, the actual literature references and the component names are stored on a disc, which also contains extra pure-component information, for example Antoine constants.

By means of external Fortran programs, it is now relatively easy to seek out specific vapor-liquid equilibrium data from the file and to read these data into another program. Sorting out the data normally requires about 20 CPU-sec. One may also seek, for example, all data for a specific binary and have the results printed out. An example of such a printout is shown

Figure 5.1

SAMPLE PRINTOUT FROM THE DORTMUND DATA BANK

SYSTEM:
1) TETRACHLORKOHLENSTOFF   CCL4
2) BENZOL   C6H6

DRUCK :   760.00   TORR

LIT.: HAUGHTON C.O., BRIT.CHEM.ENG.10,237(1965).

X1	Y1	TEMPERATUR GRAD C
0.00900	0.01100	80.30
0.03400	0.04200	80.10
0.07400	0.08800	79.80
0.11100	0.13100	79.60
0.30000	0.32900	78.60
0.40200	0.43200	78.30
0.48400	0.51200	77.90
0.56500	0.58500	77.60
0.65200	0.66800	77.60
0.72800	0.73900	77.40
0.77200	0.77900	77.20
0.90600	0.90700	77.10
0.94500	0.94500	77.00

in Figure 5.1. The data in DDB have been fitted to various models such as the Wilson equation, NRTL and UNIQUAC; the results are shown in [2], which consists of 8 volumes.

The system used in building up DDB has the advantage that new data may be readily added. Using DDB has, naturally, saved much time and effort in developing the UNIFAC method. Furthermore, punching and reading-errors are to a large extent avoided, since the data in DDB have been several times checked.

5.2 THERMODYNAMIC CONSISTENCY

Binary vapor-liquid equilibrium data are often reported as the composition of the liquid phase either at constant pressure as a function of temperature or at constant temperature as a function of pressure. According to the phase rule, this information is enough to completely characterize the system and, using the tools of classical thermodynamics, the vapor phase compositions may be calculated as shown below.

If the experimental data consist of pressure, temperature, and the composition of both phases, the extra experimental information may be used to test the data for thermodynamic consistency as shown in many textbooks, for example [3]. However, Van Ness et al. [4] have shown that the commonly applied area test is of limited value. In this test, the pressure is not accounted for at all, and errors in composition tend to cancel out. In general, the most reliable consistency test consists of calculating y from P-x (or T-x) data and then comparing the calculated y's with the experimentally obtained values. For thermodynamically consistent data, the difference between y(exp) and y(calc) will be small.

The starting point of the consistency test used in this work is the fundamental relationship given as Equation (5.1).

For an open, homogeneous system, the total Gibbs free energy depends on temperature, pressure, and the number of moles of each component, n_i:

$$d(n_T G) = n_T v dP - n_T S dT + \sum^{i} \bar{G}_i dn_i \qquad i = 1, 2 \ldots M \qquad (5.1)$$

where G is the molar Gibbs free energy, n_i the number of moles of component i, v the molar volume, S the molar entropy, and $\bar{G}_i$ the partial molar Gibbs free energy of component i in the system. By definition,

$$n_T = \sum^i n_i \quad \text{and} \quad \bar{G}_i = RT \ln \gamma_i \qquad i = 1, 2 \ldots M \qquad (5.2)$$

When equation (5.1) is applied to a system where two phases are in equilibrium, the non-isobaric, non-isothermal Gibbs-Duhem equation results (v^E and H^E are excess volume and enthalpy):

$$\sum^i x_i \, d\ln \gamma_i - \frac{v^E}{RT} dP + \frac{H^E}{RT^2} dT = 0 \qquad i = 1, 2 \ldots M \qquad (5.3)$$

In the following, this equation is applied to isothermal and to isobaric binary vapor-liquid equilibrium data.

Isothermal Data

At the relatively low pressures considered in this work, it is reasonable to neglect the term $v^E dP/RT$. For isothermal data, Equation (5.3) then becomes

$$\sum^i x_i \, d\ln \gamma_i = 0 \qquad i = 1, 2 \ldots M \qquad (5.4)$$

Applying this equation and Equation (5.5) to a binary system

$$P = \sum^i y_i P = \sum^i x_i \gamma_i f_i^\circ / \varphi_i \qquad i = 1, 2 \qquad (5.5)$$

one obtains

$$P = x_1 P_1^S \frac{\varphi_1^S}{\varphi_1} \exp\left\{ g + x_2 g' + \frac{v_1(P-P_1^S)}{RT} \right\}$$
$$+ x_2 P_2^S \frac{\varphi_2^S}{\varphi_2} \exp\left\{ g - x_1 g' + \frac{v_2(P-P_2^S)}{RT} \right\} \qquad (5.6)$$

$$g \equiv G^E/RT \quad ; \quad g' = (dg/dx_1)_\sigma \quad ; \quad g = 0 \text{ for } x_1 \text{ and } x_2 = 0$$

$$\left.\begin{array}{l} \ell n\, \gamma_1 = g + x_2 g' \\ \ell n\, \gamma_2 = g - x_1 g' \end{array}\right\} \tag{5.7}$$

The subscript σ denotes "along the saturation line".

In addition to neglecting the term $v^E dP/RT$, it is in the derivation of Equation (5.6) assumed that the pure-component liquid volumes, v_1 and v_2, are incompressible over the pressure range in question. We shall furthermore assume that the fugacities of the pure components at saturation and the fugacities of the components in the vapor mixture (φ_i^S and φ_i) may be calculated using the volume-explicit virial equation terminated after the second virial coefficient (see Chapter 2).

If Equation (5.6) can be solved for $g(x_1)$, and hence for $g'(x_1)$, then the activity coefficients follow from (5.7), and the vapor phase mole fractions are calculated from

$$y_i(\text{calc}) = x_i \gamma_i f_i^\circ / P\varphi_i \tag{5.8}$$

An iterative procedure is necessary, since φ_i in Equations (5.6) and (5.8) depend upon $y_i(\text{calc})$.

We choose to represent $g(x_1)$ by the highly flexible Legendre polynomials:

$$g = \frac{G^E}{RT} = x_1(1-x_1) \sum^k a_k L_k(x_1) \qquad k = 0,1 \ldots n$$

$$L_k(x_1) = \{(2k-1)(2x_1-1)L_{k-1}(x_1) - (k-1)L_{k-2}(x_1)\}/k \tag{5.9}$$

$$L_0(x_1) = 1 \quad ; \quad L_1(x_1) = 2x_1 - 1$$

Here n is the polynomial order used, often maximum five or six. There are now two different routes for solving Equation (5.6), i.e. for obtaining those coefficients a_k, $k = 0, \ldots n$, which best represent the P-x_i data set under consideration. One has to realize that Equation (5.6) is a first-order, ordinary, non-linear differential equation in $g(x_1)$. As discussed in

[5], this equation may then be solved using the technique of orthogonal collocation. The problem of solving the differential equation is then reduced to solving n nonlinear algebraic equations in n unknowns, i.e. the n a_k's.

The other route is to insert Equation (5.9) directly into Equation (5.6) and to find the n a_k's by a least squares regression technique, where the sum of the squared differences between P(exp) and P(calc) is minimized. This is equivalent to Barker's method [6]. Whichever of the two routes is chosen is only a secondary question. The principal question is to obtain the desired a_k's, and the best suited method should be used. For high-order approximations the method of orthogonal collocation appears to be advantageous, for low-order polynomials the method of least squares is preferable. The latter route is chosen in this work.

Although the Legendre polynomials were chosen purely from the standpoint of numerical expediency, it turns out that they are quite similar to the Redlich-Kister expansion [7]:

$$G^E/RT = g = x_1(1-x_1) \sum^k A_k(2x_1-1)^k \qquad k = 0,1 \ldots n \qquad (5.10)$$

Expressions for $L_k(x_1)$ and $(2x_1-1)^k$ for discrete values of k are given in Table 5.1. The similarity between the two expansions is striking. The constants in front of the 2'nd to 4'th order polynomials can be absorbed in the a_k and A_k's.

Table 5.1

EXPRESSIONS FOR LEGENDRE POLYNOMIALS AND THE REDLICH-KISTER EXPANSION

Polynomial order, k	$L_k(x)$, see Equation (5.9)	$(2x-1)^k$, see Equation (5.10)
0	1	1
1	$2x-1$	$2x-1$
2	$6(x^2 - x + \frac{1}{6})$	$4(x^2 - x + \frac{1}{4})$
3	$20(x^3 - \frac{3}{2}x^2 + \frac{3}{5}x - \frac{1}{20})$	$8(x^3 - \frac{3}{2}x^2 + \frac{3}{4}x - \frac{1}{8})$
4	$70(x^4 - 2x^3 + \frac{9}{7}x^2 - \frac{2}{7}x + \frac{1}{70})$	$16(x^4 - 2x^3 + \frac{3}{2}x^2 - \frac{1}{2}x + \frac{1}{16})$

The result of the calculations as outlined above is a set of calculated vapor phase mole fractions and liquid phase activity coefficients which, when fitted to the isothermal P-x data, satisfy the Gibbs-Duhem equation. Observing the cautions listed below, we consider the P-T-x-y data set consistent if the average absolute deviation between $y_i(exp)$ and $y_i(calc)$ is less than 0.01. The choice of 0.01 is arbitrary - in most cases this corresponds to a "reasonable" value for the sum of errors in the measured vapor and liquid phase mole fractions. No rule of this type is, however, generally applicable, and the following four observations are in order:

(1) A positive result of the consistency test for a first-order Legendre polynomial may be fortuitous.

(2) If the pure-component vapor pressures differ greatly, the vapor phase mole fraction of the more volatile component is nearly unity over the whole concentration range, and the deviation between $y_i(exp)$ and $y_i(calc)$ is then automatically less than 0.01. In this case, the most important condition for consistency is the quality of the fit to the total pressure.

(3) For mixtures not containing organic acids a Legendre polynomial of order 3 or 4 will normally give a good fit to the total pressure and at the same time give the lowest possible mean deviation between the experimental and calculated vapor mole fractions. Because of numerical reasons a further increase in the degree of the polynomial will tend to increase the mean deviation of the vapor mole fractions.

For mixtures containing organic acids it is normally necessary to use a higher order polynomial because the strong nonidealities in the vapor phase result in great variations in the fugacity coefficients.

(4) For mixtures containing an organic acid and a component X we have chosen to calculate the fugacity coefficients in the whole concentration range according to the chemical theory described in Chapter 2. This means that the fuga-

city coefficient for the <u>pure</u> component X will be calculated as:

$$\varphi_X = \exp[\frac{B_{free}P}{RT}] \text{ instead of}$$

$$\varphi_X = \exp[\frac{B_{total}P}{RT}]$$

It turns out that by so doing we get a faster convergence without affecting the end result significantly.

Isobaric Data

For isobaric conditions, the term $v^E dP/RT$ in Equation (5.3) vanishes. The term $H^E dT/RT^2$, is in some cases not negligible. Since experimental excess enthalpy data for systems considered in this work are scarce and no reliable methods for their prediction exist, it was chosen not to implement a procedure for isobaric data completely analogous to that used for isothermal data. Unfortunately, much of the data reported in the literature are isobaric. To deal with these, we have established an empirical procedure for testing isobaric data for consistency.

The procedure is based on first neglecting the term $H^E dT/RT^2$. In this case, Equations (5.4 - 9) still hold, and Legendre polynomials for $g(x_1)$ are fitted to the $P-T-x_i$ data via Equation (5.6) in exactly the same manner as before. The result is a set of values for $y_i(calc)$ corresponding to the experimental $T-x_i$ values. On the basis of experience - and with the same cautions as stated earlier - we consider isobaric vapor-liquid equilibrium data consistent, if the average difference between $y_i(calc)$ and $y_i(exp)$ is less than 0.01.

When $H^E dT/RT^2$ is neglected the activity coefficients calculated using Equation (5.7) are not quite correct. Therefore, when isobaric data are found consistent, the "experimental" activity coefficients are calculated from

$$\gamma_i(exp) = y_i(exp)\varphi_i P(exp)/x_i(exp)f_i^o \qquad (5.11)$$

This procedure has provided a practicable means of screening binary, isobaric vapor-liquid equilibrium data.

From a theoretical point of view, it might be more satisfactory to express H^E in terms of a chosen model for $g=G^E/RT$ using the Gibbs-Helmholz equation:

$$\left(\frac{\partial g}{\partial T}\right)_{P,x} = -H^E/RT^2 \qquad (5.12)$$

If the chosen expression for g (unlike the Legendre polynomials) is temperature-dependent, this would allow for the inclusion of the term $H^E dT/RT$ in Equations (5.4) and (5.6). A procedure completely analogous to that for isothermal data could then be developed. However, in practice it often proves difficult to simultaneously represent G^E and H^E using the same model with the same parameters. Since most often accurate prediction of equilibrium compositions (and hence G^E) is of much higher priority than accurate prediction of excess enthalpies, this procedure was not chosen.

Appendix 1 gives the current version of our consistency test program. If the data are found consistent, the resulting activity coefficients are incorporated in the data base for calculating UNIFAC parameters (see Section 5.4).

The following figures show examples of tested binary vapor-liquid equilibrium data. The isothermal data presented in Figures 5.2 and 5.3 have been accepted as thermodynamically consistent. The isobaric data on Figure 5.4 are also consistent, while the isothermal data on Figure 5.5 are discarded as not consistent.

5.3. CALCULATION OF ACTIVITY COEFFICIENTS FROM MUTUAL SOLUBILITY DATA

In some cases - for example in the case of alkane-water mixtures - it is necessary to calculate experimental activity coefficients from mutual solubility data, i.e. from liquid-liquid equilibrium compositions. Then the condition for phase equilibrium for a binary mixture becomes

$$x_1^I \gamma_1^I = x_1^{II} \gamma_1^{II}$$
$$x_2^I \gamma_2^I = x_2^{II} \gamma_2^{II} \qquad (5.13)$$

Figure 5.2
2-BUTANONE(1)-ACETIC ACID(2)
AT 351.15 K

(LYNGBY, 1976)

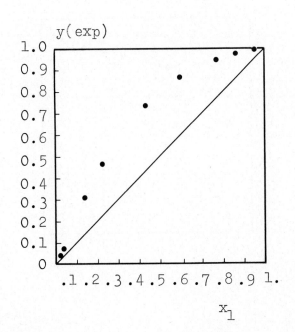

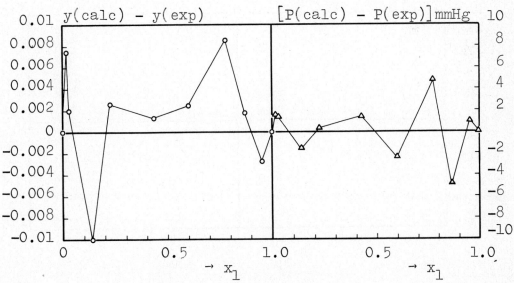

Number of data points: 11
Degree of Legendre polynomial: 5

Mean absolute deviation in:
Vapor mole fraction y: 0.0043
Pressure P: 2.15 mm Hg

Figure 5.3
BENZENE(1)-2-BUTANOL(2)
AT 318.15 K
(J.Chem.Thermo., 1(1969)273)

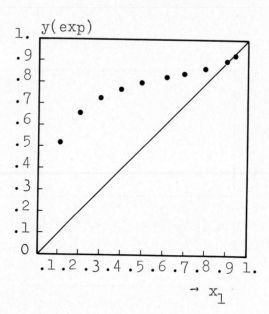

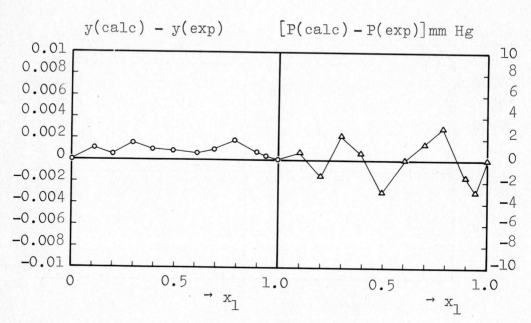

Number of data points: 12
Degree of Legendre polynomial: 5

Mean absolute deviation in:
Vapor mole fraction: 0.001
Pressure P: 0.18 mm Hg

Figure 5.4
BENZENE(1)-2-PROPANOL(2)
AT 500 mm Hg
(J.Chem.Eng. Data 10(1965)106)

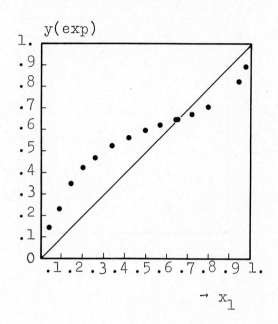

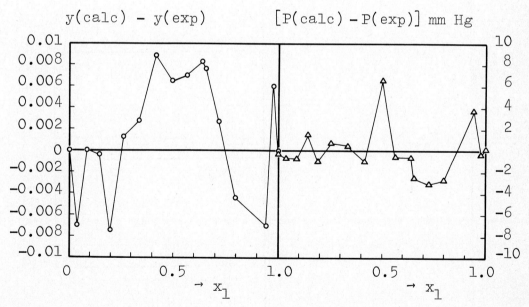

Number of data points: 17
Degree of Legendre polynomial: 3

Mean absolute deviation in:
Vapor mole fraction y: 0.0051
Pressure P: 1.82 mm Hg

Figure 5.5
TETRACHLOROMETHANE(1)-2-PROPANOL(2)
AT 338 K

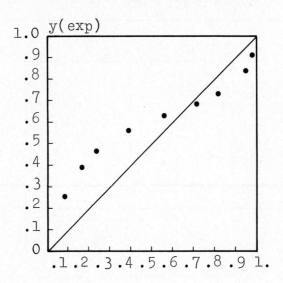

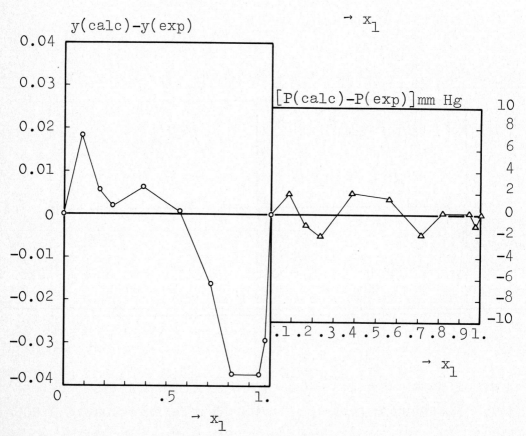

Number of data points: 11
Degree of Legendre polynomial: 3

Mean absolute deviation in:
Vapor mole fraction y: 0.017
Pressure P: 1.06 mm Hg

where the superscripts refer to phase I and phase II.

For very small mutual solubilities (in our case 5×10^{-4} mole fraction or less) the activity is assumed equal to unity.

$$x_i \gamma_i \approx 1 \tag{5.14}$$

The justification for this approximation is that the activity of component i in the phase rich in component i is very nearly unity. At equilibrium, the activity of component i in the phase poor in component i must also be unity. Thus

$$\gamma_i(\exp) \approx 1/x_i(\exp) \tag{5.15}$$

If the mutual solubilities are larger than 5×10^{-4} mole fraction, then the activity coefficients γ_1^I, γ_1^{II}, γ_2^I and γ_2^{II} are written out in terms of the UNIQUAC model.

One then obtains

$$x_1^I(\exp) F_{UNIQUAC}(x_1^I, T) = x_1^{II}(\exp) F_{UNIQUAC}(x_1^{II}, T)$$

$$\tag{5.16}$$

$$x_2^I(\exp) F_{UNIQUAC}(x_2^I, T) = x_2^{II}(\exp) F_{UNIQUAC}(x_2^{II}, T)$$

The function $F_{UNIQUAC}$ corresponds to equations (3.7 - 9) and contains two parameters, which may be calculated using an iterative technique.

The obtained UNIQUAC parameters are used in conjunction with the experimental mole fractions at the solubility limits to calculate $\gamma_1(\exp)$ and $\gamma_2(\exp)$ in phases I and II by back-substitution in Equations (3.7 - 9).

5.4 ESTIMATION OF UNIFAC PARAMETERS

In this section we describe how the UNIFAC model is fitted to the "experimental" activity coefficients. Fabries and Renon [8] discuss the possibility of eliminating the intermediate step of calculating experimental activity coefficients and instead fitting the chosen model for G^E directly to the isothermal or isobaric P-T-x-y data. At this point such an approach does not, however, appear sufficiently developed to be applied to group-contribution models.

First we will briefly discuss how the "experimental" activity coefficients are calculated:

If an isothermal set of data is accepted as consistent, the activity coefficients calculated through the consistency test (Equation 5.7) are used as the "experimental" values. The activity coefficients are those calculated for an order of the Legendre polynomial that gives a good fit to the total pressure and at the same time gives the lowest deviation between experimental and calculated vapor mole fractions (see Section 5.2). The experimental vapor mole fractions have thus only been used as a reference with which the calculated vapor mole fractions (5.8) are compared. The "experimental" activity coefficients are consistent and functions only of liquid phase mole fractions and temperature.

For a consistent isobaric set of data we have as already explained used Equation (5.11) for calculating the experimental activity coefficients.

By this procedure the experimental isothermal or isobaric P-T-x-y data for a binary mixture are converted to a set of experimental T-x_i-γ_i values.

As explained in Chapter 4 the UNIFAC model predicts the activity coefficients γ_i for given values of the temperature T, the mole fractions x_i, group sizes and surface areas R_K and Q_K, and group-interaction parameters a_{ij}. The values of R_K and Q_K are calculated from van der Waals group volumes and surface areas while the a_{ij}'s have to be estimated from experimental data. The problem of fitting the UNIFAC model to experimental activity coefficients is thus identical to finding the best values of a_{ij}, i.e. the parameters that will predict activity coefficients which are as close as possible to the experimentel values.

The objective function for this fitting procedure may be defined in many ways. We have tried different functions, and we have chosen to estimate the parameters that will minimize the following expression:

$$F = \sum^{i} \sum^{j} [\ln \gamma_i(\exp) - \ln \gamma_i(\text{UNIFAC})]_j^2 \tag{5.17}$$

where the summations are over all components (i) and data points (j) thus including all binary data sets.

We find that by using the stated objective function one will in general get a good fit to the experimental activity coefficients, both to large values and to values close to unity.

It has to be recalled that although only binary mixtures are used, we may have several different mixtures for estimating one pair of group-interaction parameters. For example, in the estimation of parameters for the interaction between the ACH group and the CCOH group, the binary mixtures included were those containing benzene and one of the following alcohols: Ethanol, 1-Propanol, 2-Propanol, 1-Butanol and 2-Butanol.

The group-interaction parameters have been estimated one pair at a time. Based on the experimental activity coefficients for mixtures containing groups i and j we have found the pair of parameters a_{ij} and a_{ji} that best fit the data. It must be recalled that in order to estimate, for example, the CCOH, CH_2CO group-interaction parameters from alcohol-ketone data it is necessary to have available the CCOH, CH_2 group-interaction parameters from alcohol-alkane data and the CH_2, CH_2CO parameters from alkane-ketone data. If all three pairs of parameters had been estimated at the same time one might have obtained a good fit to the experimental alkane-alcohol-ketone data but the parameters for the CCOH, CH_2 interaction thus estimated may not be very good when used for estimating other interaction parameters, for example those describing interaction between groups CCOH and CH_2O.

Naturally, this would not be a problem if all of the parameters were fitted at the same time. We have however refrained from doing so partly because the amount of data does not justify the implementation of a parameter estimation procedure of this magnitude and partly because it would make it more difficult to redetermine some of the parameters or to extend the parameter table.

The objective function F (Equation 5.17) is minimized by means of a sequential search procedure developed by Nelder and Mead [9]. This method is an extension of the simplex method introduced by Spendley et al. in 1962 [10] and can, for the two-parameter case, be pictured as the movement of an amoeba that slides over the surface of the objective function trying to find the deepest place to rest. During its movement it will contract and expand, thereby changing its form so as to follow the contours of the surface and to reach its goal as fast as possible.

The basic principles for the estimation of n parameters are outlined below and the estimation program is given in Appendix 3.

Step 1

The objective function F is considered to be a function of n independent variables, and F is evaluated at n + 1 points $(P_0, P_1, P_2 \ldots P_n)$ in the space of the variables. For n = 2 this means that F is calculated at the three vertices of a triangle, the "simplex".

Experience has shown that the necessary computer time for the parameter estimation is not influenced much by the chosen initial values of the parameters. The vertices on the initial simplex must on the other hand not be too far from each other or too close. We have found that it is normally a good compromise to build the initial simplex by changing each of the initial parameters one at a time by 10%. For a case with two parameters for which the initial guess is X,Y the vertices of the initial simplex would be (X, Y), (1.1·X, Y), (X, 1.1·Y).

Step 2

The values of F are calculated at all the vertices. The vertex P_h at which F takes the largest value F_h is reflected through the centroid $\bar{P}$ of the remaining vertices to give a trial point P^*, and its coordinates are defined by the relation:

$$P^* = \bar{P} + \alpha(\bar{P} - P_h) \tag{5.18}$$

Step 3

If the value F^* at P^* proves to be a new minimum, go to step 4. If a comparison of F^* with all other values of F except F_h shows that F^* is not the largest, replace P_h by P^* and go to step 6. If F^* is larger than all other values of F except F_h, replace P_h with P^* and go to step 5. If F^* also is larger than F_h, go directly to step 5.

Step 4

A new trial point P^{**} is defined by expanding the simplex by the relation:

$$P^{**} = \bar{P} + \gamma(P^* - \bar{P}) \tag{5.19}$$

If F^{**} proves to be a new minimum, replace P_h by P^{**} and go to step 6, otherwise replace P_h by P^* and go to step 6.

Step 5

A new trial point P^{**} is attempted by contraction of the simplex by the relation:

$$P^{**} = \bar{P} - \beta(\bar{P} - P_h) \tag{5.20}$$

If the value F^{**} at P^{**} is smaller than F_h, replace P_h by P^{**} and go to step 6. If F^{**} is larger than F_h, the contraction has failed. Then all the points are moved closer to the point P_ℓ at which F is lowest by the relation:

$$P_i = (P_i + P_\ell)/2 \qquad i = 0,1 \ldots n \tag{5.21}$$

After this go to step 6.

Step 6

At this step it is checked whether or not the minimum of F has been reached. The criterion chosen is to compare the "standard deviation" of the F's with a pre-set value. The standard deviation SD is defined by the relation:

$$SD^2 = \sum^{i} (F_i - \bar{F})^2/n \qquad i = 0,1 \ldots n \qquad (5.22)$$

In this expression F_i are the values of the objective function at the $n+1$ vertices P_0, P_1, ... P_n, and $\bar{F}$ is the average value of the F_i's.

If the value of SD is larger than the pre-set value go back to step 2, otherwise the procedure has converged and the best parameters are taken as the values of the coordinates at P_ℓ at which F is lowest.

The pre-set value with which we compare SD is equal to 10^{-6}. The values of α, β and γ are 1, 0.5 and 2 and identical to the values recommended by Nelder and Mead [9].

The Nelder-Mead procedure for parameter estimation has turned out to be advantageous compared with other methods that we have tried.

It is not necessary to estimate first or higher order derivatives of the objective function.

It is easy to follow the changes in F as the minimum is approached.

The method is highly reliable and we have never had any problems with convergence. It is difficult to say anything definite about the necessary computer time for the parameter estimation, because it will depend so much on the number of data points and the specific group-interactions. An indication of the speed is that the program goes through step 6 roughly 30-40 times demanding the calculation of F about twice the number of times. The CPU time on an IBM 370/165 will normally be less than 18 seconds for the estimation of a pair of parameters in the UNIFAC model.

Before we have accepted a pair of parameters, we have tried different initial guesses to be sure that we did not have false convergence at a point other than the minimum. The different initial guesses of the parameters may be quite scattered and still give the same final parameters. Two typical initial guesses would be (100., 100.) and (-100., -100.).

REFERENCES

1. I. Wichterle, J. Linek and E. Hála, Vapor-Liquid Equilibrium Data Bibliography, Elsevier, Amsterdam, 1973; Supplement I, 1976.
2. J. Gmehling and U. Onken, Vapor-Liquid Equilibrium Data Collection, DECHEMA Chemistry Data Series, Frankfurt, 1977.
3. H.C. Van Ness, Classical Thermodynamics of Nonelectrolyte Solutions, The Macmillan Company, New York, 1964.
4. H.C. Van Ness, S.M. Byer and R.E. Gibbs, AIChE Journal, 19(1973)238.
5. L.J. Christiansen and Aage Fredenslund, AIChE Journal, 21(1975)49.
6. J.A. Barker, Australian J.Chem., 6(1953)207.
7. O. Redlich and A.T. Kister, J.Chem.Phys., 15(1947)849.
8. J.F. Fabries and H. Renon, AIChE Journal, 21(1975)735.
9. J.A. Nelder and R. Mead, Computer Journal, 7(1965)308.
10. W. Spendley, G.R. Hext and F.R. Himsworth, Technometrics 4(1962)441.

CHAPTER 6

PREDICTION OF VAPOR-LIQUID EQUILIBRIA IN BINARY SYSTEMS

The purpose of this chapter is to give the reader an indication of the accuracy of the predictions when UNIFAC is applied to binary vapor-liquid equilibria. We choose to do so by showing extensive examples of predicted activity coefficients, vapor phase mole fractions, and temperatures or pressures. These are compared with the corresponding experimental values. In this chapter,

$$\Delta P = \frac{1}{N} \sum^{\ell} |P(UNIFAC) - P(exp)| \text{ mm Hg}$$

$$\Delta T = \frac{1}{N} \sum^{\ell} |T(UNIFAC) - T(exp)| \text{ K}$$

$$\Delta y = \frac{1}{2N} \sum^{\ell} \sum^{k} |y_i(UNIFAC) - y_i(exp)| \text{ mole fraction}$$

$$\ell = 1, 2 \ldots N \quad \text{and} \quad k = 1, 2$$

where N is the number of data points of an isotherm or isobar.

For isothermal P-x-y data the experimental values of temperature, $T(exp)$, and liquid mole fraction for component k, x_k, are at each data point used together with the UNIFAC method to predict the pressure $P(UNIFAC)$ and vapor mole fraction $y_k(UNIFAC)$:

$$P(UNIFAC) = \sum^{k} x_k \gamma_k f_k^o / \varphi_k$$

$$\text{and} \quad y(UNIFAC) = \frac{x_k \gamma_k f_k^o}{\varphi_k P(UNIFAC)}$$

$$k = 1, 2$$

The activity coefficients, γ_k, are calculated from the UNIFAC method. The reference fugacity, f_k^o, and the fugacity coefficient, φ_k, are calculated as explained in Chapters 2 and 3 as functions of T(exp), P(UNIFAC) and y_k(UNIFAC). This means that the above equations have to be solved by a bubble point calculation, which is an iterative procedure.

For isobaric T-x-y data the value of T(UNIFAC) and y_k(UNIFAC) are at each data point calculated by iteration from

$$P(\exp) = \sum_{}^{k} x_k \gamma_k f_k^o / \varphi_k$$

and

$$y_k(\text{UNIFAC}) = \frac{x_k \gamma_k f_k^o}{\varphi_k \; P(\exp)}$$

$$k = 1, 2$$

In obtaining γ_k, φ_k and f_k^o, the calculated values T(UNIFAC) and y_k(UNIFAC) are used together with the experimental values P(exp) and x_k.

Comparisons between experimental and predicted values are carried out both for mixtures which were included in the data base (Section 6.1) and for mixtures not included in the data base (Section 6.2). Section 6.1 thus contains results from the data correlation leading to the parameter values given in Table 4.4, while Section 6.2 contains true predictions.

In Sections 6.1 and 6.2 the UNIFAC parameters of Tables 4.3 and 4.4 are used throughout. Section 6.3 shows results for binary systems containing Ethanol when the alternative UNIFAC parameters of Section 4.5 are used.

6.1 SAMPLE RESULTS FROM THE DATA CORRELATION

Table 6.1 shows experimental and calculated activity coefficients representing most of the systems in the data base. The examples given here are from the dilute regions, because the deviations from Raoult's law are particularly large there. It should be noted, however, that the agreement between calculated and experimental activity coefficients almost always is better in the middle of the concentration range than in the dilute regions. Table 6.2 shows deviations between experimental and predicted equilibrium conditions for several entire isotherms or isobars. Figures 6.1 - 6.21 similarly show results from the data correlation for the whole concentration range. Note that the azeotropes are correlated very well using the UNIFAC method.

6.2 SAMPLE PREDICTIONS

Table 6.3 summarizes predictions of vapor-liquid equilibria for many binary mixtures. Note that none of the mixtures listed in Table 6.3 were included at all in the data base. Figures 6.22- 6.33 show similar predictions in more detail.

Most of the systems included in the data base were found thermodynamically consistent. This is not the case for the systems listed in Table 6.3, which certainly accounts for some of the reason why Δy on the average is somewhat larger in Table 6.3 than in Table 6.2.

As a whole, the results of the predictions are satisfactory. The effect of isomerization is apparently accounted for very well by UNIFAC, see for example the results for systems with branched hydrocarbons. Azeotropic conditions are also predicted very accurately, see for example Figures 6.28 and 6.31.

6.3 PREDICTIONS FOR SYSTEMS CONTAINING ETHANOL

The results of the last two sections are as a whole highly acceptable. We were, however, not completely satisfied with the results for systems containing Ethanol-Water and systems containing Ethanol-Hydrocarbons. Relatively large deviations between calculated and experimental vapor phase mole fractions resulted for these two types of systems. This was especially discouraging since these mixtures are frequently encountered in practice. We therefore developed the alternative parameters shown in Section 4.5, and as already stated, these parameters are recommended for systems containing Ethanol together with Water and Hydrocarbons. Section 4.5 already shows several examples of improved predictions when the alternative parameters are used. This section gives additional examples in Figures 6.34 - 6.40.

REFERENCES

1. L. Sieg, Chem.Ing.Techn., 22(1950)322.
2. N.D. Litvinov, Zh.Fiz.Khim., 14(1940)782.
3. S.K. Ogorodnikov, V.B. Kogan and M.S. Nemtsov, Zh.Prikl. Khim., 33(1960)2685.
4. J.D. Thornton and F.H. Garner, J.Appl.Chem. 1, Suppl. 1, 68(1951).
5. L.R. Hellwig and M. Van Winkle, Ind.Eng.Chem., 45(1953) 624.
6. W.D. Hill and M. Van Winkle, Ind.Eng.Chem., 44(1952)205.
7. L.H. Ballard and M. Van Winkle, Ind.Eng.Chem., 45(1953) 1803.
8. M. Newman, C.B. Hayworth and R.E. Treybal, Ind.Eng.Chem., 41(1949)2039.
9. I. Nagata, J.Chem.Eng. Data, 10(1965)106.
10. R.E. Treybal and L.D. Weber, Ind.Eng.Chem., 38(1946)817.
11. J.M. Harrison and L. Berg, Ind.Eng.Chem., 38(1946)117.
12. H.G. Drickamer, G.G. Brown and R.R. White, Trans.Am.Inst. Chem.Eng., 41(1945)555.
13. J.H. Weber, Ind.Eng.Chem., 48(1956)134.
14. D. Papousek and L. Pago, Coll.Czech.Chem.Commun., 24(1959) 2666.
15. F.L. Padgitt, E.S. Amis and D.W. Hughes, J.Am.Chem.Soc., 64(1942)1231.
16. D.F. Saunders and A.I.B. Spaull, Z.Phys.Chem., (Frankfurt) 28(1961)332.
17. W.D. Hill and M. Van Winkle, Ind.Eng.Chem., 44(1952)208.
18. S.K. Ogorodnikov, V.B. Kogan and M.S. Nemtsov, Zh.Prikl. Khim., 33(1960)2685.
19. R.N. Hopkins, E.S. Yerger and C.C. Lynch, J.Am.Chem.Soc., 61(1939)2460.
20. S.K. Ogorodnikov, V.B. Kogan, M.S. Nemtsov and G.V. Burova, Zh.Prikl.Khim., 34(1961)1096.
21. O. Vilim, Coll.Czech.Chem.Commun., 26(1961)2124.

22. S.R.M. Ellis and J.M. Spurr, Brit.Chem.Eng., 6(1961)92.
23. H. Michalski, Nauk.Politech.Lodz.Chem., 10(1961)73.
24. N. Isii, J.Soc.Chem.Ind. Japan, 38(1935)705.
25. E.R. Washburn and B.H. Handorf, J.Am.Chem.Soc., 57(1935) 441.
26. V.A. Kireev, Yu.N. Sheinker and E.M. Peresleni, Zh.Fiz. Khim., 26(1952)352.
27. C.P. Smith and E.W. Engel, J.Am.Chem.Soc., 51(1929)2660.
28. J.C. Chu, O.P. Kharbanda, R.F. Brooks and S.L. Wang, Ind. Eng.Chem., 46(1954)754.
29. A.I. Altsybeeva, V.P. Belousov, N.V. Ovtrakht and A.G. Morachevsky, Zh.Fiz.Khim., 38(1964)1242.
30. P. Dakshinamurti and C.V. Rao, J.Sci.Ind.Research, 17B (1958)105.
31. J. Matous, A. Zivny and J. Birds, Coll.Czech.Chem.Commun., 37(1972)3960.
32. F.G. Tenn and R.W. Missen, Can.J.Chem.Eng., 41(1963)12.
33. K.C. Chao, Ph.D. Thesis, University of Wisconsin, 1956.
34. T. Michishita, Y. Arai and S. Saito, Kagaku Kogaku 35(1971)111.
35. K. Ochi and K. Kojima, Kagaku Kogaku, 33(1969)352.
36. P. Vonka, V. Svoboda, K. Strubl and R. Holub, Coll.Czech. Chem.Commun., 36(1971)18.
37. L.A.J. Verhoeye, J.Chem.Eng.Data, 13(1968)462.
38. S.V. Vijayaraghavan, P.K. Desphande and N.R. Kuloor, J.Chem.Eng. Data, 12(1967)13.
39. J.M. Willock and M. Van Winkle, J.Chem.Eng. Data, 15(1970) 281.
40. S.M.K.A. Gurukul and B.N. Raju, J.Chem.Eng. Data, 15(1970) 361.
41. W.A. Scheller and S.V.N. Rao, J.Chem.Eng. Data, 18(1973) 223.
42. J. Linek, K. Prochazka and I. Wichterle, Coll.Czech.Chem. Commun., 37(1972)3010.
43. J. Kraus and J. Linek,Coll.Czech.Chem.Commun., 36(1971) 2547.
44. J. Gmehling, unpublished data.

45. V. Natarej and M.R. Rao, Indian J.Technol., 5(1967)212.
46. V.C. Maripuri and G.A. Ratcliff, J.Appl.Chem.Biotechnol., 22(1972)899.
47. M.P. Susarev and A.M. Toikka, Zh.Prikl.Khim. (Leningrad), 46(1973)2461.
48. V. Valent, Dissertation Tuebingen, 1976.
49. M. Maretic and V. Sirocic, Nafta, (Zagreb), 13(1962)126.
50. B. Kumar and K.S.N. Raju, J.Chem.Eng. Data, 17(1972)438.
51. C.G. McConnell and M. Van Winkle, J.Chem.Eng. Data, 12(1967)430.
52. I. Nagata, T. Ohta, M. Ogura and S. Yasuda, J.Chem.Eng. Data, 21(1976)310.
53. E.M. Gause and F.M. Ernsberger, J.Chem.Eng. Data, 2(1957)28.
54. I. Nagata, T. Ohta and S. Nakagawa, J.Chem.Eng. Japan, 9(1976)276.

Table 6.1
EXPERIMENTAL AND CALCULATED ACTIVITY COEFFICIENTS

Typical results from correlating the vapor-liquid equilibrium data included in the data base. For references see Appendix 5.

Hydrocarbon-Hydrocarbon Systems

System Component 1-Component 2	T,K	x_1	γ_1(exp)	γ_1(calc)
1-Hexene-Hexane	333.2	0.121	1.022	1.024
Hexane-1-Hexene	333.2	0.123	1.026	1.031
Benzene-Hexane	298.2	0.100	1.550	1.478
Hexane-Benzene	298.2	0.100	1.834	1.630
Heptane-Toluene	383.7	0.000	1.411	1.541
Toluene-Heptane	371.5	0.006	1.677	1.441
Octane-p-Xylene	409.7	0.064	1.246	1.244
p-Xylene-Octane	399.3	0.096	1.185	1.229

Hydrocarbon-Oxygenated Hydrocarbon Systems

Alcohols (including methanol and phenol)

	T,K	x_1	γ_1(exp)	γ_1(calc)
Ethanol-Heptane	358.2	0.025	10.63	19.78
Heptane-Ethanol	348.9	0.050	7.569	8.377
1-Propanol-Heptane	303.2	0.020	19.61	14.48
Heptane-1-Propanol	303.2	0.010	7.286	5.499
2-Propanol-Heptane	333.2	0.005	13.22	12.85
Heptane-2-Propanol	333.2	0.020	6.501	5.215
1-Butanol-Decane	373.2	0.033	7.032	6.345
Decane-1-Butanol	373.2	0.155	3.634	3.580
1-Octanol-Heptane	313.2	0.100	3.900	3.586
Heptane-1-Octanol	313.2	0.100	2.450	2.235
4-Octanol-Heptane	313.2	0.100	3.330	2.624
Heptane-4-Octanol	313.2	0.100	2.220	1.870
1-Butanol-1-Hexene	333.2	0.062	4.651	4.385
1-Hexene-1-Butanol	333.2	0.139	2.667	2.510
2-Butanol-Benzene	318.2	0.061	4.197	4.222
Benzene-2-Butanol	318.2	0.106	2.592	2.392

Table 6.1 cont.

1-Propanol-Benzene	338.2	0.090	3.627	3.852
Benzene-1-Propanol	338.2	0.032	3.082	3.025
Ethanol-Toluene	308.2	0.033	9.964	8.991
Toluene-Ethanol	308.2	0.030	5.298	5.330
1-Pentanol-Toluene	343.2	0.100	2.910	2.597
Toluene-1-Pentanol	343.2	0.100	2.436	2.148
Methanol-Benzene	328.2	0.064	7.091	7.545
Benzene-Methanol	363.2	0.144	3.729	3.713
Methanol-Toluene	376.8	0.006	7.520	7.469
Toluene-Methanol	337.2	0.034	7.453	6.412
Methanol-Hexane	318.2	0.045	15.59	12.09
Hexane-Methanol	318.2	0.072	12.45	10.43
2-Methyl-2-butene-Methanol	313.5	0.090	6.874	6.258
Methanol-2-Methyl-2-butene	306.3	0.088	9.626	9.321
Phenol-Dodecane	474.3	0.109	2.546	2.768
Dodecane-Phenol	451.8	0.065	5.034	4.864
Phenol-Benzene	345.2	0.270	1.536	1.593
Benzene-Phenol	345.2	0.032	3.971	4.197

Ethers

Dipropyl ether-Heptane	343.2	0.000	1.085	1.190
Heptane-Dipropyl ether	343.2	0.000	1.085	1.190
1,2-Dimethoxyethane-Heptane	343.2	0.027	2.312	2.081
Heptane-1,2-Dimethoxyethane	343.2	0.090	2.074	2.047
Bis(2-methoxyethyl)ether-Decane	393.2	0.130	1.931	2.319
Decane-Bis(2-methoxyethyl)ether	393.2	0.021	2.851	3.259
Tetrahydrofuran-Cyclohexane	298.2	0.050	2.290	1.873
Cyclohexane-Tetrahydrofuran	298.2	0.050	1.800	1.819
1,4-Dioxane-1-Hexene	353.2	0.115	1.797	1.800
1-Hexene-1,4-Dioxane	353.2	0.058	2.020	2.088
1,2-Dimethoxyethane-Benzene	343.2	0.000	1.010	1.108

Table 6.1 cont.

Benzene-1,2-Dimethoxy-ethane	343.2	0.000	1.010	1.140

Ketones

Acetone-Hexane	318.2	0.100	3.368	3.825
Hexane-Acetone	318.2	0.100	3.398	3.762
3-Pentanone-Heptane	353.2	0.056	2.248	1.980
Heptane-3-Pentanone	353.2	0.051	2.321	2.038
5-Nonanone-Hexane	333.2	0.100	1.676	1.690
Hexane-5-Nonanone	333.2	0.100	1.454	1.502
2-Butanone-1-Hexene	333.2	0.097	2.125	2.390
1-Hexene-2-Butanone	333.2	0.094	2.053	2.059
Acetone-Benzene	318.2	0.096	1.436	1.408
Benzene-Acetone	318.2	0.092	1.408	1.490
2-Butanone-Toluene	328.2	0.057	1.301	1.229
Toluene-2-Butanone	328.2	0.068	1.171	1.256

Organic acids (vapor-phase nonideality by dimerization equilibrium constant)

Acetic acid-Heptane	293.2	0.084	6.731	6.891
Heptane-Acetic acid	293.2	0.072	9.772	9.212
Propanoic acid-Octane	396.8	0.062	3.880	3.626
Octane-Propanoic acid	404.0	0.109	2.985	3.732
Acetic acid-Benzene	293.2	0.124	2.919	2.651
Benzene-Acetic acid	293.2	0.065	4.393	4.384
Acetic acid-p-Xylene	398.7	0.110	3.176	3.088
p-Xylene-Acetic acid	388.7	0.045	4.411	4.340
Propanoic acid-p-Xylene	412.5	0.035	2.856	3.047
p-Xylene-Propanoic acid	409.4	0.086	2.275	2.685

Esters

Butyl acetate-Heptane	347.9	0.080	1.816	1.881
Heptane-Butyl acetate	347.9	0.020	1.890	1.988
Methyl acetate-Cyclohexane	308.2	0.111	3.167	2.976
Cyclohexane-Methyl acetate	308.2	0.079	3.641	3.455
Ethyl acetate-Benzene	353.2	0.016	1.158	1.153

Table 6.1 cont.

Benzene-Ethyl acetate	350.3	0.050	1.084	1.122
Ethyl acetate-p-Xylene	406.7	0.006	1.789	1.756
p-Xylene-Ethyl acetate	351.1	0.035	1.529	1.505

<u>Aldehydes</u>

Propionaldehyde-Cyclohexane	318.2	0.000	4.630	4.606
Cyclohexane-Propionaldehyde	318.2	0.000	3.903	3.838

Mixtures of Hydrocarbons and Nitrogen Compounds

<u>Amines</u> (including Aniline)

Hexylamine-Hexane	333.2	0.123	1.546	1.493
Hexane-Hexylamine	333.2	0.094	1.486	1.423
Butylamine-Hexane	333.2	0.128	1.659	1.679
Hexane-Butylamine	333.2	0.104	1.770	1.668
Methylamine-Nonane	293.2	0.000	3.550	3.724
Nonane-Methylamine	293.2	0.000	7.901	7.974
Butylamine-Benzene	343.2	0.000	1.260	1.257
Benzene-Butylamine	343.2	0.000	1.163	1.205
Dimethylamine-Hexane	293.2	0.000	2.182	2.069
Hexane-Dimethylamine	293.2	0.000	2.849	2.898
Dipropylamine-Hexane	333.2	0.114	1.160	1.198
Hexane-Dipropylamine	333.2	0.082	1.141	1.189
Diethylamine-1-Hexene	333.2	0.108	1.120	1.113
1-Hexene-Diethylamine	333.2	0.099	1.118	1.115
Diethylamine-Benzene	328.2	0.000	1.084	1.113
Benzene-Diethylamine	328.2	0.000	1.124	1.091
Aniline-Methylcyclohexane	373.2	0.098	4.375	5.236
Methylcyclohexane-Aniline	373.2	0.076	4.950	4.772
Aniline-Benzene	343.2	0.071	1.891	2.035
Benzene-Aniline	343.2	0.202	1.558	1.505
Aniline-Toluene	373.2	0.097	1.840	1.944
Toluene-Aniline	373.2	0.102	1.815	1.808

Table 6.1 cont.

Nitriles

Acetonitrile-Heptane	293.2	0.017	27.68	26.90
Heptane-Acetonitrile	293.2	0.014	35.49	35.17
Acetonitrile-Benzene	318.2	0.094	2.275	2.417
Benzene-Acetonitrile	318.2	0.043	2.507	2.655
Acetonitrile-1-Pentene	301.9	0.100	6.278	6.401
1-Pentene-Acetonitrile	314.7	0.100	5.296	5.314
Acetonitrile-Toluene	318.2	0.021	3.267	3.306
Toluene-Acetonitrile	318.2	0.027	3.416	3.591

Others

Nitromethane-Benzene	318.2	0.045	3.041	2.862
Benzene-Nitromethane	318.2	0.040	3.038	3.065
Heptane-Nitroethane	378.6	0.037	5.423	5.023
Nitroethane-Heptane	366.7	0.062	5.339	5.531
Heptane-1-Nitropropane	398.8	0.021	3.673	3.876
1-Nitropropane-Heptane	379.1	0.058	5.073	4.834

Hydrocarbon-Chlorinated Hydrocarbon Systems

1-Chlorobutane-Heptane	323.2	0.113	1.164	1.110
Heptane-1-Chlorobutane	323.2	0.135	1.319	1.127
1,2-Dichloroethane-Toluene	381.6	0.045	1.042	1.008
Toluene-1,2-Dichloroethane	360.3	0.188	0.997	0.999
Chloroform-Hexane	308.2	0.101	1.394	1.398
Hexane-Chloroform	308.2	0.099	1.630	1.582
Chloroform-Benzene	352.4	0.060	0.851	0.853
Benzene-Chloroform	335.8	0.066	0.856	0.798
Chloroform-Toluene	372.0	0.160	0.816	0.823
Toluene-Chloroform	338.5	0.135	0.887	0.789
Tetrachloromethane-Cyclohexane	343.2	0.053	1.072	1.147
Cyclohexane-Tetrachloromethane	343.2	0.052	1.087	1.128
Tetrachloromethane-Benzene	318.2	0.099	1.105	1.084
Benzene-Tetrachloromethane	318.2	0.037	1.080	1.091

Table 6.1 cont.

Hydrocarbons with Carbon disulfide

Carbon disulfide-Benzene	303.2	0.116	1.240	1.336
Benzene-Carbon disulfide	303.2	0.080	1.549	1.415
Carbon disulfide-2-Methyl-butane	290.2	0.220	1.610	1.536
2-Methylbutane-Carbon disulfide	290.2	0.168	1.650	1.737

Systems with Water

Hydrocarbons

(These activity coefficients are based on liquid-liquid equilibria)

Water-Hexane	293.2	0.000	1880	2040
Hexane-Water	298.2	0.000	489000	402000
Water-Benzene	313.2	0.000	226.3	308.2
Benzene-Water	333.2	0.000	1730	1670
Water-Toluene	333.2	0.000	3320	391
Toluene-Water	333.2	0.000	3390	7820

Oxygenated Hydrocarbons

Water-Ethanol	313.2	0.000	4.748	3.184
Ethanol-Water	313.2	0.043	2.398	2.210
Water-1-Propanol	363.2	0.117	2.845	3.079
1-Propanol-Water	363.2	0.111	4.594	5.066
Water-2-Propanol	328.2	0.141	2.965	2.993
2-Propanol-Water	328.2	0.046	7.706	10.572
Water-1-Butanol	333.2	0.050	4.237	4.339
1-Butanol-Water	333.2	0.016	38.61	54.66
Water-Methanol	340.2	0.154	1.505	1.543
Methanol-Water	368.2	0.029	2.097	2.126
Water-Phenol	348.2	0.321	1.842	2.185
Phenol-Water	348.2	0.015	20.43	23.42
Water-Acetic acid	388.6	0.050	1.769	1.697
Acetic acid-Water	373.7	0.100	2.027	2.300
Water-Propanoic acid	394.3	0.115	2.382	2.125

Table 6.1 cont.

System	T	x	γ₁	γ₂
Propanoic acid-Water	373.1	0.108	3.462	3.809
Water-Acetaldehyde	293.2	0.100	2.982	2.688
Acetaldehyde-Water	293.2	0.100	2.653	2.606
Water-Methyl acetate	308.2	0.019	21.25	20.17
Methyl acetate-Water	298.2	0.005	21.84	24.94
Water-Ethyl acetate	343.2	0.032	8.025	8.378
Ethyl acetate-Water	343.2	0.000	21.97	14.71
Water-Acetone	333.2	0.263	2.551	2.362
Acetone-Water	333.2	0.072	5.638	6.736
Water-Tetrahydrofuran	323.2	0.078	6.800	5.690
Tetrahydrofuran-Water	323.2	0.075	10.12	14.67
Water-1,4-Dioxane	365.9	0.120	2.430	3.400
1,4-Dioxane-Water	365.1	0.073	4.670	9.840

Nitrogen compounds

System	T	x	γ₁	γ₂
Water-Aniline	433.2	0.011	4.985	4.277
Aniline-Water	371.8	0.021	34.16	50.65
Water-Butylamine	350.2	0.045	1.720	1.587
Butalamine-Water	361.3	0.020	12.58	12.95
Water-Diethylamine	311.5	0.100	3.074	1.665
Diethylamine-Water	311.5	0.100	2.485	3.398
Water-Acetonitrile	303.2	0.080	6.778	6.041
Acetonitrile-Water	303.2	0.070	9.666	8.201

Mixed Systems

System	T	x	γ₁	γ₂
Butyl acetate-Phenol	408.1	0.099	0.402	0.386
Phenol-Butyl acetate	362.1	0.106	0.283	0.293
Diethyl ether-Acetone	303.2	0.050	1.783	1.784
Acetone-Diethyl ether	303.2	0.050	1.892	1.891
Diethylamine-Ethyl acetate	350.1	0.000	1.315	1.306
Ethyl acetate-Diethylamine	328.5	0.000	1.340	1.340
Diethylamine-1,4-Dioxane	374.3	0.000	1.556	1.542
1,4-Dioxane-Diethylamine	328.5	0.000	1.514	1.514
Acetonitrile-Methyl acetate	332.3	0.156	1.100	1.166
Methyl acetate-Acetonitrile	352.5	0.055	1.224	1.189

Table 6.1 cont.

Acetonitrile-Propyl acetate	373.0	0.020	1.805	2.104
Acetonitrile-Butyl acetate	393.7	0.018	2.379	2.255
Acetone-Ethyl acetate	347.0	0.095	1.185	1.206
Ethyl acetate-Acetone	330.4	0.090	1.141	1.101
Methanol-Propanoic acid	298.2	0.043	1.353	1.426
Propanoic acid-Methanol	298.2	0.125	0.873	0.893
Methanol-Butanoic acid	318.2	0.036	1.482	1.557
Butanoic acid-Methanol	318.2	0.129	0.925	1.022
Acetic acid-Acetone	323.2	0.156	0.929	0.979
Acetone-Acetic acid	323.2	0.171	0.906	0.912
Propanoic acid-3-Pentanone	378.9	0.151	1.295	1.193
3-Pentanone-Propanoic acid	404.9	0.175	1.168	1.162
Acetic acid-Ethyl acetate	315.2	0.071	2.120	1.709
Ethyl acetate-Acetic acid	315.2	0.163	1.249	1.221
Acetic acid-Butyl acetate	398.3	0.038	1.847	1.856
Butyl acetate-Acetic acid	391.1	0.091	1.509	1.568
Acetic acid-Tetrachloromethane	382.3	0.025	4.563	4.285
Tetrachloromethane-Acetic acid	349.4	0.060	5.864	5.246
Formic acid-1,2-Dichloroethane	333.2	0.330	2.310	2.115
1,2-Dichloroethane-Formic acid	333.2	0.127	4.928	3.662
Acetic acid-1-Chloropropane	325.1	0.172	2.411	2.239
1-Chloropropane-Acetic acid	358.4	0.083	2.842	2.352
Ethyl acetate-Ethanol	343.2	0.018	2.434	2.642
Ethanol-Ethyl acetate	343.2	0.068	2.057	2.434
Ethyl acetate-1-Butanol	384.2	0.046	1.805	1.677
1-Butanol-Ethyl acetate	350.5	0.109	1.774	1.657
Diethyl ether-Ethanol	313.2	0.050	2.487	2.560
Ethanol-Diethyl ether	313.2	0.050	3.433	3.712
2-Propanol-Diisopropyl ether	340.9	0.011	5.928	4.429
Diisopropyl ether-2-Propanol	349.2	0.084	2.872	2.742
Acetone-Ethanol	321.2	0.050	2.022	1.909
Ethanol-Acetone	321.2	0.100	2.666	2.079

Table 6.1 cont.

Acetone-Methanol	335.2	0.076	1.679	1.697
Methanol-Acetone	328.2	0.083	1.629	1.678
Ethanol-2-Pentanone	309.0	0.371	1.472	1.413
2-Pentanone-Ethanol	307.2	0.058	2.091	2.104
1-Butanol-Butylamine	313.2	0.041	0.536	0.538
Butylamine-1-Butanol	313.2	0.165	0.666	0.663
Dimethylamine-Ethanol	293.2	0.659	0.948	0.921
Ethanol-Dimethylamine	293.2	0.259	0.691	0.678
Acetone-Chloroform	335.8	0.098	0.486	0.546
Chloroform-Acetone	330.2	0.051	0.553	0.533
Ethanol-1,2-Dichloroethane	313.2	0.100	4.814	4.448
1,2-Dichloroethane-Ethanol	313.2	0.100	3.874	3.362
Carbon disulfide-Chloroform	298.2	0.100	1.442	1.458
Chloroform-Carbon disulfide	298.2	0.100	1.527	1.333
Chloroethane-Diethyl ether	288.2	0.196	1.295	1.379
Diethyl ether-Chloroethane	288.2	0.080	1.684	1.637
Tetrachloromethane-Acetonitrile	318.2	0.089	4.367	3.984
Acetonitrile-Tetrachloromethane	318.2	0.104	4.903	4.955
Chloroform-Ethyl acetate	350.7	0.064	0.530	0.576
Ethyl acetate-Chloroform	336.9	0.078	0.428	0.459
Chloroform-Methyl acetate	323.2	0.064	0.509	0.460
Methyl acetate-Chloroform	323.2	0.065	0.553	0.546
Chloroform-Methanol	308.2	0.050	2.264	2.396
Methanol-Chloroform	308.2	0.050	5.167	5.172
Methanol-Ethyl acetate	323.2	0.053	2.577	2.604
Ethyl acetate-Methanol	323.2	0.075	3.032	2.429
1,2-Dichloroethane-1-Propanol	323.2	0.100	2.337	2.610
1-Propanol-1,2-Dichloroethane	323.2	0.100	3.069	3.239
Acetone-Tetrachloromethane	318.2	0.056	2.420	2.389
Tetrachloromethane-Acetone	318.2	0.036	2.066	1.983
Chlorobenzene-1-Nitropropane	348.2	0.119	1.333	1.458

Table 6.1 cont.

System				
1-Nitropropane-Chlorobenzene	348.2	0.229	1.304	1.309
Acetone-Carbon disulfide	308.3	0.062	3.441	3.235
Carbon disulfide-Acetone	308.3	0.031	5.872	5.392
Tetrachloromethane-Nitromethane	318.3	0.051	6.233	5.474
Nitromethane-Tetrachloromethane	318.3	0.046	8.899	9.249
Tetrachloromethane-1-Butanol	381.5	0.046	2.803	2.053
1-Butanol-Tetrachloromethane	350.3	0.038	5.830	3.830
Chloroform-Ethanol	318.2	0.071	1.724	1.805
Ethanol-Chloroform	318.2	0.024	4.555	4.298
Chloroform-1-Butanol	379.6	0.076	1.349	1.387
1-Butanol-Chloroform	335.8	0.046	2.319	2.093
1,2-Dichloroethane-Tetrachloromethane	354.2	0.053	1.666	1.683
Tetrachloromethane-1,2-Dichloroethane	349.2	0.062	1.491	1.553
Methanol-Ethyl acetate	328.2	0.031	2.775	2.687
Ethyl acetate-Methanol	328.2	0.100	2.312	2.306
Methanol-Methyl acetate	313.2	0.135	2.186	2.144
Methyl acetate-Methanol	313.2	0.061	2.723	2.300

Table 6.2

TYPICAL RESULTS FROM CORRELATING BINARY VAPOR-LIQUID
EQUILIBRIUM DATA INCLUDED IN THE DATA BASE

System	T or P	ΔP or ΔT	Δy
1-Chlorobutane-Heptane	328 K	10 mm Hg	0.011
Methanol-3-Methyl-1-butanol	328 K	8.5 mm Hg	0.005
Benzene-Heptane	333 K	0.7 mm Hg	0.008
1-Propanol-Ethylbenzene	760 mm Hg	0.5 K	0.006
Methanol-Toluene	760 mm Hg	0.5 K	0.009
Acetone-Chloroform	298 K	9.3 mm Hg	0.022
Ethanol-2-Butanone	760 mm Hg	0.6 K	0.022
Ethanol-1,2-Dichloroethane	313 K	4.8 mm Hg	0.018
Water-Phenol	363 K	3.8 mm Hg	0.004
1-Butanol-Butyl acetate	760 mm Hg	0.3 K	0.006
Tetrachloromethane-1,2-Dichloroethane	760 mm Hg	0.3 K	0.008
Acetonitrile-Water	760 mm Hg	1.0 K	0.029
Ethanol-Ethyl acetate	313 K	6.3 mm Hg	0.006
Acetone-Tetrachloromethane	318 K	2.3 mm Hg	0.008
Acetone-Methanol	760 mm Hg	0.2 K	0.006
Methanol-Water	313 K	1.1 mm Hg	0.002
Ethyl acetate-p-Xylene	760 mm Hg	0.3 K	0.007
Diethyl ether-Ethanol	303 K	8.0 mm Hg	0.006
Pentane-Acetone	760 mm Hg	1.1 K	0.015
Benzene-2-Propanol	500 mm Hg	0.6 K	0.011
Cyclohexane-Aniline	760 mm Hg	1.3 K	0.016
Toluene-1-Butanol	760 mm Hg	0.1 K	0.006
Carbon disulfide-Tetrachloromethane	760 mm Hg	0.1 K	0.008
Acetone-Water	298 K	18 mm Hg	0.048
Acetone-Water	760 mm Hg	1.3 K	0.029
Water-1-Propanol	363 K	4.4 mm Hg	0.008
Water-1,4-Dioxane	308 K	8.4 mm Hg	0.096
Water-Diethylamine	309 K	28 mm Hg	0.028
Tetrachloromethane-2-Butanone	760 mm Hg	0.3 K	0.012
Chloroform-Methanol	323 K	1.9 mm Hg	0.003

Table 6.2 cont.

Diethylamine-Ethyl acetate	760 mm Hg	0.2 K	0.003
Benzene-Acetonitrile	293 K	1.2 mm Hg	0.006
Tetrahydrofuran-1-Butanol	760 mm Hg	0.9 K	0.011
Acetonitrile-Benzene	318 K	1.6 mm Hg	0.008
Ethyl acetate-1-Butanol	730 mm Hg	0.7 K	0.003
Ethyl acetate-Water	343 K	2.9 mm Hg	0.008
Ethanol-Water	328 K	18 mm Hg	0.038
Toluene-2-Butanone	348 K	3.4 mm Hg	0.007
Diethylamine-Heptane	328 K	3.1 mm Hg	0.011

Table 6.3

PREDICTION OF VAPOR-LIQUID EQUILIBRIA FOR SYSTEMS
NOT INCLUDED IN THE DATA BASE

System	P or T	ΔT or ΔP	Δy	Ref.
Benzene-Methylcyclohexane	760 mm Hg	0.4 K	0.013	1
Dimethoxymethane-Chloroform	308 K	33 mm Hg	0.032	2
Methanol-Decane	760 mm Hg	2.4 K	0.005	3
Cyclohexane-2-Methoxyethanol	760 mm Hg	5.1 K	0.091	4
Benzene-2-Methoxyethanol	760 mm Hg	1.7 K	0.021	4
Water-3-Hydroxy-2-butanone	750 mm Hg	2.4 K	0.029	5
Methanol-1-Pentanol	760 mm Hg	1.8 K	0.028	6
Ethanol-1-Pentanol	760 mm Hg	0.7 K	0.018	5
2-Propanol-4-Methyl-2-pentanone	760 mm Hg	1.4 K	0.021	7
Methanol-4-Methyl-2-pentanone	760 mm Hg	2.0 K	0.010	6
2-Butanone-2-Butoxyethanol	760 mm Hg	1.7 K	0.014	8
2-Propanol-Methylcyclohexane	500 mm Hg	0.7 K	0.022	9
Acetone-1,1,2-Trichloroethane	755 mm Hg	1.0 K	0.019	10
Benzene-2,2,3-Trimethylbutane	736 mm Hg	0.1 K	0.005	11
Phenol-2,2,4-Trimethylpentane	760 mm Hg	3.3 K	0.016	12
Toluene-2,2,4-Trimethylpentane	760 mm Hg	1.3 K	0.010	12
Ethylbenzene-2,2,5-Trimethylhexane	760 mm Hg	0.2 K	0.006	13
Methanol-Tetrahydrofuran	600 mm Hg	0.2 K	0.012	14
Methanol-1,4-Dioxane	760 mm Hg	3.7 K	0.074	15
Benzene-1-Nitropropane	298 K	5.5 mm Hg	0.025	16
Methanol-1-Propanol	760 mm Hg	0.5 K	0.013	6
2-Propanol-2-Methyl-1-propanol	760 mm Hg	0.3 K	0.006	7
Methanol-2-Butanone	760 mm Hg	1.8 K	0.008	17
Methanol-1-Pentene	760 mm Hg	1.4 K	0.041	18
Ethanol-1,4-Dioxane	760 mm Hg	0.7 K	0.018	19
Methanol-2-Pentanone	760 mm Hg	1.8 K	0.024	17

Table 6.3 cont.

System				
2-Methyl-2-butene-Acetonitrile	760 mm Hg	1.8 K	0.030	20
Hexane-2-Nitropropane	298 K	15 mm Hg	0.020	16
Methanol-2-Methylpentane	745 mm Hg	1.3 K	0.042	21
Ethanol-1-Butanol	760 mm Hg	0.3 K	0.017	5
Ethanol-Ethylbenzene	760 mm Hg	5.2 K	0.073	22
Acetone-1-Butanol	746 mm Hg	0.3 K	0.014	23
Methanol-2-Methylbutane	760 mm Hg	1.3 K	0.022	18
Pentane-Ethanol	273 K	19 mm Hg	0.011	24
Cyclohexane-Ethanol	298 K	14 mm Hg	0.039	25
Toluene-2-Propanol	760 mm Hg	0.8 K	0.013	26
Heptane-1-Butanol	323 K	2.1 mm Hg	0.018	27
Aniline-p-Xylene	745 mm Hg	0.5 K	0.018	28
2-Butanol-Water	760 mm Hg	2.8 K	0.059	29
Tetrachloromethane-Nitroethane	298 K	2.0 mm Hg	0.090	16
Tetrachloromethane-1-Nitropropane	298 K	2.2 mm Hg	0.011	16
Tetrachloromethane-2-Nitropropane	298 K	4.5 mm Hg	0.014	16
Chloroform-Butyl acetate	760 mm Hg	0.5 K	0.013	30
Methanol-Tetrahydrofuran	385 K	0.2 mm Hg	0.009	31
Methanol-Pentane	750 mm Hg	2.1 K	0.042	32
Cyclohexane-Ethyl acetate	760 mm Hg	0.5 K	0.012	33
Heptane-p-Xylene	760 mm Hg	0.9 K	0.023	34
Methanol-1-Propanol	760 mm Hg	0.1 K	0.012	35
Cyclohexane-1-Propanol	343 K	7.0 mm Hg	0.010	36
Cyclohexane-2-Propanol	760 mm Hg	0.5 K	0.018	37
Heptane-1-Butanol	2205 mm Hg	1.0 K	0.027	38
Acetone-2,3-Dimethylbutane	760 mm Hg	0.1 K	0.021	39
Chloroform-2,3-Dimethylbutane	760 mm Hg	0.1 K	0.005	39
1-Propanol-1,4-Dioxane	760 mm Hg	1.1 K	0.016	40
Heptane-2-Pentanone	363 K	18 mm Hg	0.008	41
Ethyl acetate-Ethylbenzene	323 K	3.0 mm Hg	0.016	42
Ethylbenzene-2-Butanone	348 K	3.3 mm Hg	0.008	43
Ethylbenzene-Aniline	373 K	6.2 mm Hg	0.012	44
Cyclohexane-2-Methyl-2-propanol	760 mm Hg	1.7 K	0.025	45

Table 6.3 cont.

System				
Benzene-2-Methyl-2-propanol	760 mm Hg	1.0 K	0.013	45
Hexane-3-Pentanone	338 K	19 mm Hg	0.004	46
Toluene-1,4-Dioxane	373 K	2.4 mm Hg	0.010	47
Methanol-1,4-Dioxane	333 K	31 mm Hg	0.023	48
Dichloromethane-2-Butanone	750 mm Hg	2.7 K	0.033	49
1,2-Dichloroethane-p-Xylene	735 mm Hg	1.4 K	0.014	50
Acetone-2,3-Dimethylbutane	760 mm Hg	0.5 K	0.017	51
Ethyl formate-Ethanol	318 K	29 mm Hg	0.019	52
Methanol-Tetrahydrofuran	740 mm Hg	0.2 K	0.009	53
2-Butanone-2-Propanol	328 K	5.4 mm Hg	0.013	54
Methyl acetate-1-Propanol	318 K	11 mm Hg	0.017	54
Methyl acetate-2-Propanol	318 K	13 mm Hg	0.010	54

Figures 6.1-6.21

CALCULATED AND EXPERIMENTAL ACTIVITY COEFFICIENTS
AND VAPOR PHASE MOLE FRACTIONS FOR BINARY SYSTEMS
WHICH WERE IN THE DATA BASE

In the data base used for UNIFAC parameter estimation, we have not included all acceptable data for all binary systems. For some systems, for example alcohol-water mixtures, there are so many data that we have only included some isotherms and/or isobars representing a broad temperature range and different types of substances. Therefore, the results shown in Figures 6.1-6.21 represent mixtures which were included in the data base, but not necessarily precisely the data referred to in Appendix 5.

Figure 6.1

Chloroform(1)-Ethyl acetate(2) at 760 mm Hg
I. Nagata, J.Chem.Eng. Data, 7(1962)367.

Figure 6.2

Acetone(1)-Chloroform(2) at 50.0 °C
H. Roeck and W. Schroeder, Z.Phys.Chem. (Frankfurt), 11(1957) 41.
* Note the strong negative deviation from Raoult's law and the excellent representation of the azeotrope.

Figure 6.3

Acetone(1)-Benzene(2) at 45.0 °C
I. Brown and F. Smith, Austr.J.Chem., 10(1957)423.

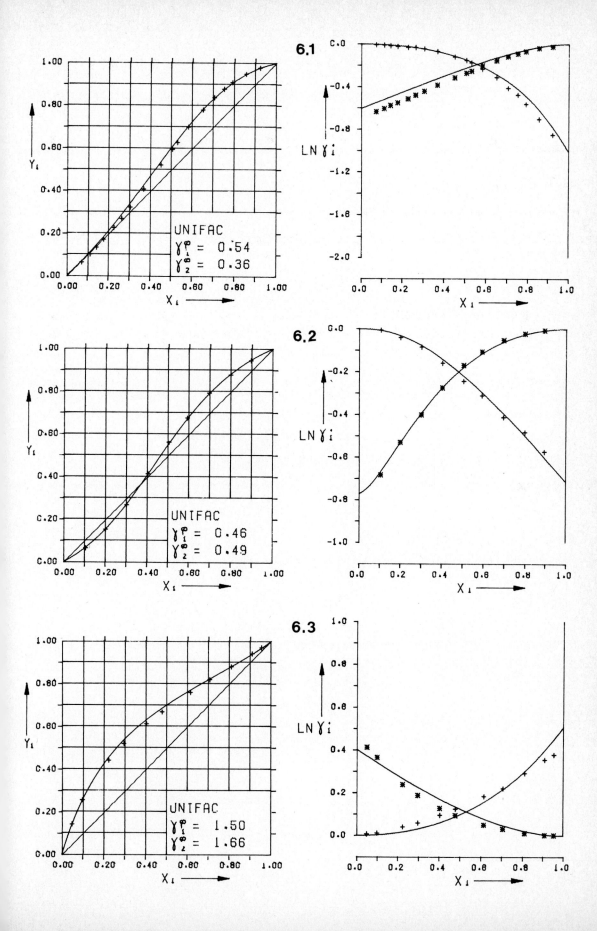

Figure 6.4

Ethanol(1)-Ethyl acetate(2) at 40.0 °C
P.S. Murti and M. Van Winkle, Chem.Eng. Data Ser., 3(1958)72.

Figure 6.5

Acetonitrile(1)-Benzene(2) at 45.0 °C
I. Brown and F. Smith, Austr.J.Chem., 8(1955)62.

Figure 6.6

Benzene(1)-1-Butanol(2) at 45.0 °C
I. Brown and F. Smith, Austr.J.Chem., 12(1959)407.

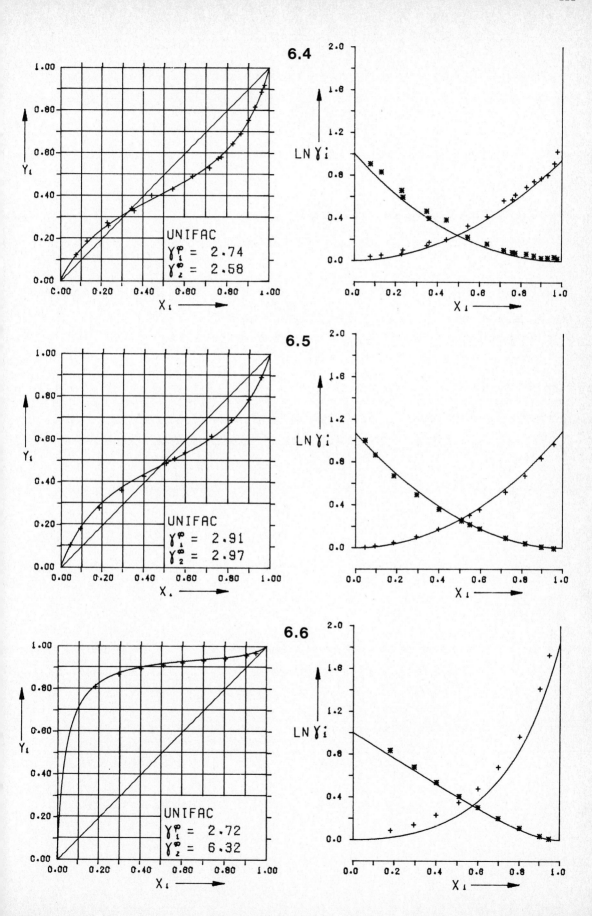

Figure 6.7

Chloroform(1)-Methanol(2) at 760 mm Hg
I. Nagata, J.Chem.Eng. Data, 7(1962)367.

Figure 6.8

Carbon disulfide(1)-Acetone(2) at 35.2 °C
V.J. Zawidski, Z.Phys.Chem., 35(1900)129.

Figure 6.9

Pentane(1)-Acetone(2) at 760 mm Hg
T.Ch. Lo, H.H. Bieber and A.E. Karr, J.Chem.Eng. Data, 7(1962)327

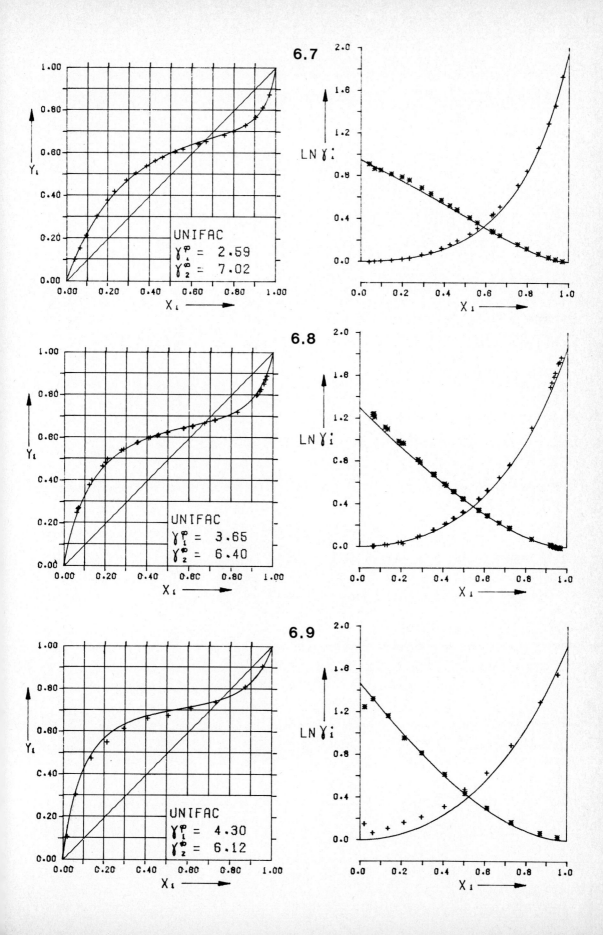

Figure 6.10

Tetrachloromethane(1)-Acetonitrile(2) at 45.0 °C
I. Brown and F. Smith, Austr.J.Chem., 7(1954)269.

Figure 6.11

Ethanol(1)-1,2-Dichloroethane(2) at 40.0 °C
V.V. Udovenko and Ts.B. Frid, Zh.Fiz.Khim., 22(1948)1263.

Figure 6.12

Methanol(1)-Toluene(2) at 760 mm Hg
B.C.-Y. Lu, Can.J.Technol., 34(1957)468.

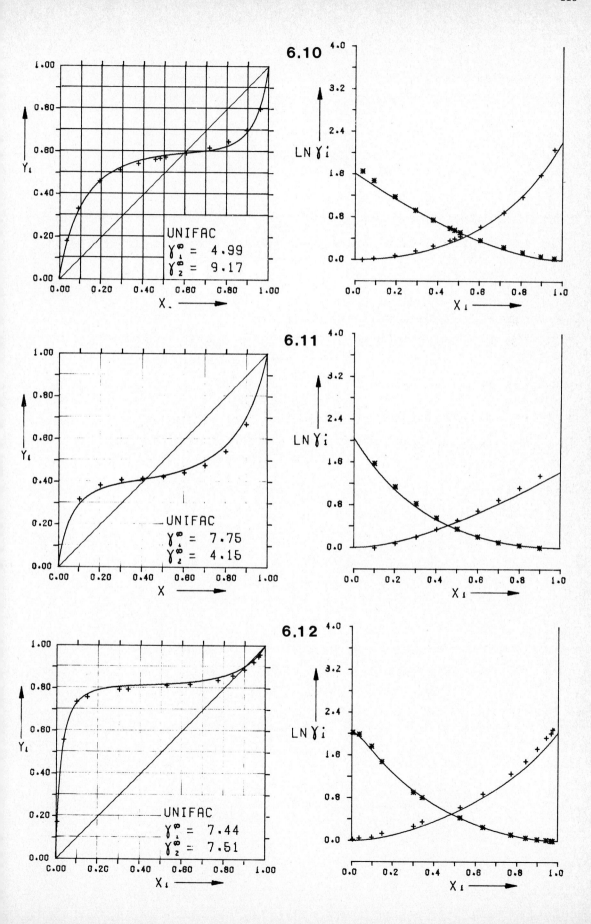

Figure 6.13

Ethanol(1)-Toluene(2) at 35.0 °C

C.B. Kretschmer and R. Wiebe, J.Am.Chem.Soc., 71(1949)1793.

Figure 6.14

Acetonitrile(1)-Water(2) at 760 mm Hg

D.S. Blackford and R. York, J.Chem.Eng. Data, 10(1965)313.

Figure 6.15

1-Propanol(1)-Water(2) at 90.0 °C

G.A. Ratcliff and K.C. Chao, Can.J.Chem.Eng., 47(1969)148.

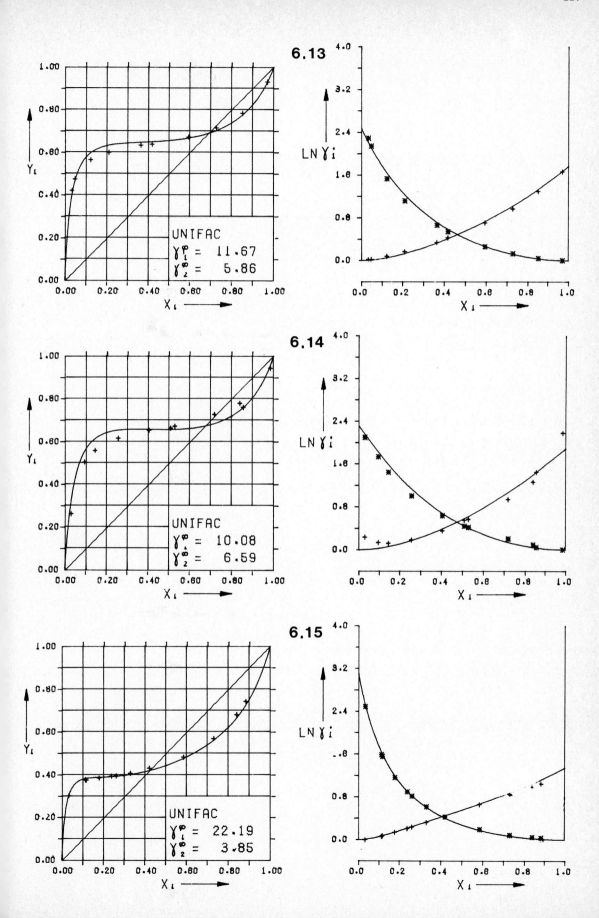

Figure 6.16

2-Propanol(1)-Water(2) at 760 mm Hg
L.A.J. Verhoeye, J.Chem.Eng. Data, 13(1968)462.

Figure 6.17

Hexane(1)-2-Butanol(2) at 60.0 °C
D.O. Hanson and M. Van Winkle, J.Chem.Eng. Data, 12(1967)319.

Figure 6.18

Tetrachloromethane(1)-Nitromethane(2) at 45.0 °C
I. Brown and F. Smith, Austr.J.Chem., 8(1955)501.

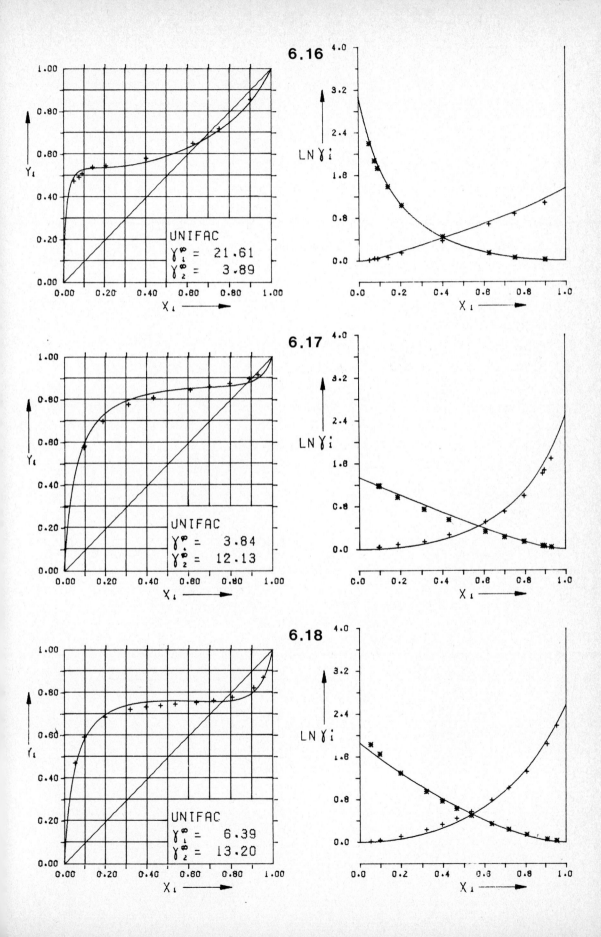

Figure 6.19

Heptane(1)-Acetonitrile(2) at 20.0 °C
G. Werner and H. Schuberth, J.Prakt.Chem., 31(1966)225.
* Note the phase splitting.

Figure 6.20

Water(1)-1-Butanol(2) at 760 mm Hg
T. Boublik, Coll.Czech.Chem.Commun., 25(1960)285.

Figure 6.21

2-Butanone(1)-Water(2) at 350 mm Hg
D.F. Othmer and R.F. Benenati, Ind.Eng.Chem., 37(1945)299.

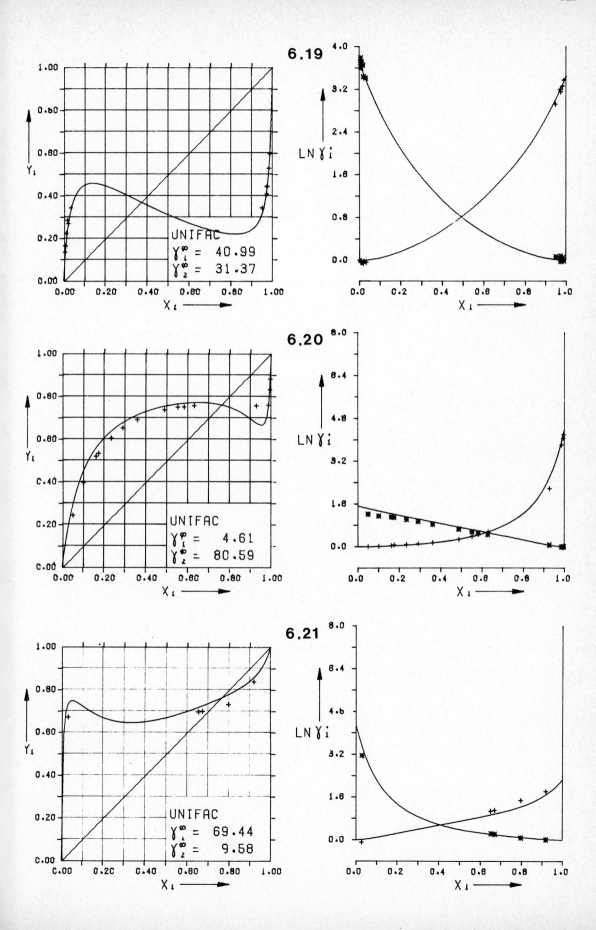

Figures 6.22-6.33

PREDICTED AND EXPERIMENTAL ACTIVITY COEFFICIENTS
AND VAPOR PHASE MOLE FRACTIONS FOR BINARY SYSTEMS
WHICH WERE NOT INCLUDED IN THE DATA BASE

Figure 6.22

Chloroform(1)-2,3-Dimethylbutane(2) at 760 mm Hg
C.G. McConnell and M. Van Winkle, J.Chem.Eng. Data,
12(1967)430.

Figure 6.23

Hexane(1)-3-Pentanone(2) at 65.0 °C
V.C. Maripuri and G.A. Ratcliff, J.Appl.Chem.Biotechnol.,
22(1972)899.

Figure 6.24

Ethanol(1)-1,4-Dioxane(2) at 760 mm Hg
Z. Gropsianu, J. Kyri and R. Gropsianu, Studii Si Cercetari
Stiintifici.Ser.Stiinte Chim., 4(1957)73.

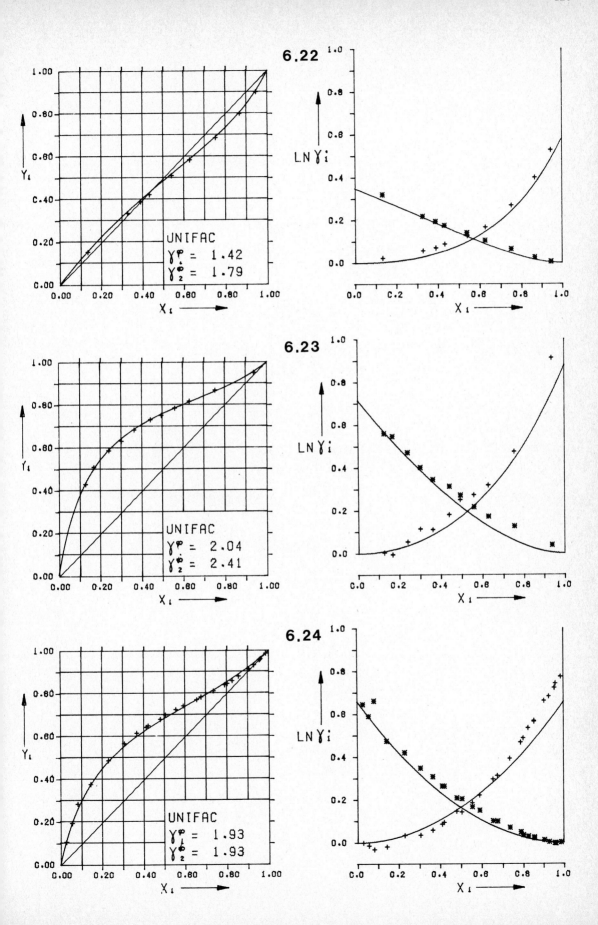

Figure 6.25

Methanol(1)-2-Butanone(2) at 760 mm Hg
W.J. Privott, D.R. Paul, K.R. Jolls and E.M. Schoenborn,
J.Chem.Eng. Data, 11(1966)331.

Figure 6.26

2,3-Dimethylbutane(1)-Acetone(2) at 760 mm Hg
J.M. Willock and M. Van Winkle, J.Chem.Eng. Data, 15(1970)281.

Figure 6.27

Tetrachloromethane(1)-2-Propanol(2) at 760 mm Hg
I. Nagata, J.Chem.Eng. Data, 10(1965)106.

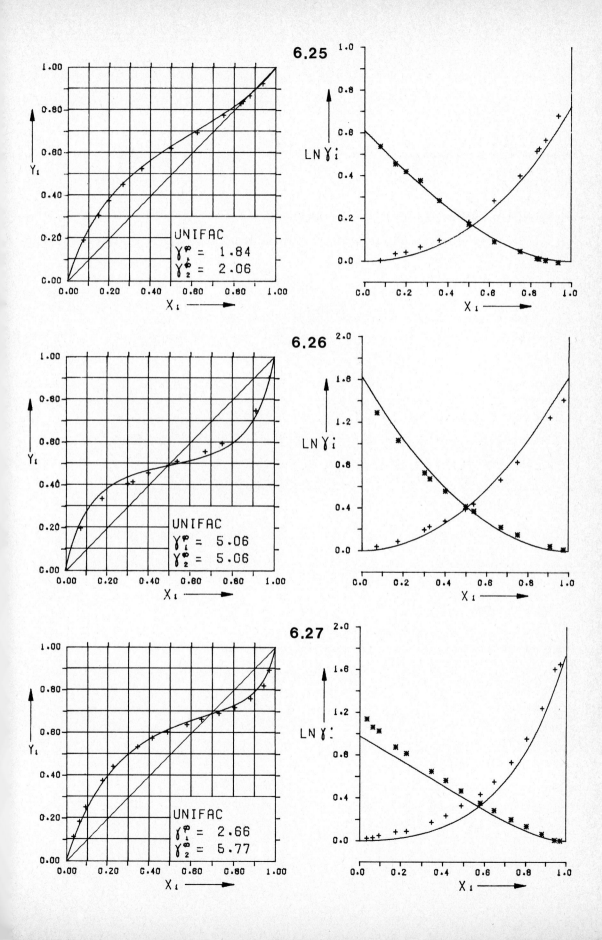

Figure 6.28

Heptane(1)-2-Pentanone(2) at 90.0 °C
W.A. Scheller and S.V.N. Rao; J.Chem.Eng. Data, 18(1973)223.

Figure 6.29

Cyclohexane(1)-Ethyl acetate(2) at 760 mm Hg
K.C. Chao, Thesis Wisconsin 1956.

Figure 6.30

Tetrahydrofuran(1)-Methanol(2) at 235.5 mm Hg
J. Matous, A. Zivny and J. Biros, Coll.Czech.Chem.Commun., 37(1972)3960.

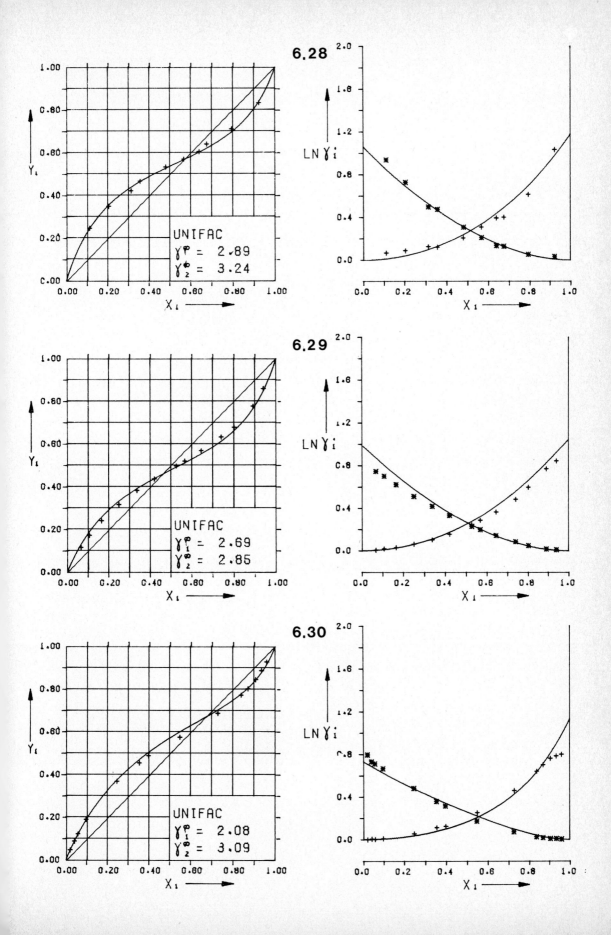

Figure 6.31

Cyclohexane(1)-1-Propanol(2) at 65.0 °C
K. Strubl, V. Svoboda, R. Holub and J. Pick, Coll.Czech.Chem.
Commun., 35(1970)3004.

Figure 6.32

Cyclohexane(1)-2-Propanol(2) at 760 mm Hg
L.A.J. Verhoeye, J.Chem.Eng. Data, 13(1968)462.

Figure 6.33

1-Chloropropane(1)-Methanol(2) at 762 mm Hg
Yu.N. Garber and V.R. Mironenko, Zh.Prikl.Khim., 41(1968)2022.

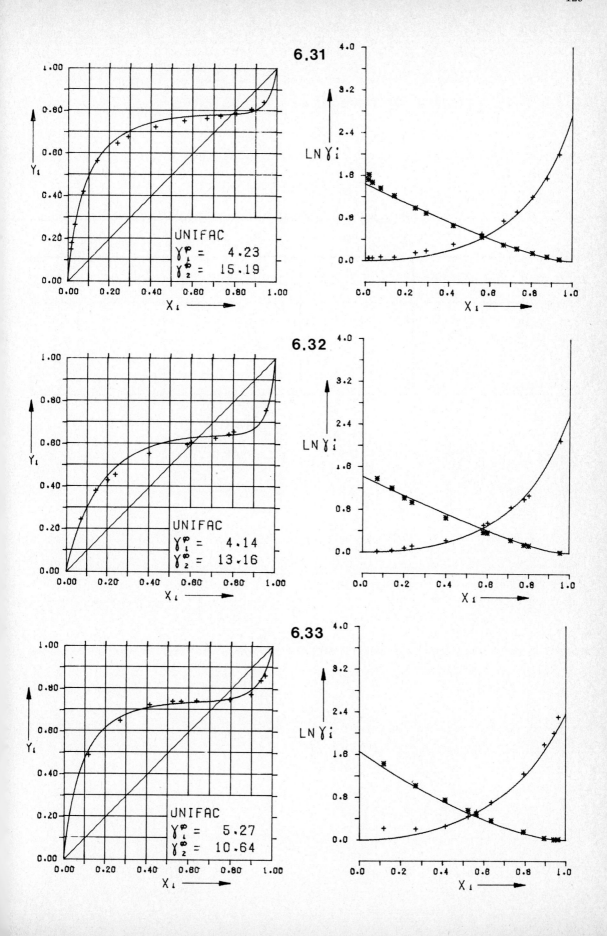

Figures 6.34-6.40

ETHANOL CONTAINING SYSTEMS

Fig.	$R_{CH_3CH_2OH}$	$Q_{CH_3CH_2OH}$	a_{H_2O/CH_3CH_2OH}	$a_{CH_3CH_2OH/H_2O}$
6.34	2.1055	1.972	-148.5	285.4
6.35	1.9	1.5	-191.6	294.7
6.36	2.1055	1.972	121.2	15.98

Ethanol(1)-Water(2) at 70.0 °C
I. Mertl, Coll.Czech.Chem.Commun., 37(1972)366.

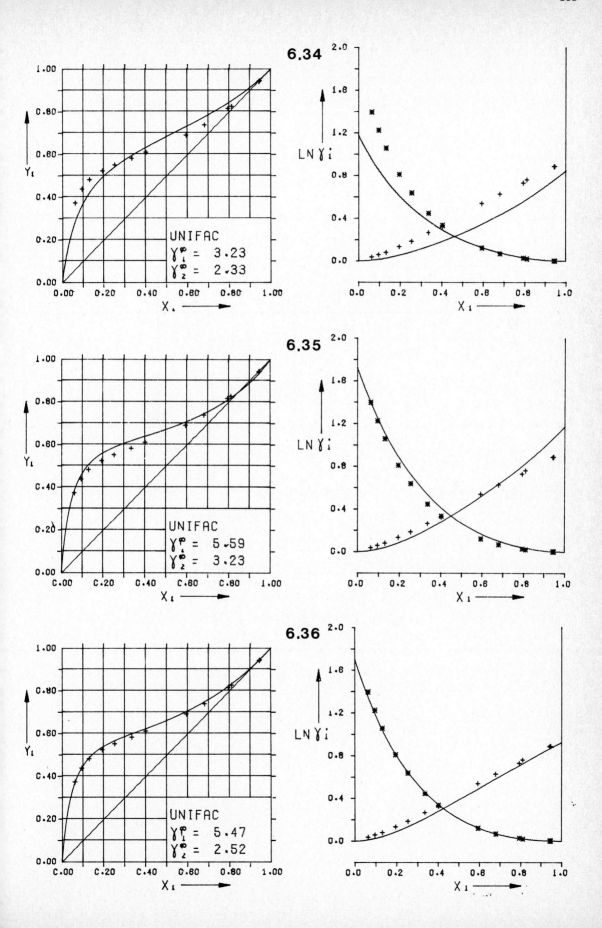

Fig.	$R_{CH_3CH_2OH}$	$Q_{CH_3CH_2OH}$	a_{CH_2/CH_3CH_2OH}	$a_{CH_3CH_2OH/CH_2}$
6.37	2.1055	1.972	737.5	−87.93
6.38	1.9	1.5	755.7	−77.70
6.39	2.1055	1.972	737.5	−87.93
6.40	1.9	1.5	755.7	−77.70

Figures 6.37 and 6.38

Ethanol(1)-Decane(2) at 760 mm Hg
S.R.M. Ellis and J.M. Spurr, Brit.Chem.Eng., 6(1961)92.

Figures 6.39 and 6.40

Ethanol(1)-Heptane(2) at 760 mm Hg
K. Katz and M. Newman, Ind.Eng.Chem., 48(1956)137.

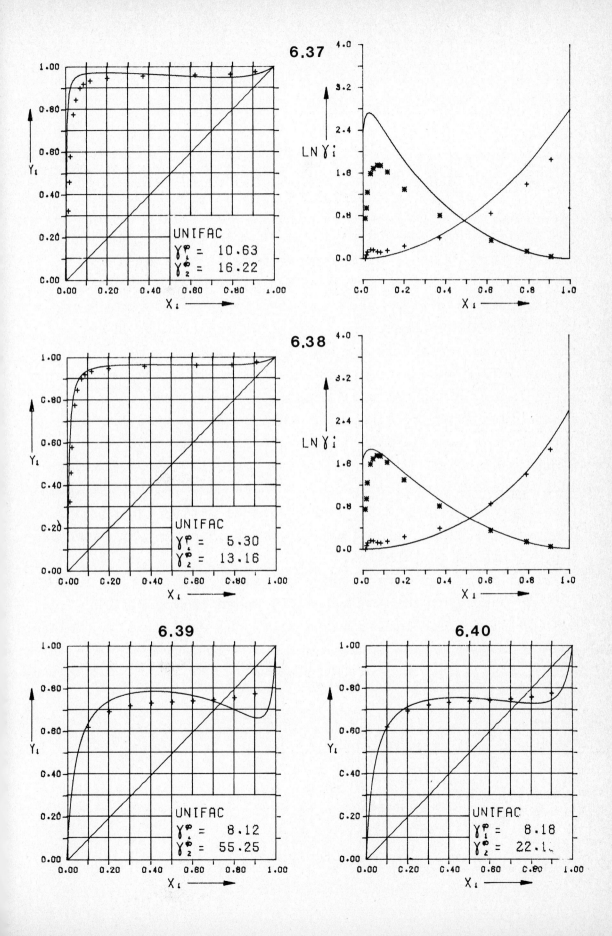

CHAPTER 7

PREDICTION OF VAPOR-LIQUID EQUILIBRIA IN MULTICOMPONENT SYSTEMS

This chapter shows comparisons between vapor-liquid equilibria in ternary and quaternary systems as predicted by the UNIFAC method and the corresponding experimental values. No multicomponent systems were included in the data base for determining UNIFAC parameters, and hence the results shown in this chapter are all true predictions.

The UNIFAC parameters of Tables 4.3 and 4.4 are used in the predictions throughout this chapter. For systems including ethanol and water or ethanol and hydrocarbons, the predictions will improve considerably if the alternative parameters of Section 4.5 were incorporated in the calculations. However, even without the incorporation of these parameters, the deviations between experimental and predicted vapor-liquid equilibrium compositions are not much greater than the estimated experimental uncertainties.

Predictions for many ternary systems are summarized in Table 7.1, and three typical examples are shown in Tables 7.2, 7.3, and 7.4. In these tables the deviations are defined as follows:

$$\Delta T = \frac{1}{N} \sum^{\ell} |T(\text{UNIFAC}) - T(\exp)| \quad K$$

$$\Delta P = \frac{1}{N} \sum^{\ell} |P(\text{UNIFAC}) - P(\exp)| \quad \text{mm Hg}$$

$$\Delta y = \frac{1}{M \cdot N} \sum^{\ell} \sum^{k} |y_k(\text{UNIFAC}) - y_k(\exp)| \quad \text{mole fraction}$$

$$\ell = 1, 2 \ldots N; \quad k = 1, 2 \ldots M$$

where N is the number of data points and M is the number of components. The calculation of ΔT, ΔP, and Δy is analogous to the method shown in Chapter 6.

Figures 7.1 - 7.12 show calculated vapor phase mole fractions vs. experimental vapor phase mole fractions for several systems. If there were complete agreement between experimental and predicted values, all the points would fall on the 45-degree line.

Tables 7.5 and 7.6 show predicted activity coefficients for two quaternary systems.

Considering the range of experimental uncertainties encountered in multicomponent vapor-liquid equilibrium measurements (Δy is usually between 0.005 and 0.025), the predictions shown in this chapter are entirely satisfactory.

REFERENCES

1. L. Stankova, F. Vesely and J. Pick, Coll.Czech.Chem. Commun., 35(1970)1.
2. H.E. Hughes and J.O. Maloney, Chem.Eng.Progr., 48(1952) 192.
3. V.V. Udovenko and L.G. Fatkulina, Zh.Fiz.Khim., 26(1952) 1438.
4. J. Griswold, P.L. Chu and W.O. Winsauer, Ind.Eng.Chem., 41(1949)2352.
5. C.H. Schneider and C.C. Lynch, J.Am.Chem.Soc., 65(1943) 1063.
6. E.M. Baker, R.E. Chaddock, R.A. Lindsay and R. Werner, Ind.Eng.Chem., 31(1939)1263.
7. B. Choffe and L. Asselineau, Rev.Inst.Franc.Petrole, 11(1956)948.
8. A.I. Altsybeeva and A.G. Morachevsky, Zh.Fiz.Khim., 38(1964)1574.
9. A.N. Campbell and W.J. Dulmage, J.Am.Chem.Soc., 70(1948) 1723.
10. N.D. Litvinov, Zh.Fiz.Khim., 26(1952)1405.
11. K.S. Yuan, B.C.-Y. Lu, J.C.K. Ho and A.K. Desphande, J. Chem.Eng.Data, 8(1963)549.

12. D.F. Othmer, M.M. Chudgar and S.L. Levy, Ind.Eng.Chem., 44(1952)1872.
13. Y. Nishi, J.Chem.Eng. Japan, 12(1972)334.
14. H.R.C. Pratt, Trans.Inst.Chem.Eng., 25(1947)43.
15. H.H. Amer, R.R. Paxton and M. Van Winkle, Ind.Eng.Chem., 48(1956)142.
16. J.C. Chu, R.J. Getty, L.F. Brennecke and R. Paul, "Distillation Equilibrium Data", Reinhold, New York, 1950.
17. I. Nagata, J.Chem.Eng. Data, 10(1965)106.
18. I. Nagata, J.Chem.Eng. Data, 7(1962)367.
19. G.O. Oakeson and J.H. Weber, J.Chem.Eng. Data, 5(1960)279.
20. I. Nagata, J.Chem.Eng. Data, 7(1962)360.
21. G. Scatchard and L.B. Ticknov, J.Am.Chem.Soc., 74(1952)3724.
22. M. Benedict, C.A. Johnson, E. Solomon and L.C. Rubin, Trans.Am.Inst.Chem.Eng., 41(1945)371.
23. R.P. Tripathi and L. Asselineau, J.Chem.Eng. Data, 20(1975)33.
24. S.E. Kharin, V.M. Perelygin and V.G. Pisarevsky, Zh.Prikl.Khim. (Leningrad), 45(1972)466.
25. S.J. Fu and B.C.-Y. Lu, J.Chem.Eng. Data, 13(1968)6.
26. L.N. Canjar, E.C. Horni and R.R. Rothfus, Ind.Eng.Chem., 48(1956)427.
27. K.C. Chao, Ph.D., Thesis, Univ. of Wisconsin, 1956.
28. K. Ochi, K. Kojima, Kagaku Kogaku, 33(1969)352.
29. S.V. Babish, R.A. Ivanchikova and L.A. Serafimov, Zh. Prikl.Khim., 42(1969)1354.
30. D.C. Freshwater and K.A. Pike, J.Chem.Eng. Data, 12(1967)179.
31. H.A. Clarke and R.W. Missen, J.Chem.Eng. Data, 19(1974)343.
32. I. Mertl, Coll.Czech.Chem.Commun., 37(1972)366.
33. J.M. Willock and M. Van Winkle, J.Chem.Eng. Data, 15(1970)281.
34. G.R. Garrett and M. Van Winkle, J.Chem.Eng. Data, 14(1969)302.
35. A.O. Delzenne, J.Chem.Eng. Data, 3(1958)224.

36. B.V.S. Rao and C.V. Rao, J.Chem.Eng. Data, 11(1966)158.
37. D.A. Palmer and D.B. Smith, J.Chem.Eng. Data, 17(1972)71.
38. I.R. Matocha Jr. and M. Van Winkle, J.Chem.Eng. Data, 16(1971)61.
39. J. Mesnage and A.A. Marsan, J.Chem.Eng. Data, 16(1974)434.
40. R.A. Dawe, D.M.T. Newman and S.B. Ng, J.Chem.Eng. Data, 18(1973)44.
41. A.N. Gorbunov, M.P. Susarev and I.M. Balashova, Zh.Prikl. Khim., 41(1968)312.
42. G.J. Rao, P. Dakshinamurti and C.V. Rao, J.Sci.Ind. Research, 18B(1959)231.
43. A.G. Morachevsky and N.P. Leontev, Zh.Fiz.Khim., 34(1960) 2347.
44. A.E. Karr, E.G. Scheibel, W.M. Bowes and D.F. Othmer, Ind. Eng.Chem., 43(1951)961.
45. V.V.G. Krishnamurty, K.V.K. Rao and C.V. Rao, J.Sci.Ind. Research, 21D(1962)312.
46. R. Nielsen and J.H. Weber, J.Chem.Eng. Data, 4(1959)145.
47. I. Nagata, J.Chem.Eng. Data, 7(1962)461.
48. M.B. Donald and K. Ridgway, J.Appl.Chem., 8(1958)408.
49. P.S. Murti and M. Van Winkle, AIChE Jornal, 3(1957)517.
50. N.D. Litvinov, Zh.Fiz.Khim., 26(1952)1405.
51. V.V. Udovenko and L.G. Fatkulina, Zh.Fiz.Khim., 26(1952) 719.
52. L.S. Kudryavtseva and M.P. Susarev, Zh.Prikl.Khim., 36(1963)1710.
53. A.G. Morachevsky and C.T. Chen, Zh.Fiz.Khim., 35(1961) 2335.
54. K.T. Ma, C. McDermott and S.R.M. Ellis, J.Chem.E.Symp.Ser. No. 32 (Inst. of Chem.Engineers, London), 3:89(1969).
55. J.H. Weber et al., J.Chem.Eng. Data, 5(1960)243, 6(1961) 485, 7(1962)344.

Table 7.1

DEVIATIONS BETWEEN EXPERIMENTAL AND PREDICTED VAPOR-LIQUID EQUILIBRIUM CONDITIONS IN TERNARY SYSTEMS

	T or P	ΔP or ΔT	Δy	No. of Data Points	Ref.
2-Propanol–Water–Toluene	760 mm Hg	2.4 K	0.045	34	1
Methanol–Ethanol–Water	760 mm Hg	0.4 K	0.010	14	2
Ethanol–1,2-Dichloroethane–Water	318 K	9.2 mm Hg	0.034	34	3
Ethanol–Ethyl acetate–Water	760 mm Hg	3.0 K	0.038	96	4
Ethanol–Water–1,4-Dioxane	760 mm Hg	1.1 K	0.022	19	5
Ethanol–2-Ethoxyethanol–Water	760 mm Hg	0.8 K	0.018	67	6
Acetone–2-Propanol–Water	760 mm Hg	0.8 K	0.019	87	7
2-Butanol–2-Butanone–Water	400 mm Hg	1.9 K	0.034	33	8
Tetrachloromethane–Ethanol–Benzene	760 mm Hg	0.9 K	0.020	51	9
Tetrachloromethane–Ethanol–Benzene	323 K	9.4 mm Hg	0.017	36	10
Tetrachloromethane–Cyclohexane–2-Propanol	760 mm Hg	1.4 K	0.027	58	11
Acetone–2-Butanone–Water	2586 mm Hg	1.9 K	0.013	37	12
Ethanol–2-Propanol–Ethyl acetate	760 mm Hg	1.0 K	0.015	6	13
Acetone–Acetonitrile–Water	760 mm Hg	1.4 K	0.026	30	14
Acetone–Methanol–Ethanol	760 mm Hg	0.7 K	0.014	83	15
Acetone–Methanol–Water	760 mm Hg	1.6 K	0.018	54	16
Tetrachloromethane–Benzene–2-Propanol	760 mm Hg	0.6 K	0.011	39	17
Chloroform–Methanol–Ethyl acetate	760 mm Hg	0.6 K	0.011	70	18
Ethanol–Benzene–Heptane	180 mm Hg	1.4 K	0.030	54	19

Table 7.1 cont.

System	P/T			
Methyl acetate–Chloroform–Benzene	760 mm Hg	0.2 K	0.007	91	20
Methanol–Tetrachloromethane–Benzene	328 K	13.3 mm Hg	0.011	8	21
Methanol–Heptane–Toluene	760 mm Hg	0.4 K	0.008	8	22
Benzene–Heptane–Acetonitrile	760 mm Hg	1.3 K	0.023	36	23
Ethanol–Water–3-Methyl-1-propanol	760 mm Hg	1.4 K	0.018	31	24
Benzene–Heptane–1-Propanol	348 K	10 mm Hg	0.018	77	25
Acetone–Benzene–1,2-Dichloroethane	760 mm Hg	0.7 K	0.017	106	26
Ethyl acetate–Benzene–Cyclohexane	760 mm Hg	0.4 K	0.005	77	27
Methanol–1-Propanol–Water	760 mm Hg	0.4 K	0.014	53	28
Methanol–2-Propanol–Water	760 mm Hg	1.0 K	0.012	56	28
Ethanol–1-Propanol–Water	760 mm Hg	0.5 K	0.015	51	28
Acetone–2-Butanone–Methyl acetate	760 mm Hg	0.4 K	0.012	16	29
Acetone–2-Butanone–Ethyl acetate	760 mm Hg	0.3 K	0.024	15	29
Ethyl acetate–Ethanol–2-Butanone	760 mm Hg	0.4 K	0.021	16	29
Acetone–Methanol–2-Propanol	760 mm Hg	0.3 K	0.010	21	30
Acetonitrile–Benzene–Tetrachloromethane	318 K	4.6 mm Hg	0.007	62	31
Ethyl acetate–Ethanol–Water	348 K	69 mm Hg	0.041	9	32
2,3-Dimethylbutane–Methanol–Acetone	760 mm Hg	0.5 K	0.016	27	33
Methanol–Acetone–Chloroform	760 mm Hg	0.2 K	0.003	5	33
Acetone–Chloroform–2,3-Dimethylbutane	760 mm Hg	1.3 K	0.020	19	34
Methanol–Ethanol–Water	760 mm Hg	1.3 K	0.029	20	35
Acetone–Tetrachloromethane–Benzene	760 mm Hg	0.7 K	0.010	58	36
Acetonitrile–Benzene–Heptane	318 K	13 mm Hg	0.020	51	37

Table 7.1 cont.

System	Pressure		Value	Ref.
2-Butanone-2-Propanol-Water	760 mm Hg	0.7 K	0.017	17	38
Cyclohexane-Cyclohexene-1,2-Dichloroethane	760 mm Hg	0.6 K	0.012	50	39
Water-1-Propanol-1-Butanol	762 mm Hg	0.3 K	0.018	16	40
Isobutyl acetate-1-Butanol-Toluene	760 mm Hg	0.8 K	0.010	21	41
Tetrachloromethane-2-Butanone-Cyclohexane	760 mm Hg	1.0 K	0.008	49	42
Chloroform-Acetone-Ethanol	760 mm Hg	0.7 K	0.013	25	43
Acetone-Chloroform-4-Methyl-2-pentanone	760 mm Hg	0.3 K	0.010	46	44
Chloroform-2-Butanone-Benzene	760 mm Hg	0.3 K	0.009	36	45
Ethanol-Benzene-Heptane	400 mm Hg	2.1 K	0.032	50	46
Methyl acetate-Benzene-Cyclohexane	760 mm Hg	0.3 K	0.011	60	47
Benzene-Cyclohexane-2-Butanone	760 mm Hg	0.2 K	0.011	86	48
Octane-Ethylbenzene-2-Ethoxyethanol	760 mm Hg	5.4 K	0.038	71	49
Methanol-Methyl acetate-Ethyl acetate	318 K	4.1 mm Hg	0.019	36	50
Ethanol-1,2-Dichloroethane-Benzene	318 K	5.4 mm Hg	0.011	63	51

Table 7.2 EXPERIMENTAL AND CALCULATED ACTIVITY COEFFICIENTS IN THE SYSTEM METHANOL(1)-2-PROPANOL(2)-WATER(3) AT 1 ATM.(16)

x_1	x_2	γ_1(exp)	γ_1(calc)	γ_2(exp)	γ_2(calc)	γ_3(exp)	γ_3(calc)
0.086	0.141	1.283	1.139	3.149	3.159	1.104	1.167
0.100	0.315	1.062	0.922	1.754	1.641	1.419	1.478
0.113	0.515	1.069	0.890	1.159	1.205	1.975	1.928
0.132	0.073	1.140	1.291	5.913	4.327	0.910	1.099
0.143	0.054	1.543	1.349	4.397	4.877	1.033	1.081
0.170	0.205	1.173	1.010	2.043	1.992	1.269	1.334
0.175	0.058	1.402	1.290	4.002	4.204	1.038	1.102
0.194	0.071	1.379	1.228	3.563	3.589	1.116	1.131
0.194	0.148	1.157	1.072	2.236	2.357	1.121	1.252
0.208	0.091	1.306	1.165	2.991	3.035	1.134	1.169
0.230	0.090	1.241	1.151	2.700	2.882	1.146	1.182
0.248	0.154	1.124	1.047	2.095	2.072	1.297	1.303
0.308	0.179	1.152	1.014	1.721	1.736	1.301	1.395
0.419	0.134	1.052	1.024	1.759	1.685	1.334	1.397
0.442	0.130	1.046	1.023	1.734	1.652	1.474	1.408
0.547	0.103	1.043	1.018	1.517	1.554	1.348	1.441

$\Delta T = 0.6$ K; $\Delta y = 0.022$

Table 7.3 EXPERIMENTAL AND PREDICTED ACTIVITY COEFFICIENTS IN THE SYSTEM ACETONE(1)-CHLOROFORM(2)-HEXANE(3) AT 318 K (52)

Note that this system exhibits large positive and negative deviations from Raoult's Law

x_1	x_2	$\gamma_1(\exp)$	$\gamma_1(\text{calc})$	$\gamma_2(\exp)$	$\gamma_2(\text{calc})$	$\gamma_3(\exp)$	$\gamma_3(\text{calc})$
0.103	0.800	0.497	0.633	0.990	0.976	1.738	1.994
0.107	0.703	0.584	0.800	1.027	0.974	1.692	1.701
0.203	0.602	0.750	0.860	0.913	0.916	1.833	1.975
0.399	0.510	0.877	0.840	0.795	0.815	2.563	3.225
0.200	0.503	0.927	1.061	0.902	0.916	1.564	1.659
0.496	0.408	0.939	0.905	0.731	0.748	2.665	3.449
0.298	0.402	1.064	1.100	0.791	0.841	1.648	1.832
0.098	0.400	1.188	1.678	1.049	1.046	1.238	1.199
0.499	0.303	1.034	1.015	0.695	0.712	2.161	2.623
0.298	0.301	1.318	1.315	0.812	0.841	1.434	1.573
0.096	0.299	1.548	2.111	1.057	1.084	1.143	1.126
0.599	0.202	1.098	1.047	0.567	0.645	2.395	2.737
0.397	0.206	1.348	1.308	0.625	0.761	1.562	1.683
0.195	0.201	1.756	1.972	0.885	0.969	1.233	1.199
0.805	0.103	1.017	1.008	0.487	0.550	3.773	3.874
0.599	0.102	1.224	1.160	0.480	0.626	2.104	2.176
0.397	0.102	1.617	1.534	0.551	0.763	1.450	1.460
0.195	0.100	2.177	2.391	0.822	0.996	1.162	1.125
0.065	0.600	0.866	1.140	1.022	1.027	1.328	1.322

$\Delta P = 18$ mm Hg; $\Delta y = 0.020$

Table 7.4 EXPERIMENTAL AND PREDICTED ACTIVITY COEFFICIENTS IN THE SYSTEM BENZENE(1)-CYCLOHEXANE(2)-1-PROPANOL(3) AT 1 ATM. (53)

x_1	x_2	$\gamma_1(\exp)$	$\gamma_1(\text{calc})$	$\gamma_2(\exp)$	$\gamma_2(\text{calc})$	$\gamma_3(\exp)$	$\gamma_3(\text{calc})$
0.084	0.155	2.049	2.205	2.561	2.783	1.197	1.055
0.166	0.077	2.000	2.157	2.677	2.868	1.181	1.054
0.089	0.331	1.639	1.783	1.929	2.050	1.392	1.218
0.177	0.247	1.599	1.737	1.945	2.108	1.440	1.208
0.350	0.081	1.641	1.664	2.133	2.269	1.306	1.204
0.181	0.341	1.511	1.560	1.658	1.796	1.636	1.372
0.095	0.532	1.375	1.446	1.389	1.482	2.062	1.717
0.190	0.440	1.391	1.402	1.459	1.521	1.870	1.680
0.374	0.261	1.268	1.333	1.570	1.632	1.880	1.632
0.554	0.086	1.273	1.287	1.684	1.798	1.984	1.625
0.197	0.548	1.254	1.279	1.232	1.291	2.559	2.312
0.293	0.453	1.224	1.242	1.312	1.332	2.402	2.238
0.482	0.268	1.158	1.182	1.428	1.444	2.361	2.150
0.574	0.178	1.135	1.161	1.416	1.520	2.442	2.130
0.204	0.663	1.657	1.209	1.518	1.115	6.827	3.918
0.402	0.467	1.090	1.130	1.136	1.184	4.163	3.543
0.689	0.183	1.029	1.057	1.313	1.375	3.836	3.229

$\Delta T = 0.6$ K; $\Delta y = 0.022$

Figures 7.1-7.12

COMPARISON BETWEEN EXPERIMENTAL AND PREDICTED VAPOR PHASE MOLE FRACTIONS IN TERNARY SYSTEMS

Figure 7.1

Acetone(1)-2-Butanone(2)-Water(3) at 2585.7 mm Hg
D.F. Othmer, M.M. Chudgar and S.L. Levy, Ind.Eng.Chem., 44(1952)1872.

Figure 7.2

Acetone(1)-Methanol(2)-Water(3) at 760 mm Hg
J.C. Chu, R.J. Getty, L.F. Brennecke and R.P. Paul, Distillation Equilibrium Data, New York 1950.

Figure 7.3

Tetrachloromethane(1)-Benzene(2)-2-Propanol(3) at 760 mm Hg
I. Nagata, J.Chem.Eng. Data, 10(1965)106.

Figure 7.4

Methyl acetate(1)-Chloroform(2)-Benzene(3) at 760 mm Hg
I. Nagata, J.Chem.Eng. Data, 7(1962)360.

Figure 7.5

Ethanol(1)-Water(2)-3-Methyl-1-butanol(3) at 760 mm Hg
S.E. Kharin, V.M. Perelygin and V.G. Pisarevsky, Zh.Prikl.Khim. (Leningrad), 45(1972)466.

Figure 7.6

Acetone(1)-Benzene(2)-1,2-Dichloroethane(3) at 760 mm Hg
L.N. Canjar, E.C. Horni and R.R. Rothfus, Ind.Eng.Chem., 48(1956)427.

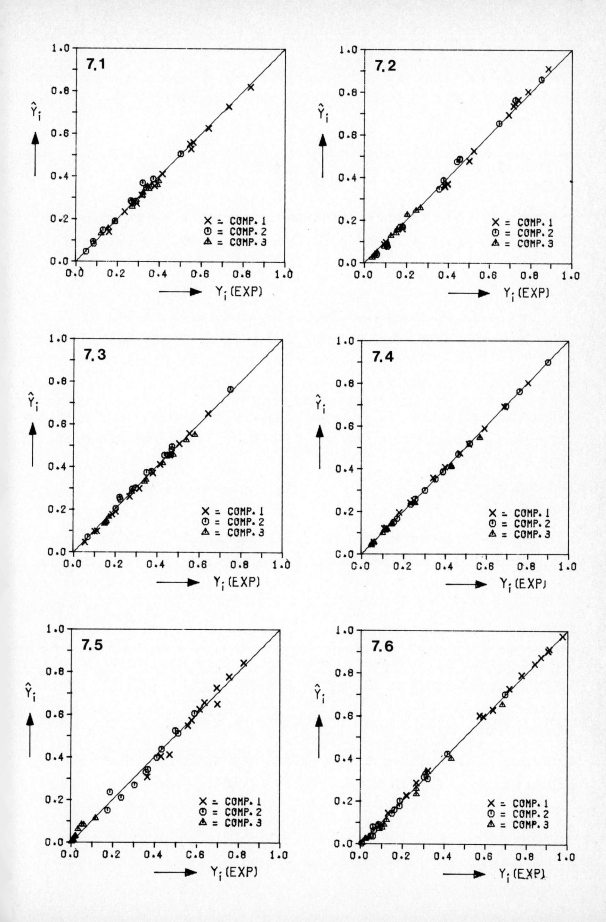

Figure 7.7

Acetonitrile(1)-Benzene(2)-Tetrachloromethane(3) at 45.0 °C
H.A. Clarke and R.W. Missen, J.Chem.Eng. Data, 19(1974)343.

Figure 7.8

Acetone(1)-Chloroform(2)-2,3-Dimethylbutane(3) at 760 mm Hg
G.R. Garrett and M. Van Winkle, J.Chem.Eng. Data, 14(1969)302.

Figure 7.9

Tetrachloromethane(1)-2-Butanone(2)-Cyclohexane(3) at 760 mm Hg
G.J. Rao, P. Dakshinamurty and C.V. Rao, J.Sci.Ind.Research, 18B(6)(1959)231.

Figure 7.10

Acetone(1)-Chloroform(2)-4-Methyl-2-pentanone(3) at 760 mm Hg
A.E. Karr, E.G. Scheibel, W.M. Bowes and D.F. Othmer, Ind.Eng. Chem., 43(1951)961.

Figure 7.11

Methyl acetate(1)-Benzene(2)-Cyclohexane(3) at 760 mm Hg
I. Nagata, J.Chem.Eng. Data, 7(1962)461.

Figure 7.12

Methanol(1)-Methyl acetate(2)-Ethyl acetate(3) at 40.0 °C
N.D. Litvinov, Zh.Fiz.Khim., 26(1952)1405.

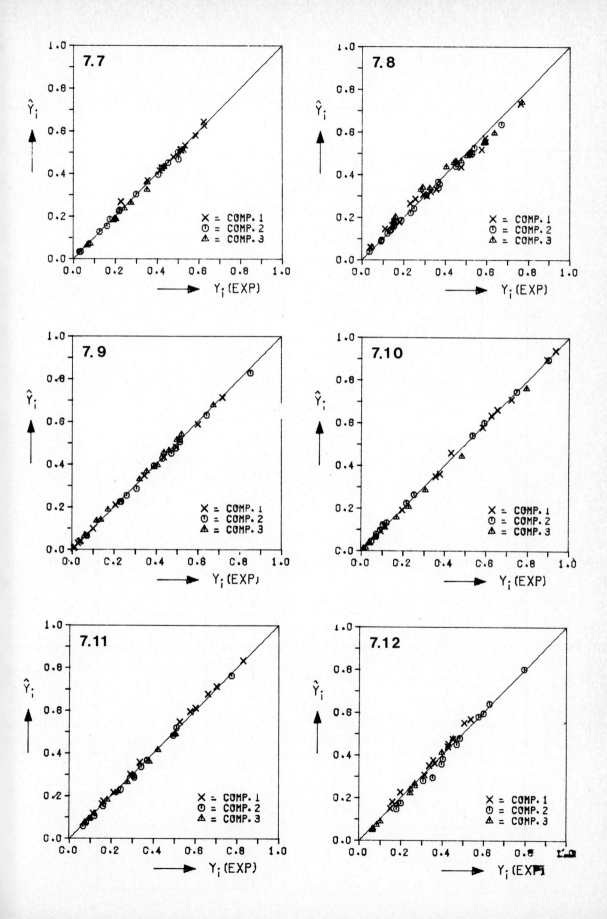

Table 7.5 EXPERIMENTAL AND PREDICTED ACTIVITY COEFFICIENTS FOR THE SYSTEM 1,2-DICHLOROETHANE(1)-1-PROPANOL(2)-TOLUENE(3)-ACETONE(4) AT 700 MM HG (54)

x_1	x_2	x_3	γ_1(exp)	γ_1(calc)	γ_2(exp)	γ_2(calc)	γ_3(exp)	γ_3(calc)	γ_4(exp)	γ_4(calc)
0.613	0.232	0.056	1.17	1.10	1.84	1.96	1.26	1.15	0.79	0.91
0.515	0.168	0.097	1.09	1.05	2.07	2.16	1.32	1.15	0.87	0.95
0.307	0.335	0.281	1.19	1.16	1.60	1.63	1.34	1.26	0.90	1.01
0.307	0.207	0.419	1.09	1.05	2.07	2.14	1.18	1.11	0.99	1.01
0.202	0.512	0.251	1.40	1.38	1.28	1.29	1.56	1.51	0.97	1.10
0.251	0.267	0.252	1.12	1.10	1.67	1.73	1.38	1.26	0.96	1.02
0.064	0.039	0.861	0.98	0.98	3.35	3.57	1.02	1.01	1.43	1.27
0.063	0.089	0.072	0.94	0.93	1.75	1.90	1.78	1.52	0.97	1.01

Mean deviations between experimental and calculated vapor-phase mole fractions:

Component	UNIFAC Method	Raoult's Law
1	0.009	0.030
2	0.006	0.087
3	0.007	0.036
4	0.016	0.022

Table 7.6 EXPERIMENTAL AND PREDICTED ACTIVITY COEFFICIENTS FOR THE SYSTEM HEXANE(1)-METHYLCYCLOPENTANE(2)-ETHANOL(3)-BENZENE(4) AT 760 MM HG (55)

x_1	x_2	x_3	γ_1(exp)	γ_1(calc)	γ_2(exp)	γ_2(calc)	γ_3(exp)	γ_3(calc)	γ_4(exp)	γ_4(calc)
0.733	0.052	0.157	1.12	1.09	1.20	1.13	4.27	4.46	1.70	1.42
0.743	0.105	0.107	1.06	1.05	1.15	1.08	5.90	5.85	1.55	1.40
0.732	0.111	0.035	1.01	1.01	1.07	1.03	12.40	9.20	1.41	1.36
0.742	0.057	0.090	1.06	1.04	1.10	1.07	6.51	6.26	1.38	1.36
0.139	0.166	0.514	2.11	2.08	1.97	1.97	1.50	1.47	1.79	1.73
0.156	0.510	0.154	1.20	1.15	1.16	1.12	4.20	4.24	1.34	1.28
0.330	0.076	0.515	1.92	1.90	1.91	1.86	1.49	1.53	1.84	1.82
0.354	0.208	0.365	1.43	1.45	1.43	1.44	2.06	2.12	1.57	1.56
0.359	0.360	0.165	1.18	1.13	1.14	1.12	4.05	4.15	1.40	1.34
0.371	0.502	0.034	1.20	1.02	1.05	1.02	12.06	9.68	1.35	1.34

Mean deviations between experimental and calculated vapor phase mole fractions:

Component	UNIFAC Method	Raoult's Law
1	0.011	0.079
2	0.012	0.056
3	0.026	0.225
4	0.010	0.040

CHAPTER 8

PREDICTION OF PHASE-SPLITTING AND EXCESS ENTHALPY

Using the tools of classical thermodynamics one finds that models for predicting vapor-liquid equilibria can, in theory, also be used for predicting liquid-liquid equilibria and excess enthalpies. It is, however, well known that models suited for vapor-liquid equilibria most often from a quantitative standpoint are inadequate for these purposes. The Wilson equation, while excellent for correlating vapor-liquid equilibria, is not able to predict phase-splitting at all. Other models, f. ex. NRTL and UNIQUAC, can predict both vapor-liquid and liquid-liquid equilibria. One often finds, however, that two different sets of model parameters should be used for the two types of equilibria. The parameters obtained for representing vapor-liquid equilibrium are not the best for predicting liquid-liquid equilibria and vice versa. This is discussed in some detail in [1].

The UNIFAC method can qualitatively predict phase-splitting but the same difficulties as described above are encountered. As already pointed out, accurate prediction of vapor-liquid equilibrium compositions is the main goal of the UNIFAC method presented in this book. Therefore, the UNIFAC parameters are as far as possible based on vapor-liquid equilibrium data only. This means that no excess enthalpy data were included in the data base, and liquid-liquid equilibrium data were only included where absolutely necessary, f.ex. in hydrocarbon-Water systems.

It follows that one can not expect quantitative prediction of excess enthalpy and phase-splitting when the UNIFAC method is used in its present form. On the other hand, attempts for order-of-magnitude estimates may be justified, f.ex. as warnings of possible phase-splitting in distillation columns and in estimating the magnitude of enthalpy of mixing effects. In the following two sections we – for better or for worse – show typical examples of predicted liquid-liquid equilibrium compositions and excess enthalpies from the UNIFAC parameters given in Chapter 4.

8.1 PHASE-SPLITTING

Consider a mixture with M components. The condition for equilibrium between liquid phase I and liquid phase II is

$$x_i^I \gamma_i^I = x_i^{II} \gamma_i^{II} \qquad i = 1,2 \ldots M \qquad (8.1)$$

The M equations of type (8.1) and M independent component material balances form a set of equations, which may be solved for the 2M unknowns: x_i^I and x_i^{II} for all i. This is described in more detail in Appendix 2, which also gives a computer program for carrying out the calculations.

Prediction of liquid-liquid equilibrium compositions has been carried out for several three-component systems exhibiting phase-splitting. The data are taken from Landolt-Börnstein [2], and the results of the predictions are summarized in Table 8.1.

Table 8.1 only includes predictions for ternary systems of type one. They have only one binary exhibiting phase-splitting and they have a "Plait point", see Figure 8.1. Ternary liquid-liquid equilibrium systems of type two have two binaries which exhibit phase-splitting, see Figure 8.2. The phase behaviour of the latter (type 2) is easier to predict than that of the former (type 1). The references given in Table 8.1 are the mixture numbers used by Landolt-Börnstein, which lists the original sources of the experimental data. In the column "remarks", the agreement between experiment and prediction is graded as follows:

  0 no phase-splitting could be predicted

  1 the agreement between the predicted and experimental liquid-liquid solubility curves is poor

  2 the predictions agree qualitatively with experimental values

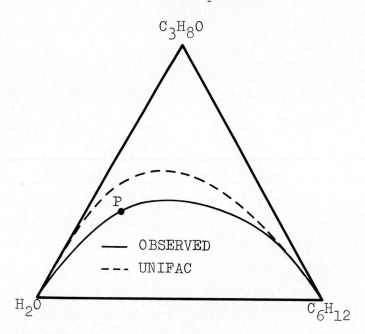

Figure 8.1
LIQUID-LIQUID EQUILIBRIA IN A TYPE ONE SYSTEM:
Water-Cyclohexane-2-Propanol
(For data see [2])
P = Plait point

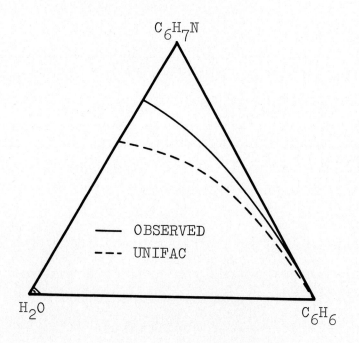

Figure 8.2
LIQUID-LIQUID EQUILIBRIA IN A TYPE TWO SYSTEM:
Water-Benzene-Aniline
(For data see [2])

3 the predictions agree quantitatively with experimental values

The UNIFAC method is seen to yield a phase split in almost all cases where such occurs. Only for systems with water plus esters and systems with aniline plus hydrocarbons the method yields no phase split. One should not, however, use the present UNIFAC parameters as a basis for routine prediction of liquid-liquid equilibrium compositions in extraction process design calculations. The agreement between predicted and experimental solubility curves is often qualitatively correct (indicated by "2" in Table 8.1), but usually not quantitatively acceptable. Several examples of predictions of liquid-liquid equilibria from Table 8.1 are shown in Figures 8.3-8.10.

8.2 EXCESS ENTHALPY

The Gibbs-Helmholz equation

$$\left(\frac{\partial (G^E/T)}{\partial T}\right)_{P,x} = -H^E/T^2 \tag{8.2}$$

where the differentiation is at constant pressure and composition, permits calculation of the excess enthalpy from a model for the excess Gibbs free energy. This is done here according to UNIFAC, i.e.

$$G^E/RT = \sum^i x_i (\ln \gamma_i^R + \ln \gamma_i^C) \qquad i = 1, 2 \ldots M \tag{8.3}$$

Since the combinatorial activity coefficients do not depend upon temperature, only the residual part enters into the calculation of the excess enthalpy. The result is

$$H^E = -RT \sum^i \sum^k x_i \nu_k^{(i)} \left[T\left(\frac{\partial \ln \Gamma_k}{\partial T}\right)_{P,x} - T\left(\frac{\partial \ln \Gamma_k^{(i)}}{\partial T}\right)_{P,x} \right] \tag{8.4}$$

where

$$T\left(\frac{\partial \ln \Gamma_k}{\partial T}\right)_{P,x} = Q_k \sum_m^m \Theta_m \left[\frac{\Psi_{mk} \ln \Psi_{mk}}{\sum_n^n \Theta_n \Psi_{nk}} + \frac{\Psi_{km} \sum_n^n \Theta_{nm} \Psi_{nm} \ln(\Psi_{km}/\Psi_{nm})}{(\sum_n^n \Theta_n \Psi_{nm})^2}\right]$$

$$i = 1, 2 \ldots M; \quad k, m, \text{ and } n = 1, 2 \ldots N$$

$$\Psi_{nm} = \exp(-a_{nm}/T)$$

A similar expression holds for $T\left(\dfrac{\partial \ln \Gamma_k^{(i)}}{\partial T}\right)_{P,x}$

Equation (8.4), with the group surface area constants Q_k from Table 4.3, and the group-interaction parameters a_{nm} in Table 4.4, were used to calculate excess enthalpies for twelve equimolar, binary mixtures, picked at random from the literature. The results are compared with experimental values in Table 8.2.

UNIFAC usually predicts large positive or large negative excess enthalpies to within a factor of two. Systems with experimental excess enthalpies near zero are usually predicted to have small values, but the sign may be wrong.

The magnitude of the excess enthalpy, up to a few thousand J/mole, is often small compared with the heat of evaporation, typically 50,000 J/mole. Since the excess enthalpy enters into the energy balances together with the heat of evaporation, the quality of predictions shown in Table 8.2 is often good enough for distillation column design.

One may conclude from the results in Table 8.2 that it is about as accurate to neglect excess enthalpy effects in distillation column design calculations as to include those from the present UNIFAC parameters. It is, however, reasonable to use UNIFAC for an approximate estimate of the influence of the excess enthalpy term on distillation calculations. If the reader desires an accurate method for predicting the excess enthalpy, he should replace Subroutine EXCES in Appendix 4 with a better method.

In [10] it is shown that the solution-of-groups concept may be used for accurate prediction of excess enthalpies, when the parameters are fitted to excess enthalpy data.

REFERENCES

1. H. Renon, L. Asselineau, G. Cohen and C. Raimbault, Calcul sur Ordinateur des Équilibres Liquide-Vapeur et Liquide-Liquide, Technip, Paris, 1971.
2. "Landolt-Börnstein Zahlenwerten und Funktionen", Volume II, Part 2c, ed. by K. Schäfer and E. Lax, Springer-Verlag, Berlin, 1964, pp. 527-697.
3. C.G. Savini, D.R. Winterhaller and H.C. Van Ness, J.Chem. Eng. Data, 10(1965)168.
4. H.D. Pflug, A.E. Pope and G.C. Benson, J.Chem.Eng. Data, 13(1968)409.
5. J.-E. A. Otterstedt and R.W. Missen, J.Chem.Eng. Data, 11(1966)361.
6. M.K. Duttachoudkury and H.B. Mathur, J.Chem.Eng. Data, 19(1974)145.
7. I. Nagata, T. Yamada and S. Nakagawa, J.Chem.Eng. Data, 20(1975)271.
8. I. Nagata, M. Nagashima, K. Kazuma and M. Nakagawa, J.Chem. Eng., Japan, 8(1975)261.
9. K.W. Morcom, Int. DATA Series [A] (1973)57.
10. T.H. Nguyen and G.A. Ratcliff, Can.J.Chem.Eng., 49(1971)120.

Table 8.1

PREDICTIONS OF TERNARY LIQUID-LIQUID EQUILIBRIA OF TYPE ONE

Summary of Results

System	Temperature °C	Reference*	Remarks
Water-Hexane-1-Propanol	38	LB 1g	1
Water-Hexane-2-Butanone	38	LB 2	2
Water-Benzene-Acetaldehyde	18	LB 4	2
Water-Benzene-1-Propanol	20	LB 5	2
Water-Benzene-2-Propanol	25	LB 8	2
Water-Benzene-1,4-Dioxane	25	LB 10	2
Water-Toluene-Acetaldehyde	17	LB 17	2
Water-Toluene-2-Propanol	25	LB 18	1
Water-Toluene-Propanoic acid	31	LB 18a	2
Water-Octane-Propanoic acid	25	LB 20	2
Water-Xylene-Acetone	25	LB 22	2
Water-Chloroform-Acetic acid	25	LB 29	2
Water-Chloroform-2-Propanol	25	LB 29a	2
Water-Chloroform-Acetone	25	LB 30	1
Water-Tetrachloromethane-1-Propanol	20	LB 31	3
Water-Tetrachloromethane-Acetone	30	LB 32	2
Water-1,2-Dichloroethane-Ethanol	30	LB 34	2
Water-Acetone-1,1,2-Trichloroethane	25	LB 49	1

Table 8.1 cont.

System			
Water-Methyl acetate-Acetone	30	LB 49a	0
Water-1-Butanol-Methanol	15	LB 50	1
Water-1-Butanol-Acetic acid	27	LB 51	1
Water-Diethyl ether-Acetone	25	LB 56	3
Water-Diethyl ether-Acetic acid	25	LB 57	1
Water-Diethyl ether-Propanoic acid	30	LB 58a	2
Water-2-Butanone-Acetone	25	LB 62	1
Water-Ethyl acetate-Methanol	20	LB 63	0
Water-1-Pentanol-Ethanol	25	LB 73	2
Water-1-Pentanol-Acetaldehyde	18	LB 74	2
Water-1-Hexanol-Methanol	28	LB 86a	2
Water-4-Methyl-2-pentanol-Acetic acid	30	LB 86b	2
Water-Diisopropyl ether-Acetic acid	23	LB 87	2
Water-4-Methyl-2-pentanone-Acetic acid	22	LB 90	2
Water-4-Methyl-2-pentanone-Acetone	25	LB 91	1
Water-p-Cresol-Methanol	35	LB 102	2
Water-Dibutyl ether-Ethanol	25	LB 106	2
Water-Diisobutyl ketone-Acetic acid	23	LB 108	2
Water-Aniline-Propanoic acid	20	LB 127	2
Hexane-Water-Acetic acid	31	LB 142a	3
Hexane-Water-Propanoic acid	31	LB 142b	3

Table 8.1 cont.

Cyclohexane-Water-Ethanol	25	LB 144	2
Benzene-Water-Ethanol	20	LB 145	2
Benzene-Water-Acetic acid	60	LB 146	3
Heptane-Water-Ethanol	30	LB 149	3
Toluene-Water-Ethanol	25	LB 151	2
Toluene-Water-Acetic acid	25	LB 152	3
2,2,4-Trimethylpentane-Water-Ethanol	25	LB 152b	1
Ethylbenzene-Water-2-Ethoxy ethanol (cellosolve)	20	LB 152f	1
Tetrachloromethane-Water-Acetic acid	28	LB 159	3
Acetonitrile-Heptane-Benzene	25	LB 239	2
Aniline-Toluene-Cyclohexane	20	LB 245	0

* The original literature references are given in [2], pp. 694-7

Figures 8.3-8.10

OBSERVED AND PREDICTED LIQUID-LIQUID EQUILIBRIA IN SOME TERNARY SYSTEMS

---- OBSERVED

——— UNIFAC

8.3   Water-Octane-Propanoic acid at 25 °C
      Ref [2]   LB20

8.4   Water-Chloroform-Acetic acid at 25 °C
      Ref [2]   LB29

8.5   Water-Chloroform-2-Propanol at 25 °C
      Ref [2]   LB29a

8.6   Water-1-Butanol-Methanol at 15 °C
      Ref [2]   LB50

8.7   Water-Diisobutyl ketone-Acetic acid at 23 °C
      Ref [2]   LB108

8.8   Cyclohexane-Water-Ethanol at 25 °C
      Ref [2]   LB144

8.9   Benzene-Water-Ethanol at 20 °C
      Ref [2]   LB145

8.10 Toluene-Water-Acetic acid at 25 °C
      Ref [2]   LB152

8.3

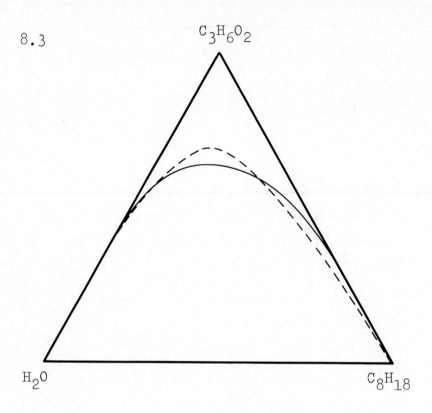

8.4

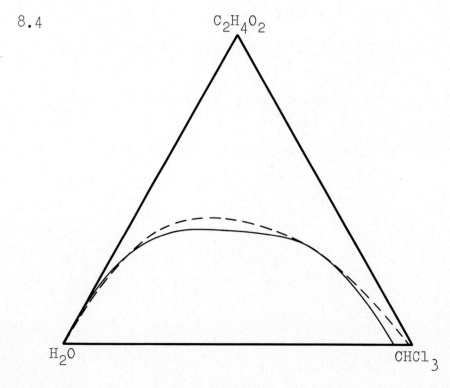

8.5

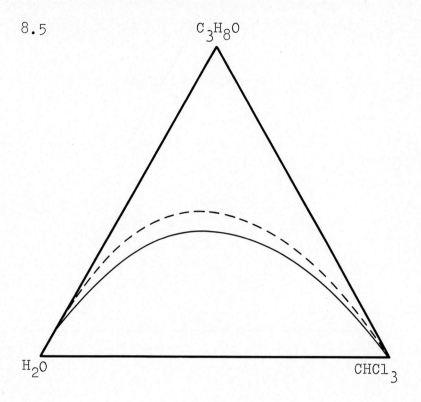

8.6

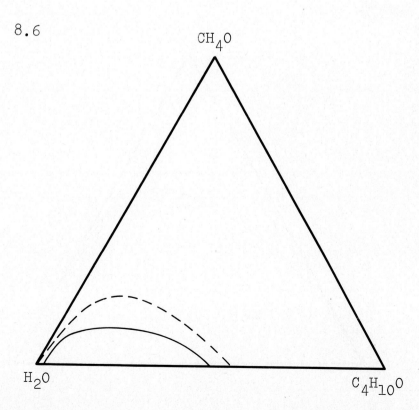

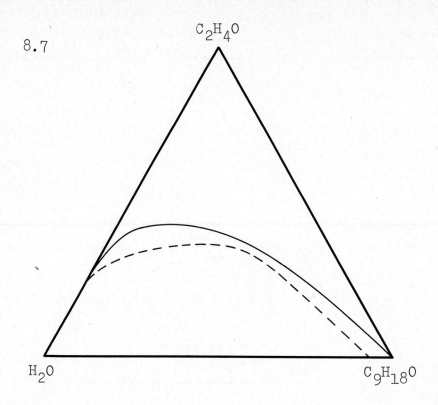

8.7

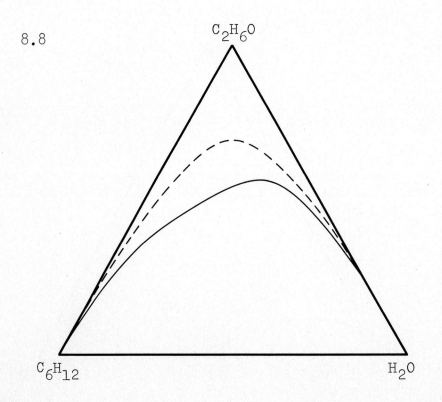

8.8

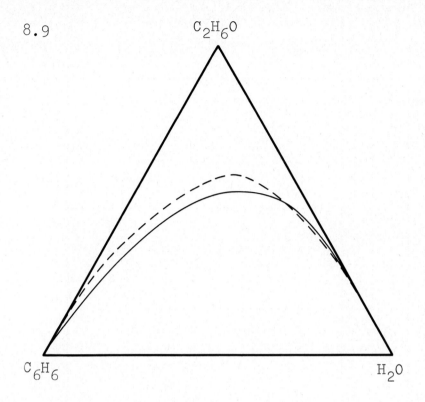

8.9

C_2H_6O

C_6H_6 H_2O

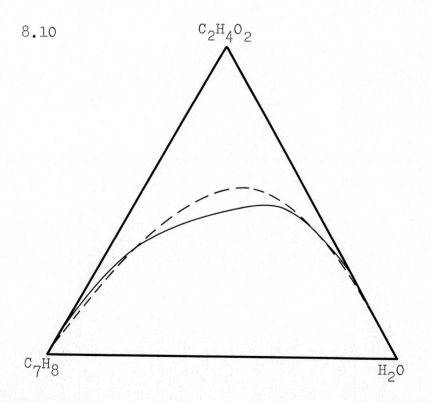

8.10

$C_2H_4O_2$

C_7H_8 H_2O

Table 8.2

EXPERIMENTAL AND PREDICTED EXCESS ENTHALPIES IN SOME BINARY SYSTEMS

System (First component is component 1)	Mole fraction of component 1 x_1	Temperature °C	Excess Enthalpy, J/mole		Ref.
			Predicted	Experimental	
Ethanol – Hexane	0.5	45	408	790	3
1-Butanol – Heptane	0.5	45	643	865	3
1-Octanol – Nonane	0.5	45	741	763	3
Ethanol – 1-Propanol	0.51	25	–35	19	4
Diethyl ether – Benzene	0.526	25	–19	–15	5
Water – Butylamine	0.52	40	–2035	–3370	6
Ethanol – Butylamine	0.5106	40	–1028	–2917	6
1-Propanol – Ethyl acetate	0.5068	25	372	1587	7
Acetic acid – Heptane	0.4841	35	1031	908	8
Propanoic acid – Heptane	0.487	35	909	625	8
Methyl acetate – Heptane	0.4862	25	885	1783	8
1,4-Dioxane – Benzene	0.5016	25	51	–32	9

CHAPTER 9

APPLICATION OF UNIFAC TO DISTILLATION COLUMN DESIGN

From the point of view of classical thermodynamics, the description of the UNIFAC method is now completed. We have seen that UNIFAC provides an accurate and reliable means of estimating phase equilibrium compositions for many mixtures of interest in chemical technology. There are, of course, numerous situations where UNIFAC may be employed in practice. In this chapter we illustrate the practical applicability of UNIFAC with an important example: interfacing UNIFAC with the design of simple distillation columns for the separation of multicomponent mixtures.

It is not the purpose of this chapter to give an "all-round" distillation column design program. The intention is to show by examples how UNIFAC may be incorporated into the reader's own column design procedures. We base our examples on the method for calculating multicomponent distillation by Naphtali and Sandholm [1].

9.1 MULTICOMPONENT DISTILLATION CALCULATIONS BY LINEARIZATION

The multicomponent column-design technique developed in [1] was chosen for our illustrations because it has the following advantages:

1. The presence of nonideal solutions and Murphree plate efficiencies are taken into account in a rigorous manner.

2. The method is flexible; the same algorithms may be used for distillation, extraction, absorption, and stripping; any number of feed- and sidestreams may be specified; and the conditions around the reboiler and condenser may be specified in any way the user wishes.

3. The method converges rapidly as the solution is approached.

The equations of conservation of mass and energy and of phase equilibrium are set up for each component and each stage. The resulting set of equations has a block-tridiagonal structure which permit rapid solution by Newton-Raphson iteration. The technique is briefly described in this section.

The distillation column configuration is shown in Figure 9.1. In the programs used here, distillation columns with up to 50 stages, up to ten components, a partial condenser, and any number of feed- and sidestreams may be considered. Changes in these limitations may be readily made. In our "operating column analysis" (see [2], p. 520), the user must specify:

1. the number of stages,
2. stage efficiencies,
3. feed- and sidestream locations,
4. feed compositions, flow rates, and thermal states,
5. column pressure,
6. sidestream phase conditions and flow rates,
7. reflux ratio, and
8. distillate flow rate.

The nomenclature for an arbitrary stage n, which includes the possibility of feed- and sidestreams, is given in Figure 9.1.

In the following, M is the total number of components and N the total number of stages including the reboiler (n=1) and the partial condenser (n=N). For stage n one can establish the following set of independent relationships, "discrepancy functions" $F_{k(n,i)}$, which must be satisfied:

Component balances (k=1):

$$F_1(n,i) = (1 + \frac{S_n^L}{L_n})\ell_{n,i} + (1 + \frac{S_n^V}{V_n})v_{n,i} - v_{n-1,i} - \ell_{n+1,i} - f_{n,i} = 0 \quad (9.1)$$

$$n = 2, 3 .. N-1$$

Figure 9.1

DISTILLATION COLUMN CONFIGURATION

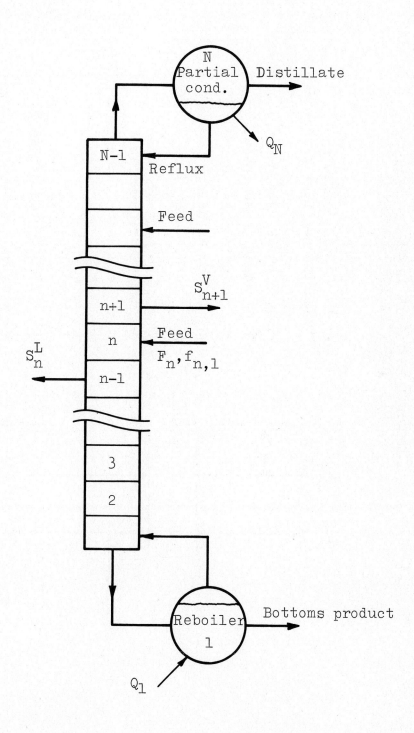

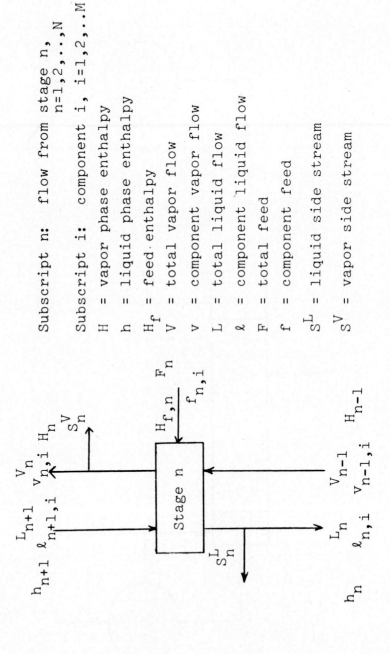

Figure 9.1 cont.

NOMENCLATURE FOR AN ARBITRARY STAGE IN A DISTILLATION COLUMN

Subscript n: flow from stage n, $n=1,2,\ldots,N$

Subscript i: component i, $i=1,2,\ldots M$

H = vapor phase enthalpy
h = liquid phase enthalpy
H_f = feed enthalpy
V = total vapor flow
v = component vapor flow
L = total liquid flow
ℓ = component liquid flow
F = total feed
f = component feed
S^L = liquid side stream
S^V = vapor side stream

$$F_{1(1,i)} = (1 + \frac{S_1^L}{L_1})\ell_{1,i} + (1 + \frac{S_1^V}{V_1})v_{1,i} - \ell_{2,i} - f_{1,i} = 0 \qquad (9.2)$$

$$F_{1(N,i)} = (1 + \frac{S_N^L}{L_N})\ell_{N,i} + (1 + \frac{S_N^V}{V_N})v_{N,i} - v_{N-1,i} - f_{N,i} = 0 \qquad (9.3)$$

V_N is the distillate and L_1 the bottoms product.

Equations (9.1-3) comprise N·M relationships.

Energy balances (k=2)   (Index i drops out):

$$F_{2(n)} = (1 + \frac{S_n^L}{L_n})h_n + (1 + \frac{S_n^V}{V_n})H_n - H_{n-1} - h_{n+1} - h_{f,n} = 0 \qquad (9.4)$$

$$n = 2, 3 \ldots N-1$$

$$Q_1 = (1 + \frac{S_1^L}{L_1})h_1 + (1 + \frac{S_1^V}{V_1})H_1 - h_2 - h_{f,1} \qquad (9.5)$$

$$-Q_N = (1 + \frac{S_N^L}{L_N})h_N + (1 + \frac{S_N^V}{V_N})H_N - H_{N-1} - h_{f,N} \qquad (9.6)$$

Equations (9.5) and (9.6) could be used to generate $F_{2(1)}$ and $F_{2(N)}$ if the heat removed from the condenser, Q_N, and the heat added to the reboiler, Q_1, were specified variables. However, in our case the reflux ratio, R, and distillate flow rate, V_N, are specified instead, see specifications 7 and 8 above. In this case we use Equations (9.5) and (9.6) to calculate Q_1 and Q_N, and the following discrepancy functions are chosen to replace them:

$$F_{2(1)} = \sum^{i} \ell_{1,i} - L_1 = 0 \qquad i = 1, 2 \ldots M \qquad (9.7)$$

where from a total column material balance

$$L_1 = \sum^n \sum^i f_{n,i} = V_N - \sum^n (S_n^L + S_n^V); \quad n = 1,2 \ldots N; \quad i = 1,2 \ldots M$$

$$F_2(N) = \sum^i \ell_{N,i} - L_N = 0 \qquad (9.8)$$

where $L_n = R \cdot V_N$

It is shown in [1] how Equations (9.7) and (9.8) may be altered for a number of other column specifications such as distillate and bottoms product temperatures, key component rates, etc.

Equations (9.4), (9.7), and (9.8) comprise N discrepancy functions.

Equilibrium condition with Murphree stage efficiency (k=3):

The Murphree stage efficiency is defined by

$$\eta_{n,i} = \frac{y_{n,i} - y_{n-1,i}}{K_{n,i} x_{n,i} - y_{n-1,i}} \qquad (9.9)$$

where $K_{n,i} \equiv \left(\frac{y_i}{x_i}\right)_n = \left(\frac{\gamma_i f_i^o}{\varphi_i P}\right)_n$

This may be rearranged to give the following discrepancy functions:

$$F_3(n,i) = \eta_{N,i} K_{n,i} V_n \ell_{n,i}/L_n - v_{n,i} + (1 - \eta_{n,i}) v_{n-1,i} V_n / V_{n-1} = 0$$

$$(9.10)$$

For the reboiler, the efficiency is 1, that is

$\eta_{1,i} = 1,$ all i

There are N·M equations of type k=3. It is in Equation (9.10) that the UNIFAC method enters into the calculations via the equilibrium ratio, $K_{n,i}$.

The discrepancy functions $F_{k(n,i)}$ are a quantitative measure of the failure of the independent variables $\ell_{n,i}$, $v_{n,i}$, and T_n to satisfy the physical conditions in the column. $F_{3(n,i)}$ is the number of heat units/time by which the enthalpy balance is unsatisfied, and for $F_{1(n,i)}$ and $F_{2(n)}$ the discrepancy is measured in moles/time. In terms of algebra, Equations (9.1-4,7,8,10) comprise a vector of discrepancy functions:

$$\underline{F}(\underline{x}) = \left\{ \begin{array}{c} \underline{F}_1 \\ \underline{F}_2 \\ \underline{F}_3 \end{array} \right\} = \underline{0} \tag{9.11}$$

which contains $N \cdot (2M+1)$ elements, and which may be solved for equally many unknowns (independent variables):

$$\underline{x} = \left\{ \begin{array}{c} \underline{\ell} \\ \underline{v} \\ \underline{T} \end{array} \right\} \tag{9.12}$$

where the vector $\underline{\ell}$ contains all the elements $\ell_{n,i}$, $\underline{v}$ all elements $v_{n,i}$, and $\underline{T}$ all elements T_n.

Once all $\ell_{n,i}$, $v_{n,i}$, and T_n's are known, the product compositions and product flow rates and the concentration- and temperature profiles in the column follow readily.

Equations (9.11-12) correspond to what we in the following will call "the full" version. An important simplification, in the following referred to as "the simplified version", may be derived.

If one assumes

$\eta_{n,i} = 1$, $\varphi_{n,i} = 1$, and $h_n^E = 0$, all n and i, and

$$\Delta H_1^{vap} = \Delta H_2^{vap} \ldots = \Delta H_M^{vap}$$

then V_n and L_n become constant in each section of the column (i.e. between neighbouring feed or product streams). Equations (9.1) and (9.10) may be combined as follows:

$$F_{4(n-1,i)} = \left[(1 + \frac{S_n^L}{L_n}) + \varsigma_{n,i}(1 + \frac{S_n^V}{V_n})\right] \ell_{n,i} \qquad (9.13)$$

$$- \varsigma_{n-1,i}\, \ell_{n-1,i} - \ell_{n+1,i} - f_{n,i} = 0$$

The relationships equivalent to Equations (9.2) and (9.3) follow from (9.13) by equating $\ell_{0,i}$ and $\ell_{N+1,i}$ to zero.

$\varsigma_{n,i}$ is the stripping factor:

$$\varsigma_{n,i} = \frac{V_n \gamma_{n,i} P_{n,i}^S}{P L_n} \qquad (9.14)$$

where $P_{n,i}^S$ is the pure-component vapor pressure of component i on stage n.

One other set of independent relations is here conveniently given by the mass balance on each stage:

$$F_{5(n)} = L_n - \ell_{n,1} - \ell_{n,2} - \cdots - \ell_{n,M} = 0 \qquad (9.15)$$

Equations (9.13) and (9.15) comprise $N \cdot (M+1)$ relations:

$$\underline{F}(\underline{x}) = \left\{ \begin{array}{c} \underline{F}_4 \\ \underline{F}_5 \end{array} \right\} = \underline{0} \qquad (9.16)$$

containing equally many independent variables

$$\underline{x} = \left\{ \begin{array}{c} \underline{\ell} \\ \underline{L} \end{array} \right\} \qquad (9.17)$$

The assumption of equality of heats of vaporization and zero excess enthalpy is tantamount to the assumption of "constant molal overflow". It is not necessary to assume ideal vapor phase at this point. However, in the evaluation of the Jacobian matrix (see below) it becomes necessary to assume that $\varsigma_{n,i}$ does not depend upon the vapor phase composition.

Equations (9.11) or (9.16) are solved by Newton-Raphson iteration, utilizing simultaneous convergence of all the independent variables, $\underline{x}$. Simultaneous convergence is more efficient than sequential convergence when the volatilities of the components depend on both temperature and composition. Solving these equations means finding the set of values of the independent variables, $\underline{x}$, which makes the set of discrepancy functions become equal to zero:

$$\underline{F}(\underline{x}) = \underline{0}$$

In Newton-Raphson iteration, a new set of values of the independent variables, $\underline{x}_r$, is generated from a previous estimate, $\underline{x}_{r-1}$, in the following fashion:

$$\underline{x}_r = \underline{x}_{r-1} - \underline{F}_{r-1}(\underline{x}_{r-1}) / \left(\frac{\partial \underline{\underline{F}}}{\partial \underline{x}}\right)\bigg|_{\underline{x}_{r-1}} \qquad (9.18)$$

When $(\underline{x}_r - \underline{x}_{r-1})$ is sufficiently small, the correct set of values of $\underline{x}$ has been found, and the iteration stops.

The variations between subsequent iterations are arbitrarily limited as follows:

(1) negative component molar flow rates are equated to zero,

(2) component flow rates exceeding L_n are equated to L_n,

(3) the maximum change in the temperature at each stage, T_n, is 10 K.

The initial guesses for the independent variables, $\underline{x}_0$, must be supplied by the user (see below). "Wild" initial guesses can make the Newton-Raphson linearization approximation invalid to such an extent that the method fails to converge. To improve convergence characteristics far from the correct solution Naphtali and Sandholm modified the Newton-Raphson method [1]. We use "straight" Newton-Raphson iteration, and in case of divergence we modify the limitations on iterations listed above or alter the initial guess, $\underline{x}_0$.

As the values of the unknowns $\underline{x}$ become more correct, the convergence to their final values accelerates. Quadratic convergence is obtained near the correct solution.

The only unexplained step in the calculation sequence given above is now the evaluation of the Jacobian correction matrix,

$(\frac{\partial \underline{\underline{F}}}{\partial \underline{x}})$, in which each element is the partial derivative of one of the discrepancy functions with respect to one of the iteration variables, with all other iteration variables held constant. For a precise description of the elements in and structure of the Jacobian, we refer the reader to [1].

The Jacobian is in our applications very large, often in the order of 200 by 200. Therefore, evaluation of its elements by ordinary Gauss elimination would be much too time consuming. Fortunately, the evaluation of this particular Jacobian is greatly facilitated by the fact that the conditions on stage n only is influenced directly by the conditions on stages n+1 and n-1. As a result, the Jacobian becomes block-tridiagonal in structure, which permits rapid solution by block elimination (see for example [3]).

The derivatives of discrepancy functions $\underline{F}$ with respect to temperature are found analytically, those with respect to component flow rates numerically.

9.2 THE COLUMN CALCULATION PROCEDURE

The distillation column configuration is that shown in Figure 9.1, and the computer programs associated with the column calculations are given in Appendices 2 and 4. The calculation procedure may be outlined as follows:

1. Read input (pure-component data, thermodynamic data for the mixture, and detailed column specifications such as feed plates, reflux ratio, etc.)

2. Guess initial temperature- and component flow profiles

3. Determine all K-factors, stream enthalpies, and their derivatives with respect to temperature and composition

4. Set up the discrepancy functions $\underline{F}(\underline{x})$

5. Determine the elements of the Jacobian matrix and solve the block-tridiagonal system of equations

6. Determine the corrections to the temperature- and component vapor and liquid flow profiles. Determine the new profiles

7. Are the corrections smaller than a specified value? If no, go to 3 with the newly determined profiles. If yes, go to 8

8. Print final output and stop.

This is shown schematically in Figure 9.2 for the simplified version.

We shall here only make detailed comments regarding points 1 and 2. Details in the column calculation procedure are indicated by comment cards in the computer programs listed in Appendix 4, which also has sample in- and outputs.

Input to the column calculations is as follows:

1. <u>Identification of components to be separated</u>

Antoine constants and enthalpies of vaporization must be known for each component. The latter are not needed for the simplified version.

2. <u>Parameters for calculation of activity coefficients</u>

In principle, the UNIFAC method may be used as is for generating activity coefficients as functions of temperature and composition (see subroutine UNIFA in Appendix 2). This is, however, much too time consuming for distillation computations. The main reason is that subroutine UNIFA uses both molecular and group compositions. Instead, as shown in Appendix 2, subroutine UNIFA is used to generate UNIQUAC (or Wilson) parameters, and the UNIQUAC (or Wilson) parameters are used in the distillation calculations. The procedure is as follows:

If all the necessary UNIFAC group-interaction parameters are available, the activity coefficient of component i, infinitely dilute in component j, is calculated for all possible i-j pairs in the multicomponent system. These calculations are carried out at two temperatures T_1 and T_2, typically the saturation temperatures of the least and most volatile components. The activity coefficients at infinite dilution form the basis for a trial-and-error estimation of the UNIQUAC or

Figure 9.2

PROCEDURE FOR DISTILLATION COLUMN CALCULATIONS
(Simplified version)

```
┌─────────────────────────────────┐
│ Define mixture to be separated: │
│ Antoine constants $A_i$, $B_i$, $C_i$ │
│ Group counts $\nu_k^{(i)}$       │
│ Temperature range $T_1$ to $T_2$ │
└─────────────────────────────────┘
              ↓
┌─────────────────────────────────┐
│ Calculate activity coefficients at infinite dilution, $\gamma_i^\infty$, │
│ at $T_1$ and $T_2$ using UNIFAC. This also gives $r_i$, $q_i$, and $\ell_i$ │
└─────────────────────────────────┘
              ↓
┌─────────────────────────────────┐
│ Fit $\gamma_i^\infty$ to UNIQUAC equation. This gives $A_{ji}^{(0)}$ and $A_{ji}^{(1)}$: │
│   $\tau_{ji} = A_{ji}^{(0)} + A_{ji}^{(1)} \cdot T$ │
└─────────────────────────────────┘
              ↓
┌─────────────────────────────────┐
│ Specify separation problem: No. of stages, feed stages, │
│ reflux ratio, feed compositions and no. of moles per hour, │
│ amount of bottoms product, sidestreams, and column pressure │
└─────────────────────────────────┘
```

(continued on next page)

177

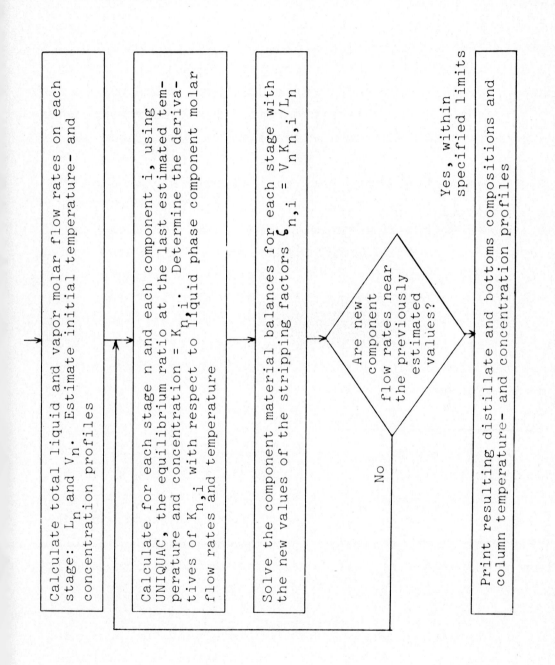

Wilson parameters:

UNIQUAC: $\tau_{ji} = A_{ji}^{(0)} + A_{ji}^{(1)} T$ (see Equation (3.9))

WILSON: $\Lambda_{ij} = A_{ij}^{(0)} + A_{ij}^{(1)} T$ (see Equation (3.5))

The values of $A^{(0)}$ and $A^{(1)}$ and, for UNIQUAC, r_i, q_i, and ℓ_i are punched out for each component; these cards may be used as input to the distillation programs. The linear temperature dependence, rather than the exponential form indicated in Equations (3.5) and (3.9), is introduced to save computer time.

Tests for systems where experimental data are available show that activity coefficients calculated from the UNIQUAC model and estimated limiting activity coefficients are, as a whole, as accurate as those calculated directly using the UNIFAC method. Table 9.1 gives an example of two sets of predicted activity coefficients. It may be seen that in a few cases there is a discrepancy between the two sets of predicted values. This discrepancy may be eliminated by fitting UNIQUAC to activity coefficients generated by UNIFAC over the whole concentration range.

3. Other mixture information

In the "full version", the vapor-phase nonideality is calculated using the virial equation of state with second virial coefficients obtained from experimental information or Hayden and O'Connell's correlation (see Chapter 2). Second virial coefficients at two different temperatures for all possible binary combinations are required. The second virial coefficients are then calculated at different temperatures by linear interpolation.

If a component dimerizes, the parameters in Table 2.3 must also be read in. At the present, the program can not handle mixtures with more than one dimerizing component.

4. Specification of the separation problem

The user must specify: the number of stages; feed- and sidestream locations; feed compositions, flow rates, and thermal

Table 9.1. PREDICTED ACTIVITY COEFFICIENTS FOR METHANOL(1)-ETHANOL(2)-WATER(3)-ACETONE(4)-1,2-DICHLOROETHANE(5) AT 330 K

A. Predictions using UNIFAC

B. Predictions using the linearized form of the UNIQUAC model with parameters calculated from UNIFAC

x_1	x_2	x_3	x_4	γ_1	γ_2	γ_3	γ_4	γ_5	UNIFAC(A) or UNIQUAC(B)
0.2	0.2	0.2	0.2	1.07	1.27	2.51	0.97	2.53	A
				1.06	1.22	2.92	1.14	2.40	B
0.1	0.1	0.1	0.1	1.78	1.97	7.93	0.70	1.33	A
				1.79	1.95	8.67	0.75	1.31	B
0.1	0.1	0.1	0.6	1.21	1.62	2.64	1.05	1.45	A
				1.29	1.63	4.08	1.06	1.36	B
0.1	0.1	0.6	0.1	0.92	1.36	1.51	1.31	9.64	A
				0.93	1.38	1.52	1.97	9.31	B
0.1	0.6	0.1	0.1	1.05	1.05	2.40	1.35	3.15	A
				1.05	1.04	2.67	1.49	3.07	B
0.6	0.1	0.1	0.1	1.03	1.15	1.90	1.23	3.35	A
				1.02	1.13	2.14	1.39	3.27	B
0.425	0.425	0.05	0.05	1.02	1.05	2.05	1.52	3.74	A
				1.02	1.04	2.19	1.62	3.70	B
0.425	0.05	0.425	0.05	1.01	1.34	1.45	1.40	9.13	A
				1.00	1.33	1.47	2.13	8.93	B
0.425	0.05	0.05	0.425	1.17	1.36	1.67	1.15	1.88	A
				1.15	1.32	2.74	1.18	1.80	B
0.425	0.05	0.05	0.05	1.33	1.43	4.18	0.78	1.76	A
				1.33	1.42	4.48	0.82	1.75	B
0.05	0.425	0.05	0.05	1.39	1.31	5.08	0.90	1.73	A
				1.39	1.31	5.39	0.92	1.73	B
0.05	0.05	0.425	0.05	0.93	1.23	3.36	0.54	2.56	A
				0.95	1.24	3.40	0.70	2.53	B
0.05	0.05	0.05	0.425	1.90	2.45	6.92	0.91	1.15	A
				2.01	2.46	9.18	0.92	1.13	B

179

states; total distillate flow rate; sidestream phase conditions and flow rates; reflux ratio; and column pressure.

An initial estimate of the temperature in the top and the bottom of the column must also be furnished.

5. Murphree tray efficiencies

For the "full version", the tray efficiencies may be read in for each stage, or they may be calculated using a subroutine supplied by the user.

This completes the input to the column calculations. The use of the UNIQUAC (or Wilson) model makes it possible to combine the binary parameters $A_{ji}^{(0)}$ and $A_{ji}^{(1)}$ obtained from UNIFAC with those obtained from experimental information on some of the constituent binaries.

To carry out the energy balances rigorously, the enthalpy of mixing should be included. This is implicitly included in Equation (9.4) if we define the liquid phase enthalpy on stage n by

$$h_n = h_{n,ideal} + h_n^E \qquad (9.19)$$

The full version can, if the user so specifies, take the excess enthalpy into account. It is then calculated using the UNIQUAC parameters obtained from UNIFAC:

$$h^E = -RT \sum^i \frac{q_i x_i \sum_j^j \theta_j \tau_{ji} \ell n \tau_{ji}}{\sum_j^j \theta_j \tau_{ji}} \qquad i \text{ and } j = 1,2 \ .. \ M \qquad (9.20)$$

$$\tau_{ji} = A_{ji}^{(0)} + A_{ji}^{(1)} T$$

Equation (9.20) provides only a <u>crude</u> estimate when, as done here, the molecular energy interaction parameters are determined from vapor-liquid equilibrium data and are assumed to be independent of temperature. In the illustrations given below, the magnitude of h^E is fortunately small compared with the heat of vaporization. Typically, the inclusion of h^E alters the total flow rates 3-5% as compared with equating h^E

to zero. The product compositions are nearly unaffected. However, if strong heat of mixing effects are expected, the user should provide a better procedure for calculating h^E.

Step two of the calculation procedure consists of establishing an initial estimate of the column temperature- and concentration profiles. In both the full and simplified versions, the temperature profile is initially guessed to be linear with respect to the top and bottom temperatures. For the simplified version, the total molar flow rates on each stage can be calculated on the basis of the input information - for the full version, these values serve as an initial guess. For the zeroth estimate of $\ell_{n,i}$ and $v_{n,i}$, the phase compositions on each stage are equated to the nearest feed composition.

Although this initial estimate of the variables $\underline{x}$ is built into the computer programs listed in Appendix 4, one can easily alter the guesses by changing the programs where indicated by comment cards.

The actual column calculations can now proceed. Convergence is normally attained within 4-9 iterations. The computer time per iteration is proportional to the number of stages and to the second power of the number of components. In the simplified version (constant molal overflow), four components and 20 stages typically require 0.4 sec. per iteration on an IBM 370/165 computer using a Fortran G compiler. About 50% of this time is used for algebraic manipulations, and 50% is used for evaluating activity coefficients. The full Naphtali-Sandholm procedure requires the same number of iterations and typically 2.0 sec. per iteration for the above example.

Similar calculations, where the Wilson equation is used instead of the UNIQUAC equation, require slightly less computer time.

9.3 EXAMPLES OF DISTILLATION CALCULATIONS (SIMPLIFIED VERSION)

In this section we show results illustrating multicomponent distillation, including azeotropic and extractive distillation. In the examples shown here, the assumption of constant molal overflow is justifiable, and hence the simplified version of the distillation program is used. Section 9.4 shows examples of results from the full version distillation program.

The System Cyclohexane(1)-Ethanol(2)-1-Propanol(3)-Toluene(4)

Hydrocarbon-alcohol systems are known to deviate greatly from ideality; minimum-boiling azeotropes are found in many cases. In the above system, no fewer than four binary azeotropes exist:

Toluene(1)-Ethanol(2) $x_1 = 0.19$ $T = 77\ ^{\circ}C$

Toluene(1)-1-Propanol(2) $x_1 = 0.37$ $T = 92\ ^{\circ}C$

Cyclohexane(1)-Ethanol(2) $x_1 = 0.57$ $T = 65\ ^{\circ}C$

Cyclohexane(1)-1-Propanol(2) $x_1 = 0.79$ $T = 77\ ^{\circ}C$

For this system, we choose as feed composition $x_1 = 0.1$, $x_2 = 0.4$, $x_3 = 0.4$, $x_4 = 0.1$. The reflux ratio is 2.25, and 60 moles out of 100 moles in the feed are specified as distillate.

There are 20 stages, no sidestreams, and feed is introduced on stage 10.

The main results of the column calculations are shown in Table 9.2. The Toluene-Ethanol azeotrope has a marked effect on the results. Although Toluene is the heaviest-boiling component, it goes into the distillate almost quantitatively. The relative content of Toluene and Ethanol in the distillate corresponds closely to that of the azeotrope. It boils at 77 $^{\circ}C$, i.e. 20 $^{\circ}C$ lower than the normal boiling point of 1-Propanol, which forms 99% of the bottoms product. A similar calculation, where the UNIQUAC model is replaced by the Wilson equation gives essentially the same results. Raoult's Law, on the other hand, gives completely erroneous results.

Computations performed with the full version of the column design procedure indicate that the assumptions of constant molal overflow and ideal vapor phase are justified in this case.

Results of similar calculations, but at different feed compositions, are shown in Figure 9.3. For the three-component system Ethanol(2)-1-Propanol(3)-Toluene(4), Figure 9.3 shows

Figure 9.3

EFFECT OF TOLUENE IN THE FEED ON THE ALCOHOL SEPARATION

Feed: $x_3 = 0.4$, 100 moles total (No Cyclohexane)

For Ethanol, F_2 = moles in distillate/moles in bottoms

For 1-Propanol, F_3 = moles in bottoms/moles in distillate

Reflux ratio   = 2.25
No of stages   = 20
Feed stage   = 10
Total feed: 100 moles, saturated liquid
1-Propanol in feed: 40 moles
Bottoms product : 60 moles
Column pressure : 1 atm.

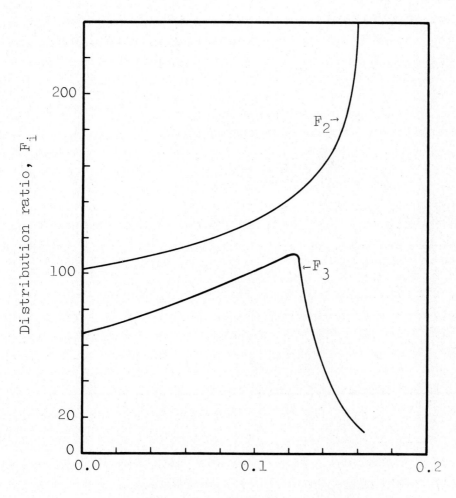

that addition of Toluene to the alcohols improves the alcohol separation to the point where so much Toluene is added to the feed, that all of it no longer can be used to form the Toluene-Ethanol azeotrope. In that case some of the Toluene must emerge as bottoms product displacing some of the 1-Propanol. In all these calculations, the amount of bottoms product specified equals the number of moles of 1-Propanol in the feed.

The System 1,2-Dichloroethane(1)-1-Propanol(2)-Toluene(3)-Acetone(4)

Essential results of the column calculations are given in Table 9.2 together with those based on Raoult's Law and those using the Wilson equation instead of the UNIQUAC equation. Experimental vapor-liquid equilibrium data are available for this quaternary system. The agreement between experimental activity coefficients and those obtained from UNIFAC are shown in Table 7.5.

In the example given here, the feed contains mainly components 1 and 2, and the reflux ratio is 4. Separation between these two components is poor due to the 1,2-Dichloroethane - 1-Propanol azeotrope (boiling point 81 $^\circ$C) going into the distillate.

In both this and the previous example, the agreement between the results obtained using the UNIQUAC model (with parameters obtained from UNIFAC) and those using Wilson's equation is especially comforting, since the Wilson parameters were obtained directly from experimental data. Raoult's Law predicts a different (and erroneous) distribution of the components.

The System Acetone(1)-Methanol(2)-2-Butanone(3)-Water(4)

The distillation problem illustrated in Figure 9.4 represents a realistic industrial problem encountered in the manufacture of 2-Butanone. Water forms a minimum-boiling azeotrope with 2-Butanone ($x_3 \approx 0.64$), and hence water leaves from the top of the distillation column.

The concentration profiles in the column are shown in Figure 9.4. The profiles are believed to be in good agreement with

Table 9.2. COMPONENT FLOW RATES OF DISTILLATE AND BOTTOMS. COMPARISON OF CALCULATIONS BASED ON UNIFAC, RAOULT'S LAW AND WILSON'S EQUATION

Simplified Version

20 stages; feed stage: 10; pressure: 1 atm.; no sidestreams

Feed, Mol of comp. i	Distillate Mol Total[*]	Reflux Ratio[*]	UNIQUAC[**]		RAOULT'S LAW		WILSON[+]	
			Distillate Mol of comp. i	Bottoms Mol of comp. i	Distillate Mol of comp. i	Bottoms Mol of comp. i	Distillate Mol of comp. i	Bottoms Mol of comp. i
Cyclohexane(1)-Ethanol(2)-1-Propanol(3)-Toluene(4)								
(1): 10	60	2.25	(1): 10.0	(1): 0.0	(1): 9.7	(1): 0.3	(1): 10.0	(1): 0.0
(2): 40			(2): 39.7	(2): 0.3	(2): 39.9	(2): 0.1	(2): 39.9	(2): 0.1
(3): 40			(3): 0.3	(3): 39.7	(3): 10.3	(3): 29.7	(3): 0.7	(3): 39.3
(4): 10			(4): 9.9	(4): 0.1	(4): 0.1	(4): 9.9	(4): 9.3	(4): 0.7
1,2-Dichloroethane(1)-1-Propanol(2)-Toluene(3)-Acetone(4)								
(1): 45	50	4	(1): 34.2	(1): 10.8	(1): 42.3	(1): 2.7	(1): 35.5	(1): 9.5
(2): 45			(2): 10.8	(2): 34.2	(2): 2.6	(2): 42.4	(2): 9.6	(2): 35.3
(3): 5			(3): 0.0	(3): 5.0	(3): 0.0	(3): 5.0	(3): 0.0	(3): 5.0
(4): 5			(4): 5.0	(4): 0.0	(4): 5.0	(4): 0.0	(4): 4.9	(4): 0.1

[*]   Specified by user
[**] Parameters estimated with UNIFAC method
[+]   Parameters fitted with binary data

Figure 9.4

COLUMN CONFIGURATION AND LIQUID PHASE CONCENTRATION PROFILES FOR DISTILLATION OF Acetone(1), Methanol(2), 2-Butanone(3), and Water(4)

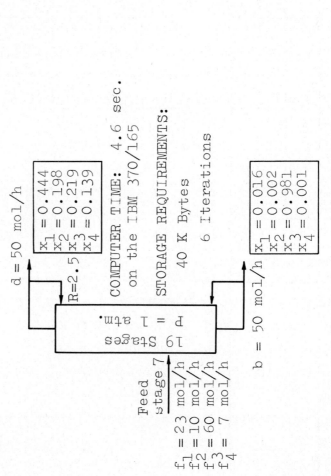

what one would measure in practice, since UNIFAC is found to represent vapor-liquid equilibria in the systems Acetone-Methanol-Water and Acetone-2-Butanone-Water very well (see Table 7.1, lines 12 and 16).

Extractive distillation of Methylcyclohexane(1) and Toluene(2) with Phenol(3)

Methylcyclohexane and Toluene are difficult to separate by simple distillation. Smith [4, pp. 424-8] presents a stage-to-stage solution of an extractive-distillation process for the above two components with Phenol as a solvent. Using UNIFAC coupled with the simplified version of the distillation program, we obtain the same results as those presented by Smith. The process and resulting concentration profiles are shown in Figure 9.5. The solution of this problem required seven iterations and 3.4 sec. on the IBM 370/165 computer.

A similar calculation was performed for the quaternary system Octane(1)-Ethylcyclohexane(2)-Ethylbenzene(3)-Phenol(4), and the detailed results (INPUT and OUTPUT) are shown in Appendix 4. This example corresponds to Example 11.1 in [5], and the results are in agreement with those shown there.

The convergence is summarized in Table 9.3. This distillation is particularly difficult to converge. It is evident from the first 5-6 iterations that a change of the profiles would make the computation converge faster. Note the rapid convergence obtained once the correct solution is approached.

Azeotropic distillation of Heptane(1) and Toluene(2) with 2-Butanone(3)

Toluene and Heptane may be separated by azeotropic distillation using 2-Butanone as the entrainer. The ketone forms a minimum boiling, binary azeotrope with Heptane at $x_3 \approx 0.78$. No phase-splitting occurs in the distillate. A detailed stage-to-stage calculation is shown in [4, p. 408], and we have recalculated the example shown there. The process and resulting profiles are shown in Figure 9.6. The results are in agreement with those given in [5].

Figure 9.5

COLUMN CONFIGURATION AND LIQUID PHASE CONCENTRATION PROFILES FOR EXTRACTIVE DISTILLATION OF Methylcyclohexane(1) and Toluene(2) with Phenol(3)

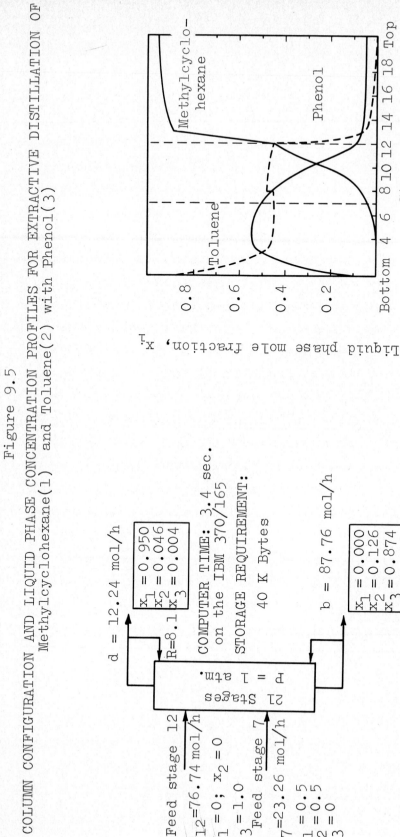

Figure 9.6
AZEOTROPIC DISTILLATION OF
Heptane(1) and Toluene(2) with 2-Butanone(3)

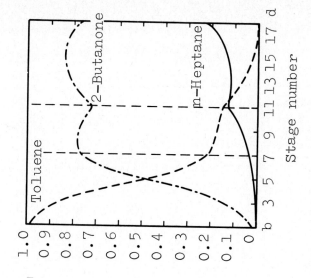

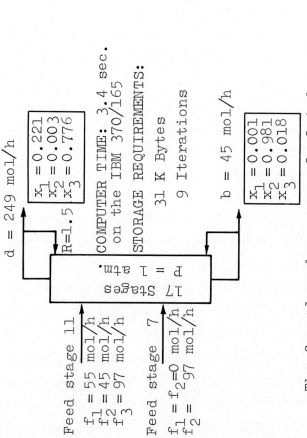

The framed numbers are calculated results

Table 9.3

CONVERGENCE FOR THE EXTRACTIVE DISTILLATION OF
OCTANE-ETHYLCYCLOHEXANE-ETHYLBENZENE WITH PHENOL
(For more details, see Appendix 4)

Iteration No., r	$\sum_i \sum_n \sum_k (x_r - x_{r-1})$ (see Equation (9.18))	Temperature, °C	
		Stage 1	Stage 30
1	$0.11 \cdot 10^8$	162.5	106.5
2	$0.39 \cdot 10^8$	165.3	116.5
3	$0.13 \cdot 10^8$	161.0	124.5
4	$0.18 \cdot 10^7$	165.5	125.6
5	$0.92 \cdot 10^6$	165.6	125.4
6	$0.65 \cdot 10^7$	165.6	125.3
7	$0.59 \cdot 10^6$	165.4	125.5
8	$0.75 \cdot 10^5$	165.4	125.5
9	$0.53 \cdot 10^4$	165.4	125.5
10	$0.18 \cdot 10^2$	165.4	125.5
11	0.12	165.4	125.5

9.4 EXAMPLES OF DISTILLATION CALCULATIONS (FULL VERSION)

If enthalpy effects cannot be neglected, if stage efficiencies are to be included, or if the vapor phase is strongly nonideal, it is necessary to use the full Naphtali-Sandholm procedure. (Slight vapor phase nonidealities can be included in the simplified version by inserting the vapor phase fugacity coefficient in the stripping factor (see Equation (9.14)) and subsequently neglecting derivatives af the discrepancy functions with respect to vapor phase composition). Systems containing organic acids are examples of cases where the full Naphtali-Sandholm procedure must be used. Liquid phase nonidealities are, as before, accounted for via the UNIFAC correlation, and vapor phase nonidealities, including organic acid dimerization, are determined as explained in Chapter 2. The results shown in Tables 9.4 and 9.5 are for the case of a simple distillation column with one feed stream at stage 7, no

Table 9.4

COMPONENT FLOW RATES OF DISTILLATE AND BOTTOMS
COMPARISON OF CALCULATIONS BASED ON UNIFAC AND RAOULT'S LAW

Full Version

Reflux ratio: 2.0; Distillate: 50 moles/h
Number of stages: 15; Feed stage: number 7
Feed: (1): 25 moles/h; (2): 25 moles/h
       (3): 25 moles/h; (4): 25 moles/h

Ethanol(1)-1-Propanol(2)-Water(3)-Acetic acid(4)

Component	Moles in distillate	Moles in bottoms product	Method
1	24.8	0.2	Raoult's Law
2	14.3	10.7	
3	10.4	14.5	
4	0.4	24.6	
1	22.4	2.6	Full Naphtali-Sandholm with UNIFAC
2	9.2	15.7	
3	18.4	6.6	
4	0.0	25.0	

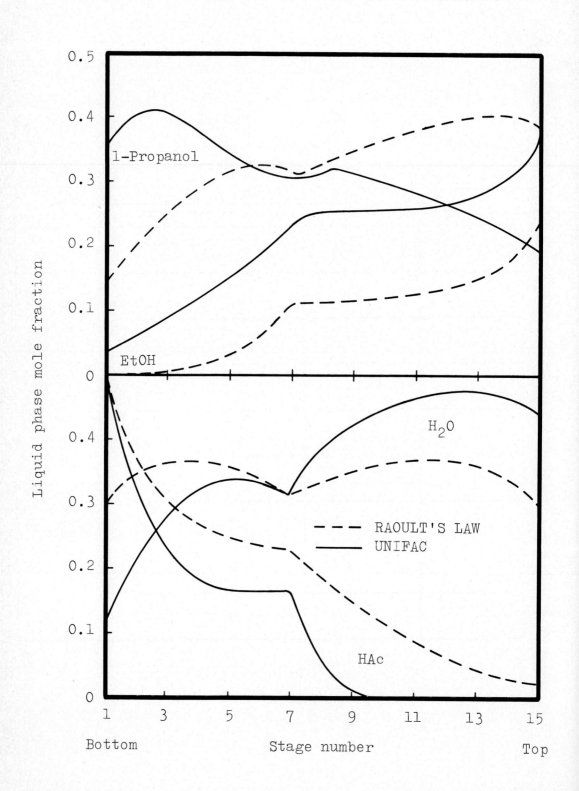

Figure 9.7
LIQUID PHASE CONCENTRATION PROFILES IN THE SYSTEM
Ethanol(1), 1-Propanol(2), Water(3), Acetic acid(4)
(See Table 9.4)

Figure 9.8
DISTILLATION RESULTS FOR THE SYSTEM
Ethanol(1), 1-Propanol(2), Water(3), Acetic acid(4)

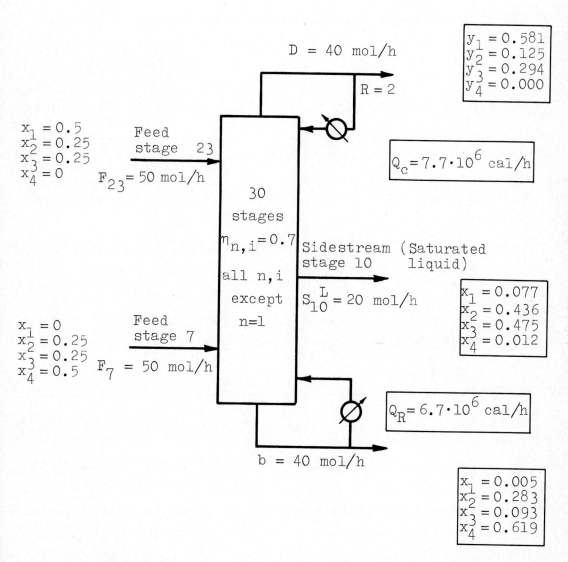

COMPUTER TIME: 17.4 sec. on the IBM 370/165
STORAGE REQUIREMENT: 88 K Bytes

The framed numbers are calculated results

sidestreams, and 15 stages. Figure 9.8 shows results for a more complex column.

The System Ethanol(1)-1-Propanol(2)-Water(3)-Acetic acid(4)

For a feed of 25 moles/h of each component to a column as specified in Table 9.4, one might expect a split between 1-Propanol and Water, such that most of the 1-Propanol goes into the distillate, most of the Water into the bottoms product. Raoult's Law gives this result. However, as the UNIFAC correlation correctly predicts, Water has a much higher activity in Acetic acid than 1-Propanol has, and hence most of the 1-Propanol goes into the bottoms product. Figure 9.7 shows the calculated concentration profiles. In these calculations, we have neglected the effect of the relatively slow esterification of the alcohols with Acetic acid.

Figure 9.8 shows a summary of results for a complex column, where Murphree stage efficiencies of 0.7 were included. This required five iterations and 17.4 sec. computer time on the IBM 370/165 computer. This example corresponds to the sample in- and output given in Appendix 4.

The System Benzene(1)-1-Propanol(2)-Toluene(3)-Acetic acid(4)

Table 9.5 shows results for the case of the feed: 25 moles/h Benzene, 10 moles/h each of 1-Propanol and Toluene, and 55 moles/h of Acetic acid. These results show the effects of successively including corrections for liquid phase nonideality (II), vapor phase nonideality (III), and enthalpy effects (IV). It is essential to include corrections for both liquid and vapor phase nonideality, whereas the inclusion of enthalpy balances has only limited effect on the calculations. For most of the systems we have studied, the inclusion of enthalpy balances has little effect on the product compositions. They may, however, be of importance in the determination of internal column flow rates.

Some of the above examples of distillation calculations are discussed in [6].

Table 9.5

COMPONENT FLOW RATES OF DISTILLATE AND BOTTOMS
COMPARISON OF CALCULATIONS BASED ON UNIFAC AND RAOULT'S LAW

Full Version

Benzene(1)-1-Propanol(2)-Toluene(3)-Acetic acid(4)

Reflux ratio: 2.0; Distillate: 50 moles/h
Number of stages: 15; Feed stage: number 7
Feed: (1): 25 moles/h; (2): 10 moles/h
      (3): 10 moles/h; (4): 55 moles/h

	Component	Distillate	Bottoms	Method
I	1	24.95	0.05	Raoult's Law $\varphi_i = 1$ $\gamma_i = 1$
	2	9.77	0.23	
	3	5.39	4.61	
	4	9.89	45.11	
II	1	25.00	0.00	Constant molal flow $\varphi_i = 1$ γ_i from UNIFAC
	2	9.43	0.57	
	3	10.00	0.00	
	4	5.57	49.43	
III	1	25.00	0.00	Constant molal flow $\varphi_i \neq 1$ γ_i from UNIFAC
	2	6.95	3.05	
	3	9.99	0.01	
	4	8.06	46.94	
IV	1	25.00	0.00	Full Naphtali-Sandholm $\varphi_i \neq 1$ γ_i from UNIFAC
	2	6.96	3.04	
	3	9.99	0.01	
	4	8.05	46.95	

REFERENCES

1. L.M. Naphtali and D.P. Sandholm, AIChE Journal 17(1971)148.
2. C.J. King, Separation Processes, McGraw-Hill Book Company, New York, 1971.
3. D.W. Birmingham and F.D. Otto, Hydrocarbon Process.,46(10)(1967)163.
4. B.D. Smith, Design of Equilibrium Stage Processes, McGraw-Hill Book Company, New York, 1963.
5. M. Van Winkle, Distillation, McGraw-Hill Book Company, New York, 1967.
6. Aa. Fredenslund, J. Gmehling, M.L. Michelsen, P. Rasmussen and J.M. Prausnitz, Ind.Eng.Chem., Process Design and Development (1977)(in print).

APPENDICES

Appendices 1 and 2 contain the current versions of the subroutines for calculating fugacity coefficients, activity coefficients, and other thermodynamic properties according to the methods shown in this book. These programs have undergone significant changes during our work with UNIFAC. Some of the subroutines in Appendices 3 and 4 differ from the current versions shown in Appendices 1 and 2, because the parameter estimation and distillation programs were developed at an early stage of our work with UNIFAC.

In sample inputs the column numbers are sometimes indicated by hand written numbers, f.ex. $\overset{24}{1|}$

APPENDIX 1

CALCULATION OF FUGACITY AND ACTIVITY COEFFICIENTS

THE CONSISTENCY TEST

This appendix contains the following programs:

1.1 SVIR, subroutine for calculating second virial coefficients (pure substances and binary mixtures).

1.2 PHIB, subroutine for calculating fugacity coefficients for pure substances and in binary mixtures (calls SVIR).

1.3 YENWO, subroutine for calculating saturated liquid molar volumes.

1.4 GAUSL, subroutine for solving a set of N linear algebraic equations.

1.5 MLMEN, subroutine for calculating the terms of Legendre polynomials.

1.6 BIJ, main program for calculating second virial coefficients (multicomponent mixture)(calls SVIR).

An example of the data input and of the printout is presented.

1.7 THERMODYNAMIC CONSISTENCY TEST, this main program tests binary vapor-liquid equilibrium data for thermodynamic consistency (calls SVIR, PHIB, YENWO, GAUSL, MLMEN).

An example of the data input and of the printout is presented.

The comment cards included in the programs are assumed to make the programs selfexplanatory.

```
C     *********************************************************************
C     *
C     *   SUBROUTINE SVIR(N,TEMP,BF,BB)
C     *
C     *********************************************************************
C
C
C
C     ****************************
C     CALCULATION OF PURE COMPONENT AND CROSS VIRIAL COEFFICIENTS
C     FOR TWO COMPONENTS AT TEMPERATURE TEMP
C     FROM HAYDEN AND O'CONNELL
C     IEC PROC.DES.DEV.14(3)209(1975)
C
C     N  = NUMBER OF COMPONENTS (1 OR 2)
C     BF = BFREE
C     BB = BTOTAL
C     COMMON BLOCK /VIRDAT/ CONTAINS
C           NC = NUMBER OF COMPONENT
C           PC = CRITICAL PRESSURE IN ATM
C           RD = MEAN RADIUS OF GYRATION IN A
C           DMU = DIPOLE MOMENT IN DEBYE
C           ETA(1) AND ETA(2) = ASSOCIATION PARAMETERS (PURE COMPONENTS)
C           ETA(3) = SOLVATION PARAMETER (CROSS INTERACTION)
C           TC = CRITICAL TEMPERATURE IN K
C           ZC = CRITICAL COMPRESSIBILITY FACTOR
C
C     FOR GIVEN COMMON VALUES OF VIRDAT THE SUBROUTINE WILL
C     RETURN VALUES OF BFREE AND BTOTAL
C     ****************************
C
      SUBROUTINE SVIR (N,TEMP,BF,BB)
C
      DIMENSION BF(3),BB(3),W(3),EPSI(3),SIGM3(3),RDMU(3),RDMM(3),A(3),
     1DELH(3),D(3),BO(3)
      COMMON/VIRDAT/NC(2),PC(2),RD(2),DMU(2),ETA(3),TC(2),ZC(2)
C
C           CALCULATION OF COMPONENT PARAMETERS
C           EQ'S 15,30,17,24,25,23,21,22,10
C     THE EQUATION NUMBERS REFER TO THE ARTICLE BY HAYDEN AND O'CONNELL
C
      DO 101 I=1,N
      W(I)=0.006*RD(I)+0.02087*RD(I)**2-0.00136*RD(I)**3
      EPSI(I)=TC(I)*(0.748+0.91*W(I)-0.4*ETA(I)/(2.+20.*W(I)))
      SIGM3(I)=(2.44-W(I))**3*(TC(I)/PC(I))
      IF(DMU(I)-1.45)101,101,103
  103 PN=16.+400.*W(I)
      C=2.882-1.882*W(I)/(0.03+W(I))
      XI=DMU(I)**4/(C*EPSI(I)*(SIGM3(I)**2)*TC(I)*5.723E-8)
      PPN=PN/(PN-6.)
      EPSI(I)=EPSI(I)*(1.-XI*PPN+PPN*(PPN+1.)*(XI**2)/2.)
      SIGM3(I)=SIGM3(I)*(1.+3.*XI/(PN-6.))
  101 RDMU(I)=(DMU(I)**2)*7243.8/(EPSI(I)*SIGM3(I))
      IF(N-1)300,300,400
```

```
      300 J=1
          GO TO 301
      400 J=3
          GO TO 401
C
C         PARAMETERS FOR MIXTURE CALCULATION
C         NONPOLAR-NONPOLAR, EQ'S 32,33,34
C
      401 EPSI(3)=0.7*SQRT(EPSI(1)*EPSI(2))+0.60/(1./EPSI(1)+1./EPSI(2))
          SIGM3(3)=SQRT(SIGM3(1)*SIGM3(2))
          W(3)=0.5*(W(1)+W(2))
          IF(DMU(1)*DMU(2))500,501,500
C
C         POLAR-NONPOLAR, EQ'S 38,24,36,37
C
      501 IF(DMU(1)+DMU(2)-2.)500,500,19
       19 XI38=(DMU(1)**2*(EPSI(2)**2*SIGM3(2))**(1./3.)*SIGM3(2)+DMU(2)**2*
         1(EPSI(1)**2*SIGM3(1))**(1./3.)*SIGM3(1))/(EPSI(3)*SIGM3(3)**2)
          PN=16.+400.*W(3)
          EPSI(3)=EPSI(3)*(1.+XI38*PN/(PN-6.))
          SIGM3(3)=SIGM3(3)*(1.-3.*XI38/(PN-6.))
C
C         POLAR-POLAR, EQ'S 35,27
C
      500 RDMU(3)=7243.8*DMU(1)*DMU(2)/(EPSI(3)*SIGM3(3))
      301 DO 600 I=1,J
          IF(RDMU(I)-0.04)14,15,15
       14 RDMM(I)=RDMU(I)
          GO TO 600
       15 IF(RDMU(I)-0.25)16,17,17
       16 RDMM(I)=0.
          GO TO 600
       17 RDMM(I)=RDMU(I)-0.25
      600 CONTINUE
C
C         LAST PARAMETERS, EQ'S 7,8,9,29
C
          DO 609 I=1,J
          BO(I)=1.2618*SIGM3(I)
          A(I)=-0.3-0.05*RDMU(I)
          DELH(I)=1.99+0.2*RDMU(I)**2
          IF(ETA(I)-4.)604,604,605
      604 D(I)=650./(EPSI(I)+300.)
          GO TO 609
      605 D(I)=42800./(EPSI(I)+22400.)
      609 CONTINUE
C
C         CALCULATION OF VIRIAL COEFFICIENTS, EQ'S 14,13,26,6,29
C
          DO 651 I=1,J
          TSTR=EPSI(I)/TEMP-1.6*W(I)
          BFN=0.94-1.47*TSTR-0.85*TSTR**2+1.015*TSTR**3
          BFP=(0.75-3.*TSTR+2.1*TSTR**2+2.1*TSTR**3)*RDMM(I)
          BF(I)=(BFN-BFP)*BO(I)
          BB(I)=BF(I)+BO(I)*A(I)*EXP(DELH(I)*EPSI(I)/TEMP)
          IF(ETA(I))651,651,653
      653 BCHEM=BO(I)*EXP(ETA(I)*(D(I)-4.27))*(1.-EXP(1500.*ETA(I)/TEMP))
          BB(I)=BB(I)+BCHEM
```

```
651 CONTINUE
    RETURN
    END
```

```
C     ****************************************************************
C     *
C     *   SUBROUTINE PHIB(N,P,TEMP,Y,FUG,NACID)
C     *
C     ****************************************************************
C
C
C     *********************************
C     CALCULATION OF FUGACITY COEFFICIENTS FOR PURE SUBSTANCES AND
C     BINARY MIXTURES WITH COMPOSITION GIVEN BY THE MOLEFRACTIONS Y
C     AT PRESSURE P(IN ATMOSPHERES) AND TEMPERATURE TEMP(IN K)
C
C     N = NUMBER OF COMPONENTS(1 OR 2)
C     THE PROGRAM CALLS SUBROUTINE SVIR WHERE THE CONTENTS OF VIRDAT
C     IS EXPLAINED
C     NACID REPRESENTS THE NUMBER OF CARBONATOMS IN
C     AN ORGANIC ACID. FOR ACETIC ACID NACID IS 2.
C
C     THE SUBROUTINE WILL RETURN VALUES OF FUGACITY COEFFICIENTS FUG
C     *********************************
C
      SUBROUTINE PHIB(N,P,TEMP,Y,FUG,NACID)
C
      DIMENSION Y(2),FUG(2),A(4),B(4),BF(3),BB(3)
      COMMON/VIRDAT/NC(2),PC(2),RD(2),DMU(2),ETA(3),TC(2),ZC(2)
      DATA A/10.743,10.4205,10.843,10.136/,B/3083.,3166.,3316.,3079./
      RG=82.06
      PR=P/(RG*TEMP)
C
C     PURE COMPONENT OR BINARY MIXTURE
C     NO ACIDS INVOLVED
C     EQ'S (2.6) AND(2.7)
C
  100 CALL SVIR(N,TEMP,BF,BB)
      DO 101 I=1,N
      IF(ETA(I).EQ.4.5) GO TO 102
  101 CONTINUE
      IF(N.EQ.2) GO TO 103
      FUG(1)=PR*BB(1)
      FUG(1)=EXP(FUG(1))
      RETURN
  103 BMIX=Y(1)**2*BB(1)+Y(2)**2*BB(2)+2.*Y(1)*Y(2)*BB(3)
      FUG(1)=PR*(2.*Y(1)*BB(1)+2.*Y(2)*BB(3)-BMIX)
      FUG(2)=PR*(2.*Y(1)*BB(3)+2.*Y(2)*BB(2)-BMIX)
      DO 104 I=1,2
  104 FUG(I)=EXP(FUG(I))
      RETURN
C
C     PURE COMPONENT OR BINARY MIXTURE
C     ONE COMPONENT IS AN ORGANIC ACID
C     EQ'S (2.14),(2.15) AND (2.18)
C
  102 IF(I.EQ.1) GO TO 106
```

```
      NA=2
      NB=1
      GO TO 107
  106 NA=1
      NB=2
  107 IF (NACID.GT.4) GO TO 105
C
C  EXPERIMENTAL VALUE OF K
C
      AKA=A(NACID)-B(NACID)/TEMP
      AKA=EXP(-AKA*2.30259)*760
      GO TO 108
C
C  K PREDICTED FROM SECOND VIRIAL COEFFICIENTS
C  EQ (2.7)
C
  105 AKA=-(BB(NA)-BF(NA))/(RG*TEMP)
  108 AKT=AKA*P*EXP(BF(NA)*PR)
      SQ=SQRT(1.+4.*AKT*Y(NA)*(2.-Y(NA)))
      ZA=(SQ-1.)/(2.*AKT*(2.-Y(NA)))
      FUG(NA)=(ZA/Y(NA))*EXP(BF(NA)*PR)
      IF(N.EQ.1)GO TO 109
      ZB=Y(NB)*(1.+4.*AKT*(2.-Y(NA))-SQ)
      ZB=ZB/(2.*AKT*(2.-Y(NA))**2)
      FUG(NB)=(ZB/Y(NB))*EXP(BF(NB)*PR)
  109 CONTINUE
      RETURN
      END
```

```
C     ************************************************************************
C     *
C     *   SUBROUTINE YENWO(T,VLIQ)
C     *
C     ************************************************************************
C
C
C     ****************
C        COMPUTATION OF SATURATED LIQUID MOLAR VOLUMES BY THE METHOD
C        OF L.C.YEN AND S.S.WOODS, AICHE JOURNAL 12(1),95,(1966)
C
C     FOR GIVEN TEMPERATURE T(IN K) AND COMMON VALUES
C     OF VIRDAT, THE SUBROUTINE WILL RETURN MOLAR LIQUID
C     VOLUMES VLIQ IN ML/MOL
C     THE CONTENTS OF VIRDAT IS EXPLAINED IN SUBROUTINE SVIR
C     ****************
C
      SUBROUTINE YENWO(T,VLIQ)
C
      DIMENSION VLIQ(2)
      COMMON/VIRDAT/NC(2),PC(2),RD(2),DMU(2),ETA(3),TC(2),ZC(2)
      DO 10 I=1,2
      A=17.4425-214.578*ZC(I)+989.625*(ZC(I)**2)-1522.06*ZC(I)**3
      IF(ZC(I)-0.26)100,100,101
  100 B=-3.28257+13.6377*ZC(I)+107.4844*ZC(I)**2-384.211*ZC(I)**3
      GO TO 102
  101 B=60.2091-402.063*ZC(I)+501.*ZC(I)**2+641.*ZC(I)**3
  102 D=0.93-B
      TRED=T/TC(I)
      TR1=(1.-TRED)**(1./3.)
      TR2=(1.-TRED)**(2./3.)
      TR4=(1.-TRED)**(4./3.)
      RHO=1.+A*TR1+B*TR2+D*TR4
   10 VLIQ(I)=82.06*ZC(I)*TC(I)/(PC(I)*RHO)
      RETURN
      END
```

```fortran
C     ********************************************************************
C     *
C     *   SUBROUTINE GAUSL(ND,NCOL,N,NS,A)
C     *
C     ********************************************************************
C
C
C     *****************
C     SUBROUTINE GAUSL SOLVES N LINEAR ALGEBRAIC EQUATIONS BY GAUSS
C     ELIMINATION WITH ROW PIVOTING
C     TO SOLVE THE PROBLEM QX=U, WHERE Q IS A NXN MATRIX AND U IS NXNS,
C     ONE PLACES Q IN THE FIRST N COLUMNS OF A AND U IS PLACED IN THE
C     FOLLOWING NS COLUMNS.
C     THE PROGRAM RETURNS X=Q**(-1)*U AT THE PREVIOUS POSITION OF U.
C     *
C     ND IS THE ROW DIMENSION AND NCOL IS THE COLUMN DIMENSION OF A.
C     BOTH MUST BE TRANSFERRED TO THE SUBROUTINE.
C     *****************
C
      SUBROUTINE GAUSL(ND,NCOL,N,NS,A)
C
      IMPLICIT REAL*8 (A-H,O-Z)
      DIMENSION A(ND,NCOL)
      N1=N+1
      NT=N+NS
      IF (N .EQ. 1) GO TO 50
C
C        START ELIMINATION
C
      DO 10 I=2,N
      IP=I-1
      I1=IP
      X=DABS(A(I1,I1))
      DO 11 J=I,N
      IF (DABS(A(J,I1)) .LT. X) GO TO 11
      X=DABS(A(J,I1))
      IP=J
   11 CONTINUE
      IF (IP .EQ. I1) GO TO 13
C
C        ROW INTERCHANGE
C
      DO 12 J=I1,NT
      X=A(I1,J)
      A(I1,J)=A(IP,J)
   12 A(IP,J)=X
   13 DO 10 J=I,N
      X=A(J,I1)/A(I1,I1)
      DO 10 K=I,NT
   10 A(J,K)=A(J,K) - X*A(I1,K)
C
C        ELIMINATION FINISHED, NOW BACKSUBSTITUTION
C
```

```
   50 DO 20 IP=1,N
      I=N1-IP
      DO 20 K=N1,NT
      A(I,K) = A(I,K)/A(I,I)
      IF (I .EQ. 1) GO TO 20
      I1=I-1
      DO 25 J=1,I1
   25 A(J,K) = A(J,K) - A(I,K)*A(J,I)
   20 CONTINUE
      RETURN
      END
```

```
C     ***********************************************************************
C     *
C     *   SUBROUTINE MLMEN(N,X,POL,DPOL)
C     *
C     ***********************************************************************
C
C
C     *****************
C     CALCULATION OF THE TERMS OF THE LEGENDRE
C     POLYNOMIAL AND OF THE DERIVATIVES
C     N IS THE NUMBER OF TERMS
C     X IS THE VALUE OF THE VARIABLE
C     POL(I)=L(I-1)(X)
C     DPOL(I)=D(POL(I))/D(X)
C     SEE EQUATION 5.8
C     *****************
C
      SUBROUTINE MLMEN(N,X,POL,DPOL)
C
      DIMENSION POL(15),DPOL(15)
      POL(1)=1.
      POL(2)=2.*X-1.
      DPOL(1)=0.
      DPOL(2)=2.
      FX=POL(2)
      DO 100 I=3,N
      XX=(2*I-3)*POL(I-1)
      POL(I)=(FX*XX-(I-2)*POL(I-2))/(I-1)
  100 DPOL(I)=((2*I-3)*FX*DPOL(I-1)-(I-2)*DPOL(I-2)+2.*XX)/(I-1)
      RETURN
      END
```

```
C     ******************************************************************
C     *
C     *  BIJ (SECOND VIRIAL COEFFICIENTS)
C     *
C     ******************************************************************
C
C
C
C
C           ***********
C     THIS PROGRAM BUILDS A MATRIX OF PURE COMPONENT AND
C     CROSS VIRIAL COEFFICIENTS, B(I,J), FOR UP TO 10
C     COMPONENTS
C           ***********
C
C
C     THE PROGRAM CALLS SUBROUTINE SVIR
C
C     DATA CARDS
C     1. 5 CARDS,ANY TEXT(40A2)
C     2. 1 CARD, NN,NT(2I2)
C           NN   = NUMBER OF COMPONENTS
C           NT   = NUMBER OF TEMPERATURES
C     3. NN CARDS, ONE CARD FOR EACH COPONENT
C           NNC,PPC,TTC,RRD,DDMU,EETA,ZC(I2,6F10.3)
C           NNC  = NUMBER OF COMPONENT
C           PPC  = CRITICAL PRESSURE IN ATM
C           TTC  = CRITICAL TEMPERATURE IN K
C           RRD  = MEAN RADIUS OF GYRATION IN A
C           DDMU = DIPOLE MONENT IN DEBYE
C           EETA = ASSOCIATION PARAMETER FOR COMPONENT NNC
C           ZC   = CRITICAL COMPRESSIBILITY FACTOR
C     4. (NN-1) CARDS, ONE CARD FOR EACH COMPONENT EXEPT THE LAST
C           ETM(8F10.3)
C           ETM  = SOLVATION PARAMETERS. EACH CARD CONTAINS
C           THE SOLVATION PARAMETERS BETWEEN A
C           GIVEN I=1,NN-1 AND ALL J=I+1,NN
C     5. 1 CARD, T(8F10.3)
C           T    = TEMPERATURES IN K, THE CARD CONTAINS NT TEMPERATURES
C
C
C     INPUT
C
      DIMENSION NNC(10),PPC(10),RRD(10),DDMU(10),EETA(10),TTC(10),ETM(10
     1,10),B(10,10),T(10),NTEXT(40),BB(3),BF(3),ZZC(10)
      COMMON/VIRDAT/NC(2),PC(2),RD(2),DMU(2),ETA(3),TC(2),ZC(2)
      WRITE(6,21)
   21 FORMAT(1H1,'    CALCULATION OF VIRIAL COEFFICIENTS B(I,J) '//)
      DO 1 I=1,5
      READ(5,20)NTEXT
   20 FORMAT(40A2)
      WRITE(6,20)NTEXT
    1 CONTINUE
```

```
      READ(5,22)NN,NT
   22 FORMAT(2I2)
      DO 17 I=1,NN
   17 READ(5,23)NNC(I),PPC(I),TTC(I),RRD(I),DDMU(I),EETA(I),ZZC(I)
   23 FORMAT(I2,6F10.3)
      WRITE(6,24)
   24 FORMAT(1H0,' COMPONENT  PC IN ATM   RD IN A    DMU IN DEB.    ETA         TC
     1 IN K')
      WRITE(6,25)(NNC(I),PPC(I),RRD(I),DDMU(I),EETA(I),TTC(I),I=1,NN)
   25 FORMAT(5X,I2,3X,5F10.3)
      IN=NN-1
      DO 2 I=1,IN
      II=I+1
    2 READ(5,26)(ETM(I,J),J=II,NN)
   26 FORMAT(8F10.3)
      WRITE(6,27)
   27 FORMAT(/1H0,'   SOLVATION PARAMETERS')
      IN=NN-1
      DO 11 I=1,IN
      II=I+1
      DO 11 J=II,NN
   11 ETM(J,I)=ETM(I,J)
      DO 12 I=1,NN
   12 ETM(I,I)=EETA(I)
      WRITE(6,28)(NNC(I),I=1,NN)
   28 FORMAT(10X,I2,9(8X,I2))
      DO 13 I=1,NN
   13 WRITE(6,29)NNC(I),(ETM(I,J),J=1,NN)
   29 FORMAT(2X,I2,10F10.3)
      READ(5,26)(T(I),I=1,NT)
C
C
C  MAIN PROGRAM
C
C
      N=2
      WRITE(6,40)
   40 FORMAT(/1H0,' VIRIAL COEFFICIENTS')
      DO 650 KK=1,NT
      TEMP=T(KK)
      WRITE(6,41)TEMP
   41 FORMAT(/10X,'TEMPERATURE=',F10.3)
      IN=NN-1
      DO 32 I=1,IN
      II=I+1
      DO 32 J=II,NN
      K=1
      M=I
   33 NC(K)=NNC(M)
      PC(K)=PPC(M)
      RD(K)=RRD(M)
      DMU(K)=DDMU(M)
      ETA(K)=EETA(M)
      TC(K)=TTC(M)
      ZC(K)=ZZC(M)
      IF(K.EQ.2) GO TO 34
      K=2
      M=J
```

```
      GO TO 33
   34 ETA(3)=ETM(I,J)
      CALL SVIR(N,TEMP,BF,BB)
      B(I,I)=BB(1)
      B(J,J)=BB(2)
      B(I,J)=BB(3)
   32 B(J,I)=B(I,J)
      WRITE(6,28)(NNC(I),I=1,NN)
      DO 19 I=1,NN
   19 WRITE(6,29)NNC(I),(B(I,J),J=1,NN)
  650 CONTINUE
      STOP
      END
```

```
****************************************************************
*
*   SAMPLE INPUT (BIJ)
*
****************************************************************


 MIXTURE OF 4 COMPONENTS
  1 BENZENE,   2 CHLOROFORM
  3 ACETONE,   4 ETHYL ACETATE
  2  4
  4  4
  1 48.6       562.1       3.004       0.          0.          0.274
  2 54.        536.6       3.178       1.02        0.          0.294
  3 47.        509.1       2.74        2.88        0.9         0.237
  4 37.8       523.3       3.348       1.78        0.53        0.252
 0.12        0.5         0.6
 1.26        1.65
 1.1
 323.2       333.2       353.2       368.
```

```
***********************************************************************
*
*   SAMPLE OUTPUT (BIJ)
*
***********************************************************************


    CALCULATION OF VIRIAL COEFFICIENTS B(I,J)



MIXTURE OF  4 COMPONENTS
1 BENZENE,    2 CHLOROFORM
3 ACETONE,    4 ETHYL ACETATE

COMPONENT  PC IN ATM    RD IN A     DMU IN DEB.     ETA      TC IN K
    1        48.600      3.004         0.0          0.0      562.100
    2        54.000      3.178         1.020        0.0      536.600
    3        47.000      2.740         2.880        0.900    509.100
    4        37.800      3.348         1.780        0.530    523.300

    SOLVATION PARAMETERS
               1          2            3            4
    1        0.0        0.120        0.500        0.600
    2        0.120      0.0          1.260        1.650
    3        0.500      1.260        0.900        1.100
    4        0.600      1.650        1.100        0.530
```

VIRIAL COEFFICIENTS

```
        TEMPERATURE=    323.200
            1         2         3         4
1   -1194.375 -1131.534 -1047.051 -1473.408
2   -1131.534  -956.183 -1509.186 -2268.721
3   -1047.051 -1509.186 -1437.744 -1654.402
4   -1473.408 -2268.721 -1654.402 -1521.767

        TEMPERATURE=    333.200
            1         2         3         4
1   -1100.822 -1046.662  -973.294 -1358.302
2   -1046.662  -882.104 -1327.730 -1936.503
3    -973.294 -1327.730 -1292.786 -1473.516
4   -1358.302 -1936.503 -1473.516 -1403.742

        TEMPERATURE=    353.200
            1         2         3         4
1    -947.395  -906.571  -849.074 -1168.373
2    -906.571  -759.921 -1054.309 -1468.944
3    -849.074 -1054.309 -1061.544 -1192.322
4   -1168.373 -1468.944 -1192.322 -1207.682

        TEMPERATURE=    368.000
            1         2         3         4
1    -855.719  -822.274  -772.813 -1054.294
2    -822.274  -686.483  -905.285 -1231.634
3    -772.813  -905.285  -928.262 -1034.120
4   -1054.294 -1231.634 -1034.120 -1089.175
```

```
C     ****************************************************************
C     *
C     *   THERMODYNAMIC CONSISTENCY TEST
C     *
C     ****************************************************************
C
C
C     *****************
C     THIS PROGRAM TESTS BINARY VLE DATA FOR THERMODYNAMIC
C     CONSISTENCY AS EXPLAINED IN SECTION 5.2
C
C     THE PROGRAM CALLS SUBROUTINES
C     PHIB AND SVIR FOR CALCULATION OF FUGACITY COEFFICIENTS
C
C     YENWO FOR CALCULATION OF THE MOLAR VOLUMES USED
C     IN THE POYNTING CORRECTION FACTOR FOR THE FUGACITIES
C     OF PURE SUBSTANCES
C
C     MLMEN AND GAUSL FOR CALCULATION OF THE COEFFICIENTS
C     IN THE LEGENDRE POLYNOMIALS
C
C     ****************
C     DATA CARDS
C     1. 1 CARD, ISO(I5)
C            ISO IS LESS THAN ZERO FOR ISOTHERMAL DATA
C            ISO IS EQUAL TO ZERO AT THE END OF THE DATA SET
C            ISO IS GREATER THAN ZERO FOR ISOBARIC DATA
C     2. 1 CARD, ANY TEXT(40A2)
C     3. 2 CARDS, ONE CARD FOR EACH COMPONENT WITH PURE COMPONENT DATA
C            TC,PC,RD,ETA,DMU,ZC(6F10.5)
C
C            TC  = CRITICAL TEMPERATURE IN K
C            PC  = CRITICAL PRESSURE IN ATM
C            RD  = MEAN RADIUS OF GYRATION IN A
C            ETA = ASSOCIATION PARAMETER
C            DMU = DIPOLE MOMENT IN DEBYE
C            ZC  = CRITICAL COMPRESSIBILITY FACTOR
C     4. 1 CARD, THIS CARD IS ONLY INCLUDED IF ONE OF THE COMPONENTS
C            IS AN ORGANIC ACID (ETA IS EQUAL TO 4.5)
C            NACID(I2) NUMBER OF CARBONATOMS IN THE ACID
C     5. 2 CARDS, THESE CARDS ARE ONLY INCLUDED IF THE DATA ARE ISOBARIC
C            A,B,C(3F10.5)
C            ONE CARD FOR EACH COMPONENT, ANTOINE CONSTANTS FOR
C            CALCULATION OF VAPOR PRESSURES ACCORDING TO HALA
C     6. 1 CARD, ETA(3) (F10.2)
C            ETA(3) = SOLVATION PARAMETER
C     7. 1 CARD, NT,NPOL(2I5)
C            NT = NUMBER OF DATA SETS
C            NPOL = ORDER OF LEGENDRE POLYNOMIAL
C     8. 1 CARD, TEMP(F10.2) = TEMPERATURE IN K FOR ISOTHERMAL DATA
C            OR
C            PRES(F10.2) = PRESSURE IN MM HG FOR ISOBARIC DATA
C     9. NT CARDS, *ISOTHERMAL DATA* X,Y,P(3F10.5)
```

```
C               OR
C       NT CARDS, *ISOBARIC DATA* X,Y,T(3F10.5)
C           X = LIQUID MOLE FRACTION OF COMPONENT ONE, THE MORE VOLATILE
C               COMPONENT
C           Y = VAPOR MOLE FRACTION OF COMPONENT ONE
C           X(1) AND Y(1) SHOULD BE ZERO
C           X(NT) AND Y(NT) SHOULD BE ONE
C           P = PRESSURE IN MM HG
C           T = TEMPERATURE IN K
C
C   ****************
C
C
C   ****************
C   OUTPUT INFORMATION
C
C           ONLY SYMBOLS NOT USED IN DATA CARDS ARE EXPLAINED
C
C           PEXP,PEST AND YEXP,YEST ARE EXPERIMENTAL AND ESTIMATED
C           VALUES OF PRESSURES IN ATM AND OF VAPOR MOLE FRACTIONS
C           GAMMA = LIQUID PHASE ACTIVITY COEFFICIENTS FOR COMPONENT
C           ONE AND TWO
C           G = EXCESS GIBBS ENERGY
C           FUG10 AND FUG20 ARE PURE COMPONENT FUGACITIES
C           PHI1 AND PHI2 ARE FUGACITY COEFFICIENTS
C           OF THE COMPONENTS IN THE MIXTURE
C
C   *****************
C
C
C
C
      DIMENSION YE(25),PHI1(25),PHI2(25),PHI(2),PHIR1(25),PHIR2(25)
      DIMENSION P1(25),P2(25),V1(25),V2(25),A(2),B(2),C(2)
      DIMENSION YY(2),Z(25)
      DIMENSION BB(15),FA(25),FB(25),X(25),PEXP(25),PEST(25),PAR(15,25)
      DIMENSION Y(25),PP1(25),PP2(25),NTEXT(40)
      DIMENSION DPAR(15,25),DPEST(25,25),TEXP(25)
      REAL*8 SS(10,11)
      DIMENSION G1(25),G2(25),G(25)
      DIMENSION V(2)
      COMMON/VIRDAT/NC(2),PC(2),RD(2),DMU(2),ETA(3),TC(2),ZC(2)
      M=5
      N=6
      NK=2
   96 READ(M,5) ISO
      IF (ISO) 99,199,99
   99 READ(M,2) NTEXT
    2 FORMAT (40A2)
      WRITE(N,3) NTEXT
    3 FORMAT(1H1,/40A2//)
      DO 8 I=1,NK
      NC(I)=I
    8 READ(M,1) TC(I),PC(I),RD(I),ETA(I),DMU(I),ZC(I)
      IF(ETA(1).EQ.4.5)GO TO 411
      IF(ETA(2).EQ.4.5)GO TO 411
      NACID=0
      GO TO 412
  411 READ(M,801)NACID
```

```
      412 WRITE(N,910)
      910 FORMAT (25X,'PURE COMPONENT DATA'/)
          IF(ISO)417,199,414
      414 WRITE(N,911)
      911 FORMAT (1X,'NO',9X,            'TC',8X,'PC',8X,'RD',8X,'ETA',8X,
         1        'DMU',7X,'ZC',8X,'A',8X,'B',8X,'C'/)
          DO 88 I=1,NK
          READ(M,1)A(I),B(I),C(I)
          WRITE(N,912) NC(I),TC(I),PC(I),RD(I),ETA(I),DMU(I),ZC(I),A(I),B(I)
         1,C(I)
   88     CONTINUE
          GO TO 415
      417 WRITE(N,913)
      913 FORMAT (1X,'NO',9X,            'TC',8X,'PC',8X,'RD',8X,'ETA',8X,
         1'DMU',7X,'ZC'/)
          DO 89 I=1,NK
   89     WRITE(N,912)NC(I),TC(I),PC(I),RD(I),ETA(I),DMU(I),ZC(I)
      415 CONTINUE
      912 FORMAT (1X,I2,3X,9F10.3)
      801 FORMAT (I2)
      803 FORMAT (8F10.2)
          READ(M,803) ETA(3)
          WRITE(N,135) ETA(3)
      135 FORMAT(//,'       SOLVATION PARAMETER=',F10.2)
          RG=82.06
          NH=1
          READ(M,5)NT,NPOL
          IF (ISO.GT.0.) GOTO 313
          WRITE (N,6)
        6 FORMAT(/'  NT    NPOL NACID P1SAT      P2SAT      V1       V2'/)
          READ(M,803) TEMP
          DO 7 I=1,NT
          TEXP(I) = TEMP
          READ(M,1) X(I),Y(I),PEXP(I)
    7     YE(I)=Y(I)
          DO 727 I=1,NT
      727 PEXP(I)=PEXP(I)/760.
          CALL YENWO(TEMP,V)
          WRITE(N,666)NT,NPOL,NACID,PEXP(NT),PEXP(1),V(1),V(2)
    5     FORMAT(3I5,F10.5)
      666 FORMAT(3I5,4F10.5)
        1 FORMAT(8F10.5)
          DO 701 I=1,NT
          YY(1)=Y(I)
          IF(YY(1).EQ.0.) YY(1)=1.E-06
          YY(2)=1.-Y(I)
          IF(YY(2).EQ.0.) YY(2)=1.E-06
          P=PEXP(I)
          CALL PHIB(NK,P,TEMP,YY,PHI,NACID)
          PHI1(I)=PHI(1)
      701 PHI2(I)=PHI(2)
          DO 9 I=1,NT
          PP1(I)=PEXP(NT)*PHI1(NT)*EXP(V(1)*(PEXP(I)-PEXP(NT))/(RG*TEMP))
    9     PP2(I)=PEXP(1)*PHI2(1)*EXP(V(2)*(PEXP(I)-PEXP(1))/(RG*TEMP))
          GOTO 314
      313 CONTINUE
          WRITE(N,667)
      667 FORMAT(/'   NT    NPOL NACID')
```

```
      READ(M,803) PRES
      PRES = PRES/760.
      DO 107 I=1,NT
      PEXP(I)=PRES
      READ(M,1) X(I),YE(I),TEXP(I)
  107 Y(I)=YE(I)
      WRITE(N,5)NT,NPOL,NACID
      DO 413 I=1,NT
      T=TEXP(I)
      P1(I)=10.**(A(1)-B(1)/(C(1)+T-273.15))/760.
      YY(1)=1.
      YY(2)=1.E-24
      CALL PHIB(NK,PRES,T,YY,PHI,NACID)
      PHIR1(I)=PHI(1)
      P2(I)=10.**(A(2)-B(2)/(C(2)+T-273.15))/760.
      YY(1)=1.E-24
      YY(2)=1.
      CALL PHIB(NK,PRES,T,YY,PHI,NACID)
      PHIR2(I)=PHI(2)
      CALL YENWO(T,V)
      PP1(I)=P1(I)*PHIR1(I)*EXP(V(1)*(PRES-P1(I))/(RG*TEXP(I)))
  413 PP2(I)=P2(I)*PHIR2(I)*EXP(V(2)*(PRES-P1(I))/(RG*TEXP(I)))
      DO 1701 I=1,NT
      YY(1)=Y(I)
      IF(YY(1).EQ.0.) YY(1)=1.E-06
      YY(2)=1.-Y(I)
      IF(YY(2).EQ.0.) YY(2)=1.E-06
      T=TEXP(I)
      CALL PHIB(NK,PRES,T,YY,PHI,NACID)
      PHI1(I)=PHI(1)
 1701 PHI2(I)=PHI(2)
  314 CONTINUE
   97 DO 730 I=1,NT
      CALL MLMEN(15,X(I),FA,FB)
      DO 11 J=1,15
      PAR(J,I)=FA(J)
   11 DPAR(J,I)=FB(J)
  730 CONTINUE
      DO 30 K=1,NPOL
   30 BB(K)=0.
      DO 40 IK=1,NPOL,NH
      NDG=NT-IK-2
  706 IT=0
      IF (NDG) 200,200,45
   45 DO 50 I=1,NT
      FA(I)=0.
      FB(I)=0.
      DO 51 J=1,IK
      FA(I)=FA(I)+(PAR(J,I)+X(I)*DPAR(J,I))*BB(J)
   51 FB(I)=FB(I)+(PAR(J,I)-(1.-X(I))*DPAR(J,I))*BB(J)
      C1=PP1(I)
      C2=PP2(I)
      IF (PHI1(I).EQ.0) GO TO 780
      FA(I)=X(I)*C1*EXP((1.-X(I))**2*FA(I))/PHI1(I)
      GO TO 781
  780 FA(I)=0.
  781 IF (PHI2(I).EQ.0) GO TO 782
      FB(I)=(1.-X(I))*C2*EXP(X(I)**2*FB(I))/PHI2(I)
```

```
          GO TO 783
782   FB(I)=0.
783   CONTINUE
      PEST(I)=FA(I)+FB(I)
      Z(I)=FA(I)/PEST(I)
      DO 50 J=1,IK
      DPEST(I,J)=FA(I)*(PAR(J,I)+X(I)*DPAR(J,I))*(1.-X(I))**2+
     1 FB(I)*(PAR(J,I)-(1.-X(I))*DPAR(J,I))*X(I)**2
 50   CONTINUE
      IF (20-IT) 661,661,662
661   WRITE(N,663)
663   FORMAT(/'   NO CONVERGENCE IN 20 ITERATIONS   '/)
      GO TO 40
662   CONTINUE
      IT=IT+1
      SSQ=0.
      DO 61 I=1,NT
 61   SSQ=SSQ+(PEST(I)-PEXP(I))**2
      DO 62 I=1,IK
      SS(I,IK+1)=0.
      DO 62 J=1,NT
 62   SS(I,IK+1)=SS(I,IK+1)-DPEST(J,I)*(PEST(J)-PEXP(J))
      DO 63 I=1,IK
      DO 63 J=1,IK
      SS(I,J)=0.
      DO 63 K=1,NT
 63   SS(I,J)=SS(I,J)+      DPEST(K,I)*DPEST(K,J)
      CALL GAUSL (10,11,IK,1,SS)
      R=0.
      RR=0.
      DO 70 I=1,IK
      BB(I)=BB(I)+SS(I,IK+1)
      R=R+BB(I)**2
 70   RR=RR+SS(I,IK+1)**2
      D=ABS(RR/R)
      IF (D-1.E-10) 42,42,45
42    YDEV=0.
      PDEV=0.
      DO 43 I=1,NT
      IF ((I-1)*(I-NT)) 795,557,795
795   G1(I)=FA(I)*PHI1(I)/X(I)/PP1(I)
      G2(I)=FB(I)*PHI2(I)/(1.-X(I))/PP2(I)
      GO TO 556
557   GG1=0.
      GG2=0.
      DO 558 J=1,IK
      GG1=GG1+(PAR(J,I)+X(I)*DPAR(J,I))*BB(J)
558   GG2=GG2+(PAR(J,I)-(1.-X(I))*DPAR(J,I))*BB(J)
      G2(I)=EXP(X(I)**2*GG2)
      G1(I)=EXP((1.-X(I))**2*GG1)
556   GG=0.
      DO 555 JJ=1,IK
555   GG=GG+BB(JJ)*PAR(JJ,I)
      G(I)=GG*X(I)*(1.-X(I))
      YDEV=YDEV+ ABS((Z(I)-Y(I))/(NT-2))
      PDEV=PDEV+ ABS((PEXP(I)-PEST(I))/(NT-2))
43    CONTINUE
      IF (YDEV-.0001)702,702,703
```

```
 703    IF (ISO.GT.0.) GOTO 1703
        DO 704 I=1,NT
        Y(I)=Z(I)
        YY(1)=Y(I)
        IF(YY(1).EQ.0.) YY(1)=1.E-06
        YY(2)=1.-Y(I)
        IF(YY(2).EQ.0.) YY(2)=1.E-06
        P=PEXP(I)
        CALL PHIB(NK,P,TEMP,YY,PHI,NACID)
        PHI1(I)=PHI(1)
 704    PHI2(I)=PHI(2)
        GOTO 706
1703    DO 1706 I=1,NT
        Y(I)=Z(I)
        YY(1)=Y(I)
        IF(YY(1).EQ.0.) YY(1)=1.E-06
        YY(2)=1.-Y(I)
        IF(YY(2).EQ.0.) YY(2)=1.E-06
        T=TEXP(I)
        CALL PHIB(NK,PRES,T,YY,PHI,NACID)
        PHI1(I)=PHI(1)
1706    PHI2(I)=PHI(2)
        GO TO 706
 702    YDEV=0.
        WRITE(N,48)
  48    FORMAT(//'     X        PEXP      PEST      YEXP      YEST     GAMMA1
       1 GAMMA2    G/RT   '/)
        DO 711 I=1,NT
        YDEV=YDEV + ABS((Z(I)-YE(I))/(NT-2))
 711    WRITE(N,44) X(I),PEXP(I),PEST(I),YE(I),Z(I),G1(I),G2(I),G(I)
  44    FORMAT(8F9.4,F9.2,4F9.4)
        WRITE(N,49)
  49    FORMAT(//'     TEMP      FUG10     FUG20     PHI1     PHI2'/)
        DO 712 I=1,NT
 712    WRITE(N,145) TEXP(I),PP1(I),PP2(I),PHI1(I),PHI2(I)
 145    FORMAT(F9.2,4F9.4)
        WRITE(N,82) NDG,IT,PDEV,YDEV
  82    FORMAT(/,'  DEGREES OF FREEDOM ',I3,' NUMBER OF ITERATIONS ',I3,/,
       1         '  MEAN DEV IN P           ',F10.5/,
       2         '  MEAN DEV IN Y           ', F10.5,/)
        WRITE(N,83)(BB(K),K=1,IK)
  83    FORMAT(/'  LEGENDRE COEFFICIENTS ',/, 15(F15.5/))
  40    CONTINUE
 200    GO TO 96
 199    STOP
        END
```

```
************************************************************************
*
*   SAMPLE INPUT (THERMODYNAMIC CONSISTENCY TEST)
*
************************************************************************

     5
    -1
    2-BUTANONE(1)-ACETIC ACID(2) AT 351.15 K,LYNGBY 1976
535.        41.        3.139     0.9       2.7       0.249
594.8       57.1       2.595     4.5       1.74      0.2
   2
1.8   5       10
     11        5
351.15
0.          0.         192.12
0.0177      0.0373     198.58
0.0315      0.0749     205.83
0.1397      0.3112     260.09
0.2262      0.4653     304.17
0.4334      0.7356     414.98
0.6006      0.8622     505.55
0.7754      0.9438     600.94
0.8674      0.9728     654.24
0.9519      0.9915     676.04
1.          1.         696.5
```

```
*************************************************************************
*
*   SAMPLE OUTPUT (THERMODYNAMIC CONSISTENCY TEST)
*
*************************************************************************


2-BUTANONE(1)-ACETIC ACID(2) AT 351.15 K,LYNGBY 1976

                        PURE COMPONENT DATA
NO        TC         PC        RD        ETA        DMU         ZC

1      535.000    41.000     3.139     0.900      2.700       0.249
2      594.800    57.100     2.595     4.500      1.740       0.200


     SOLVATION PARAMETER=       1.80

   NT    NPOL  NACID P1SAT     P2SAT      V1         V2

   11     5     2    0.91645   0.25279   97.07004   59.23389
```

X	PEXP	PEST	YEXP	YEST	GAMMA1	GAMMA2	G/RT
0.0	0.2528	0.2528	0.0	0.0	1.4172	1.0000	0.0
0.0177	0.2613	0.2653	0.0373	0.0522	1.3999	1.0001	0.0061
0.0315	0.2708	0.2762	0.0749	0.0908	1.3868	1.0003	0.0106
0.1397	0.3422	0.3577	0.3112	0.3404	1.2944	1.0068	0.0419
0.2262	0.4002	0.4191	0.4653	0.4855	1.2321	1.0180	0.0610
0.4334	0.5460	0.5528	0.7356	0.7181	1.1184	1.0677	0.0856
0.6006	0.6652	0.6522	0.8622	0.8426	1.0572	1.1340	0.0836
0.7754	0.7907	0.7578	0.9438	0.9359	1.0177	1.2332	0.0607
0.8674	0.8608	0.8182	0.9728	0.9705	1.0061	1.2999	0.0401
0.9519	0.8895	0.8787	0.9915	0.9925	1.0008	1.3715	0.0160
1.0000	0.9164	0.9164	1.0000	1.0000	1.0000	1.4172	0.0

TEMP	FUG10	FUG20	PHI1	PHI2
351.15	0.9094	0.0767	1.6941	0.3033
351.15	0.9094	0.0767	1.6284	0.2996
351.15	0.9094	0.0767	1.5843	0.2958
351.15	0.9097	0.0767	1.3508	0.2816
351.15	0.9098	0.0767	1.2463	0.2803
351.15	0.9103	0.0767	1.1115	0.2978
351.15	0.9107	0.0767	1.0522	0.3385
351.15	0.9110	0.0768	1.0138	0.4376
351.15	0.9113	0.0768	1.0016	0.5490
351.15	0.9113	0.0768	0.9956	0.7656
351.15	0.9114	0.0768	0.9945	0.9989

DEGREES OF FREEDOM   8 NUMBER OF ITERATIONS   2
MEAN DEV IN P   0.01666
MEAN DEV IN Y   0.01428

LEGENDRE COEFFICIENTS
      0.34865

X	PEXP	PEST	YEXP	YEST	GAMMA1	GAMMA2	G/RT
0.0	0.2528	0.2528	0.0	0.0	0.9267	1.0000	0.0
0.0177	0.2613	0.2608	0.0373	0.0355	0.9503	0.9998	-0.0011
0.0315	0.2708	0.2684	0.0749	0.0639	0.9680	0.9993	-0.0017
0.1397	0.3422	0.3340	0.3112	0.2967	1.0828	0.9893	0.0019
0.2262	0.4002	0.3956	0.4653	0.4732	1.1416	0.9779	0.0127
0.4334	0.5460	0.5580	0.7356	0.7600	1.1717	0.9697	0.0513
0.6006	0.6652	0.6755	0.8622	0.8735	1.1205	1.0205	0.0764
0.7754	0.7907	0.7776	0.9438	0.9418	1.0487	1.1876	0.0755
0.8674	0.8608	0.8293	0.9728	0.9692	1.0189	1.3581	0.0568
0.9519	0.8895	0.8815	0.9915	0.9909	1.0027	1.5989	0.0252
1.0000	0.9164	0.9164	1.0000	1.0000	1.0000	1.7881	0.0

TEMP	FUG10	FUG20	PHI1	PHI2
351.15	0.9094	0.0767	1.6941	0.3033
351.15	0.9094	0.0767	1.6499	0.2994
351.15	0.9094	0.0767	1.6174	0.2953
351.15	0.9097	0.0767	1.3887	0.2779
351.15	0.9098	0.0767	1.2549	0.2785
351.15	0.9103	0.0767	1.0900	0.3147
351.15	0.9107	0.0767	1.0388	0.3660
351.15	0.9110	0.0767	1.0116	0.4521
351.15	0.9113	0.0768	1.0020	0.5416
351.15	0.9113	0.0768	0.9959	0.7378
351.15	0.9114	0.0768	0.9945	0.9989

DEGREES OF FREEDOM   7 NUMBER OF ITERATIONS   2
MEAN DEV IN P   0.01008
MEAN DEV IN Y   0.00856

LEGENDRE COEFFICIENTS
     0.25253
     0.32864

X	PEXP	PEST	YEXP	YEST	GAMMA1	GAMMA2	G/RT
0.0	0.2528	0.2528	0.0	0.0	1.4759	1.0000	0.0
0.0177	0.2613	0.2654	0.0373	0.0523	1.4054	1.0004	0.0064
0.0315	0.2708	0.2759	0.0749	0.0889	1.3580	1.0013	0.0109
0.1397	0.3422	0.3457	0.3112	0.3033	1.1409	1.0163	0.0323
0.2262	0.4002	0.3970	0.4653	0.4395	1.0848	1.0272	0.0392
0.4334	0.5460	0.5402	0.7356	0.7302	1.1049	1.0140	0.0511
0.6006	0.6652	0.6756	0.8622	0.8780	1.1244	0.9978	0.0695
0.7754	0.7907	0.7939	0.9438	0.9487	1.0759	1.1147	0.0811
0.8674	0.8608	0.8414	0.9728	0.9704	1.0346	1.3411	0.0685
0.9519	0.8895	0.8851	0.9915	0.9895	1.0057	1.8022	0.0337
1.0000	0.9164	0.9164	1.0000	1.0000	1.0000	2.2862	0.0

TEMP	FUG10	FUG20	PHI1	PHI2
351.15	0.9094	0.0767	1.6941	0.3033
351.15	0.9094	0.0767	1.6282	0.2996
351.15	0.9094	0.0767	1.5867	0.2958
351.15	0.9097	0.0767	1.3829	0.2784
351.15	0.9098	0.0767	1.2795	0.2740
351.15	0.9103	0.0767	1.1050	0.3024
351.15	0.9107	0.0767	1.0367	0.3711
351.15	0.9110	0.0768	1.0091	0.4720
351.15	0.9113	0.0768	1.0016	0.5480
351.15	0.9113	0.0768	0.9962	0.7150
351.15	0.9114	0.0768	0.9945	0.9989

```
DEGREES OF FREEDOM    6 NUMBER OF ITERATIONS    2
MEAN DEV IN P              0.00657
MEAN DEV IN Y              0.01036


LEGENDRE COEFFICIENTS
       0.35645
       0.21882
       0.25163
```

X	PEXP	PEST	YEXP	YEST	GAMMA1	GAMMA2	G/RT
0.0	0.2528	0.2528	0.0	0.0	0.8247	1.0000	0.0
0.0177	0.2613	0.2601	0.0373	0.0335	0.8949	0.9993	-0.0027
0.0315	0.2708	0.2677	0.0749	0.0625	0.9450	0.9979	-0.0038
0.1397	0.3422	0.3404	0.3112	0.3207	1.1749	0.9808	0.0059
0.2262	0.4002	0.4045	0.4653	0.4889	1.1954	0.9779	0.0231
0.4334	0.5460	0.5427	0.7356	0.7320	1.1120	1.0137	0.0537
0.6006	0.6652	0.6658	0.8622	0.8707	1.1020	1.0210	0.0667
0.7754	0.7907	0.7985	0.9438	0.9523	1.0849	1.0683	0.0780
0.8674	0.8608	0.8496	0.9728	0.9730	1.0466	1.2712	0.0714
0.9519	0.8895	0.8883	0.9915	0.9889	1.0089	1.8805	0.0388
1.0000	0.9164	0.9164	1.0000	1.0000	1.0000	2.7522	0.0

TEMP	FUG10	FUG20	PHI1	PHI2
351.15	0.9094	0.0767	1.6941	0.3033
351.15	0.9094	0.0767	1.6526	0.2994
351.15	0.9094	0.0767	1.6192	0.2953
351.15	0.9097	0.0767	1.3676	0.2798
351.15	0.9098	0.0767	1.2440	0.2808
351.15	0.9103	0.0767	1.1042	0.3030
351.15	0.9107	0.0767	1.0398	0.3635
351.15	0.9110	0.0768	1.0078	0.4837
351.15	0.9113	0.0768	1.0008	0.5632
351.15	0.9113	0.0768	0.9963	0.7070
351.15	0.9114	0.0768	0.9945	0.9989

DEGREES OF FREEDOM   5 NUMBER OF ITERATIONS   2
MEAN DEV IN P   0.00383
MEAN DEV IN Y   0.00807

LEGENDRE COEFFICIENTS
    0.29444
    0.42205
    0.11539
    0.18054

X	PEXP	PEST	YEXP	YEST	GAMMA1	GAMMA2	G/RT
0.0	0.2528	0.2528	0.0	0.0	1.2558	1.0000	0.0
0.0177	0.2613	0.2634	0.0373	0.0448	1.2014	1.0004	0.0036
0.0315	0.2708	0.2725	0.0749	0.0769	1.1709	1.0010	0.0060
0.1397	0.3422	0.3404	0.3112	0.3011	1.1170	1.0034	0.0184
0.2262	0.4002	0.4006	0.4653	0.4679	1.1461	0.9976	0.0290
0.4334	0.5460	0.5477	0.7356	0.7369	1.1272	1.0107	0.0579
0.6006	0.6652	0.6618	0.8622	0.8647	1.0907	1.0450	0.0697
0.7754	0.7907	0.7970	0.9438	0.9524	1.0829	1.0654	0.0760
0.8674	0.8608	0.8541	0.9728	0.9746	1.0534	1.2219	0.0717
0.9519	0.8895	0.8909	0.9915	0.9888	1.0118	1.8994	0.0420
1.0000	0.9164	0.9164	1.0000	1.0000	1.0000	3.1779	0.0

TEMP	FUG10	FUG20	PHI1	PHI2
351.15	0.9094	0.0767	1.6941	0.3033
351.15	0.9094	0.0767	1.6378	0.2995
351.15	0.9094	0.0767	1.6013	0.2956
351.15	0.9097	0.0767	1.3847	0.2783
351.15	0.9098	0.0767	1.2586	0.2778
351.15	0.9103	0.0767	1.1017	0.3049
351.15	0.9107	0.0767	1.0424	0.3577
351.15	0.9110	0.0768	1.0078	0.4840
351.15	0.9113	0.0768	1.0003	0.5739
351.15	0.9113	0.0768	0.9963	0.7056
351.15	0.9114	0.0768	0.9945	0.9989

DEGREES OF FREEDOM   4 NUMBER OF ITERATIONS   2
MEAN DEV IN P   0.00283
MEAN DEV IN Y   0.00433

LEGENDRE COEFFICIENTS
     0.34614
     0.33896
     0.25104
     0.12525
     0.09482

APPENDIX 2

UNIFAC PROGRAMS

Computer programs which use directly the UNIFAC method for calculating liquid phase activity coefficients are given in the following.

The "heart" of this appendix is Subroutine UNIFA, which for specified temperature and composition calculates activity coefficients according to the UNIFAC method as outlined in Chapter 4.

A main program which calls Subroutine UNIFA must define the following variables in the call to the subroutine:

NK = the number of components

NG = the number of different functional groups in the mixture

ITAB(I,K) = the number of groups of type K (Table 4.3) in molecule I (corresponds to $\nu_k^{(i)}$ in Equation (3.7))

T = temperature, K

X(I) = mole fraction of each component I

K = an integer between 1 and 56 according to the group definitions given in Table 4.3.

The main program must contain a COMMON statement as specified by the comment cards in the listing of Subroutine UNIFA.

In the present COMMON statement, NK is maximum 5 and NG is maximum 8. These may readily be changed when appropriate DIMENSION statements in Subroutine UNIFA are changed accordingly.

Subroutine UNIFA uses Subroutine GCOMB for calculating the combinatorial activity coefficients (according to Equation (3.7)), Subroutine GRES for calculating the term $\Sigma \nu_k^{(i)} \ln \Gamma_k$ in

Equations (3.20)$_k$ and (3.24) and GREF for calculating the corresponding term $\Sigma \nu_k^{(i)} \ln \Gamma_k^{(i)}$. Subroutine SYSTM is used to extract from Tables 4.3 and 4.4 precisely the parameters needed to solve a given problem. The large parameter matrix and the matrix ITAB are "compacted" so that they only contain the needed elements.

An important <u>warning</u> must be emphasized in connection with the use of Subroutine SYSTM. If an off-diagonal element in the "compacted" matrix PARA (corresponding to a_{mn} in Equation (3.24)) is zero it could be because:

> 1. Subgroups within the same main group have group interaction parameters equal to zero.
>
> or
>
> 2. The group interaction parameter in question is not available (corresponding to "n.a." in Table 4.4).

The first case gives no reason for concern. In the second case, however, the calculated activity coefficients are invalid. The only alternative to discontinuing the calculations is then to attempt to estimate the missing parameters as discussed in Chapter 5.

The UNIFA subroutine package described above is used in the following main programs:

1. Prediction of activity coefficients.
2. Prediction of liquid-liquid equilibrium compositions.
3. Calculation of UNIQUAC or Wilson parameters from UNIFAC.

The first main program merely serves to read in the compositions and temperatures where the activity coefficients are to be predicted and to write the needed output. The third main program, which in addition to the regular UNIFA Subroutine package uses Subroutine HCON, contains a trial-and-error procedure for calculating UNIQUAC (or Wilson) parameters from limiting activity coefficients predicted by the UNIFAC method. With the comment cards included in the main programs and the associated subroutines, these two programs are assumed selfexplanatory.

The second program needs further clarification. Consider a liquid mixture with M components containing Z_1 moles of component 1, Z_2 moles of component 2 .... and Z_M moles of component M. Component 1 is a solvent, components 2,3, ... M-1 solutes, and component M is the other solvent. The task of the second main program is to find out whether two liquid phases coexist at the given temperature and overall composition and, if so, what the compositions of the two liquid phases denoted by (') and (") are. Let (Z_1', Z_2',...... Z_M') and (Z_1'', Z_2'', ... Z_M'') be the numbers of moles each component in, respectively, phase ' and phase ". Then the following relationships hold for i = 1,2, ... M:

$$Z_i' + Z_i'' = Z_i \qquad (A2.1)$$

$$x_i' \gamma_i' = x_i'' \gamma_i'' \qquad (A2.2)$$

$$x_i' = Z_i'/\sum_i Z_i' \qquad (A2.3)$$

$$x_i'' = Z_i''/\sum_i Z_i'' \qquad i = 1,2 \ldots M \qquad (A2.4)$$

In Equation (A2.2), γ_i' and γ_i'' are given by the UNIFAC method. Equations (A2.1-4) constitute 2M equations in 2M unknowns, Z_i' and Z_i'', i = 1,2 .... M.

The program finds the Z_i' and Z_i'' for all i from a given set of Z_i by trial-and-error solution of the 2M equations. The amounts of each phase, $\sum_i Z_i'$ and $\sum_i Z_i''$, and the mole fractions given by equations (A2.3) and (A2.4) are printed out.

The above programs are listed as follows:

		Page
Subroutines:	UNIFA	231
	SYSTM	232
	GRES	236
	GREF	237
	GCOMB	237
	HCON	238
Main program:	Calculation of activity coefficients	239
	Sample INPUT	241
	Sample OUTPUT	242

Main program: Prediction of liquid-liquid equilibrium
                compositions   243
                Sample INPUT   246
                Sample OUTPUT   247

Main program: UNIQUAC Parameters from UNIFAC   250
                Sample INPUT   253
                Sample OUTPUT   254

```
C     ***********************************************************************
C     *
C     *   SUBROUTINES FOR THE MAIN PROGRAMS OF THIS APPENDIX
C     *
C     ***********************************************************************
C
C
C     ***********************************************************************
C     *
C     *   SUBROUTINE UNIFA(NK,NG,ITAB,T,X,XA)
C     *
C     ***********************************************************************
C
C
C     UNIFA GIVES THE ACTIVITY COEFFICIENT (THE VECTOR XA)  FOR GIVEN
C     VALUES OF TEMPERATURE T (IN K) AND COMPOSITION X (MOLE FRACTION)
C
C
C     PARA CONTAINS THE A(I,J) GROUP INTERACTION PARAMETERS NEEDED FOR
C     THE PARTICULAR MIXTURE.  PARB = EXP(-PARA/T)
C
C     GAMC IS THE COMBINATORIAL ACTIVITY COEFFICIENT
C     GAMRF IS THE RESIDUAL ACTIVITY COEFFICIENT STEMMING FROM THE PURE
C     COMPONENT (THE SECOND TERM ON THE RIGHT HAND SIDE OF EQ. (3.11))
C     GAMR IS THE RESIDUAL ACTIVITY COEFFICIENT MINUS GAMRF
C     ITAB IS EXPLAINED IN THE FIRST MAIN PROGRAM OF THIS APPENDIX
C     NY(I,J) IS THE COMPACTED MATRIX, WHICH ON THE BASIS OF THE INFOR-
C     MATION STORED IN THE MATRIX ITAB GIVES THE NUMBER OF GROUPS OF
C     KIND I IN MOLECULE J, I = 1,NG AND J = 1,NK
C     R AND Q ARE THE GROUP VOLUMES AND SURFACE AREAS
C     RS AND QS ARE THE MOLECULAR VOLUMES AND SURFACE AREAS
C     XL IS LOWER CASE L, A COMBINATION OF RS AND QS
C
C     NR MUST BE ZERO THE FIRST TIME UNIFA IS CALLED FOR A MIXTURE OF
C     GIVEN COMPONENTS, IN SUBSEQUENT CALLS FOR MIXTURES WITH THE SAME
C     COMPONENTS, NR SHOULD BE GREATER THAN ZERO
C
      SUBROUTINE UNIFA(NK,NG,ITAB,T,X,XA)
C
      DIMENSION PARA(8,8),PARB(8,8),GAMC(5),GAMRF(5),GAMR(8),X(5),
     *XA(5)
      DIMENSION ITAB(5,56)
      COMMON NY(5,8),R(8),Q(8),RS(5),QS(5),XL(5),NR
      IF(NR.NE.0) GOTO 10
      CALL SYSTM(PARA,NK,NG,ITAB)
   10 CONTINUE
      CALL GREF(GAMRF,PARA,PARB,T,NK,NG)
      CALL GRES(PARB,X,GAMR,NK,NG)
      CALL GCOMP(X,GAMC,NK,NG)
      DO 20 J=1,NK
   20 XA(J)=GAMC(J)*GAMR(J)/GAMRF(J)
      RETURN
      END
```

```
C     ****************************************************************
C     *
C     *    SUBROUTINE SYSTM(PARA,NK,NG,ITAB)
C     *
C     ****************************************************************
C
C
C     FORMATION OF 'THE COMPACTED PARAMETER MATRIX' FROM THE LARGE
C     UNIFAC PARAMETER TABLE ON THE BASIS OF INFORMATION STORED IN ITAB
C
C     THE SUBGROUPS ARE
C     1 = CH3, 2 = CH2, 3 = CH, 4 = C, 5 = CH=CH2, 6 = CH=CH, 7 = CH=C,
C     8 = CH2=C, 9 = ACH, 10 = AC, 11 = ACCH3, 12 = ACCH2, 13 = ACCH,
C     14 = CH2CH2OH, 15 = CHOHCH3, 16 = CHOHCH2, 17 = CH3CH2OH,
C     18 = CHCH2OH, 19 = CH3OH, 20 = H2O, 21 = ACOH, 22 = CH3CO,
C     23 = CH2CO, 24 = CHO, 25 = COOCH3, 26 = COOCH2, 27 = CH3O,
C     28 = CH2O, 29 = CH-O, 30 = FCH2O, 31 = CH3NH2, 32 = CH2NH2,
C     33 = CHNH2, 34 = CH3NH, 35 = CH2NH, 36 = CHNH, 37 = ACNH2,
C     38 = CH3CN, 39 = CH2CN, 40 = COOH, 41 = HCOOH, 42 = CH2CL,
C     43 = CHCL, 44 = CCL, 45 = CH2CL2, 46 = CHCL2, 47 = CCL2,
C     48 = CHCL3, 49 = CCL3, 50 = CCL4, 51 = ACCL, 52 = CH3NO2,
C     53 = CH2NO2, 54 = CHNO2, 55 = ACNO2, 56 = CS2
C
      SUBROUTINE SYSTM(PARA,NK,NG,ITAB)
C
      DIMENSION ARR(25,25)
      DIMENSION ITAB(5,56),RR(56),QQ(56),PARA(8,8),KTAB(8),NKTAB(56)
      DIMENSION AA1(25), AA2(25),AA3(25),AA4(25),AA5(25),AA6(25),AA7(25)
     *,AA8(25), AA9(25),AA10(25),AA11(25),AA12(25),AA13(25),AA14(25),AA1
     *5(25),AA16(25),AA17(25),AA18(25),AA19(25),AA20(25),AA21(25),AA22(2
     *5),AA23(25),AA24(25),AA25(25)
      COMMON NY(5,8),R(8),Q(8),RS(5),QS(5),XL(5),NR
      DATA NKTAB/4*1,4*2,2*3,3*4,5*5,6,7,8,2*9,10,2*11,4*12,3*13,3*14,15
     *,2*16,2*17,3*18,3*19,2*20,21,22,3*23,24,25/
      DATA RR/.9011,.6744,.4469,.2195,1.3454,1.1167,.8886,1.1173,.5313,.
     *3652,1.2663,1.0396,.8121,1.8788,1.8780,1.6513,2.1055,1.6513,1.4311
     *,.92,.8952,1.6724,1.4457,.998,1.9031,1.6764,1.1450,.9183,.6908,.91
     *83,1.5959,1.3692,1.1417,1.4337,1.2070,.9795,1.0600,1.8701,1.6434,1
     *.3013,1.528,1.4654,1.2380,.791,2.2564,2.0606,1.8016,2.8700,2.6401,
     *3.390,1.1562,2.0086,1.7818,1.5544,1.4199,2.057/
      DATA QQ/.848,.540,.228,.000,1.176,.867,.676,.988,.400,.120,.968,.6
     *6,.348,1.664,1.66,1.352,1.972,1.352,1.432,1.4,.68,1.488,1.18,.948,
     *1.728,1.420,1.088,.780,.468,1.100,1.544,1.236,.924,1.244,.936,.624
     *,.816,1.724,1.416,1.224,1.532,1.264,.952,0.724,1.988,1.684,1.448,2
     *.410,2.184,2.91,.844,1.868,1.56,1.248,1.104,1.65/
C
C     THE MAIN GROUPS ARE
C     1 = CH2, 2 = C, 3 = ACH, 4 = ACCH2, 5 = CCOH, 6 = CH3OH, 7 = H2O,
C     8 = ACOH, 9 = CH2CO, 10 = CHO, 11 = COOC, 12 = CH2O, 13 = CNH2,
C     14 = CNH, 15 = ACNH2, 16 = CCN, 17 = COOH, 18 = CCL, 19 = CCL2,
C     20 = CCL3, 21 = CCL4, 22 = ACCL, 23 = CNO2, 24 = ACNO2, 25 = CS2,
C
C     ************* W A R N I N G ******************
```

```
C      AN OFF-DIAGONAL ZERO IN THE A-PARAMETER MATRIX MEANS THAT THE
C      PARAMETER IN QUESTION IS NOT AVAILABLE
C      ****************************************************************
C
       DATA            AA1/0.,-200.0,61.1300,76.500,737.50,697.20,1318
      *.0,2789.0,476.40,677.00,232.10,251.50,391.50,255.70,1245.0,612.00,
      *663.50,35.930,53.760,24.900,104.30,321.50,661.50,543.00,114.10/
       DATA            AA2/2520.0,0.,340.70,4102.0,535.20,1509.0,599.6
      *0,000.00,524.50,000.00,000.00,289.30,396.00,273.60,000.00,370.90,7
      *30.40,99.610,337.10,4583.00,5831.0,000.00,542.10,2*000.00/
       DATA            AA3/-11.12,-94.78,0.,167.00,477.00,637.40,903.8
      *0,1397.0,25.770,000.00,5.9940,32.140,161.70,122.80,668.20,212.50,5
      *37.40,-18.81,000.00,-231.9,3.0000,538.20,168.10,194.9,97.53/
       DATA            AA4/-69.70,-269.7,-146.8,0.,469.00,603.30,5695.
      *00,726.30,-52.10,000.00,5688.0,213.10,000.00,-49.29,612.50,6096.0,
      *603.80,-114.1,000.00,-12.14,-141.3,000.00,3629.0,000.00,000.00/
       DATA            AA5/-87.93,121.50,-64.13,-99.38,0.,127.40,285.4
      *0,257.30,48.160,000.00,76.200,70.000,110.80,188.30,412.00,000.00,7
      *7.610,-38.23,-185.9,-170.9,-98.66,290.00,000.00,000.00,73.520/
       DATA            AA6/16.510,-52.39,-50.00,-44.50,-80.78,0.,-181.
      *0,000.00,23.390,000.00,-10.72,-180.6,359.30,266.0,000.00,45.540,00
      *0.00,-38.32,-102.5,-139.4,-67.80,000.00,75.140,000.00,-31.09/
       DATA            AA7/580.60,511.70,362.30,377.60,-148.5,289.60,0
      *00.00,442.00,-280.8,649.1,-455.4,-400.6,357.50,287.00,213.00,112.6
      *0,225.40,325.40,370.40,353.70,497.50,678.20,-19.44,399.50,887.10/
       DATA            AA8/311.00,000.00,2043.0,6245.0,-455.4,000.00,-
      *540.6,0.,000.00,000.00,-713.2,9*000.00,4894.0,4*000.00/
       DATA            AA9/26.760,-82.92,140.10,365.80,129.20,108.70,6
      *05.60,000.00,0.,-37.36,-213.7,5.2020,000.00,000.00,000.00,428.50,6
      *69.40,-191.7,-284.0,-354.6,-39.20,000.00,137.50,000.00,162.3/
       DATA            AA10/505.70,000.00,000.00,000.00,000.00,000.00,
      *-155.7,000.00,128.00,0.,15*000.00/
       DATA            AA11/114.80,000.00,85.840,-170.0,109.90,249.60,
      *1135.0,853.60,372.20,000.00,0.,-235.7,000.00,-73.50,000.00,533.60,
      *660.20,000.00,108.90,-209.7,54.470,3*000.00,162.7/
       DATA            AA12/83.360,76.440,52.130,65.690,42.000,339.70,
      *634.20,000.00,52.380,000.00,461.30,0.,000.00,141.70,000.00,000.00,
      *664.60,301.10,137.80,-154.3,000.00,000.00,95.180,000.00,151.1/
       DATA            AA13/-30.48,79.400,-44.85,000.00,-217.2,-481.7,
      *-507.1,5*000.00,0.,63.720,7*000.00,68.810,3*0.00/
       DATA            AA14/65.330,-41.32,-22.31,223.00,-243.3,-500.4,
      *-547.7,000.00,000.00,000.00,136.00,-49.30,108.80,0.,6*000.00,71.23
      *0,4350.0,3*0.00/
       DATA            AA15/5339.0,000.00,650.40,3399.0,-245.0,000.00,
      *-339.5,7*000.00,0.,5*000.00,8455.0,2*000.00,-62.73,000.00/
       DATA            AA16/35.760,26.090,-22.97,-138.4,000.00,168.80,
      *242.80,000.00,-275.1,000.00,-297.3,4*000.00,0.,3*000.00,-15.62,-54
      *.86,4*000.00/
       DATA            AA17/315.30,349.20,62.320,268.20,-17.59,000.00,
      *-292.0,000.00,-297.8,000.00,-256.3,-338.5,4*000.00,0.,44.420,-183.
      *4,000.00,217.70,000.00,3*0.00/
       DATA            AA18/91.460,-24.36,4.6800,122.90,368.60,529.00,
      *698.20,000.00,286.30,000.00,000.00,225.40,4*000.00,326.40,0.,108.3
      *0,249.20,62.420,4*000.00/
       DATA            AA19/34.010,-52.71,000.00,000.00,601.60,669.90,
      *708.70,000.00,423.20,000.00,-132.9,-197.7,000.00,000.00,000.00,000
      *.00,1821.0,-84.53,0.,000.00,56.330,4*000.00/
       DATA            AA20/36.700,-185.1,288.50,33.610,491.10,649.10,
```

```
      *826.80,000.00,552.10,000.00,176.50,-20.93,000.00,000.00,000.00,74.
      *040,000.00,-157.1,000.00,0.,-30.100,3*000.00,256.5/
            DATA              AA21/-78.45,-293.7,-4.700,134.70,570.70,860.10,
      *1201.0,1616.0,372.00,000.00,129.50,000.00,000.00,91.130,1302.0,492
      *.00,689.00,11.800,17.970,51.900,0.,475.80,490.9,534.7,132.2/
            DATA              AA22/-141.3,000.00,-237.7,0.,134.1,0.,920.4,5*0
      *.00,203.50,-108.40,6*000.00,-255.4,0.,-154.5,2*0.0/
            DATA              AA23/-32.69,-49.92,10.380,-97.05,000.00,252.60,
      *614.20,000.00,-142.6,000.00,000.00,-94.49,8*000.00,-34.68,794.4,0.
      *,2*0.0/
            DATA              AA24/5541.0,000.00,1825.0,3*000.00,360.7,7*000.
      *00,5250.0,5*000.00,514.6,4*000.00/
            DATA              AA25/11.460,000.00,-18.99,000.00,442.80,914.20,
      *1081.0,000.00,298.70,000.00,233.70,79.790,7*000.00,-125.8,-60.71,4
      **000.00/
         DO 99 J=1,25
         ARR(1,J)=AA1(J)
         ARR(2,J)=AA2(J)
         ARR(3,J)=AA3(J)
         ARR(4,J)=AA4(J)
         ARR(5,J)=AA5(J)
         ARR(6,J)=AA6(J)
         ARR(7,J)=AA7(J)
         ARR(8,J)=AA8(J)
         ARR(9,J)=AA9(J)
         ARR(10,J)=AA10(J)
         ARR(11,J)=AA11(J)
         ARR(12,J)=AA12(J)
         ARR(13,J)=AA13(J)
         ARR(14,J)=AA14(J)
         ARR(15,J)=AA15(J)
         ARR(16,J)=AA16(J)
         ARR(17,J)=AA17(J)
         ARR(18,J)=AA18(J)
         ARR(19,J)=AA19(J)
         ARR(20,J)=AA20(J)
         ARR(21,J)=AA21(J)
         ARR(22,J)=AA22(J)
         ARR(23,J)=AA23(J)
         ARR(24,J)=AA24(J)
         ARR(25,J)=AA25(J)
      99 CONTINUE
         NT=56
         NM=25
         M=0
         DO 15 J=1,NT
         JJ=0
         DO 16 I=1,NK
      16 JJ=JJ+ITAB(I,J)
         IF (JJ) 15,15,17
      17 M=M+1
         KTAB(M)=J
      15 CONTINUE
         NG=M
         DO 20 J=1,NG
         JJ=KTAB(J)
         R(J)=RR(JJ)
         Q(J)=QQ(JJ)
```

```
      DO 20 I=1,NK
20    NY(I,J)=ITAB(I,JJ)
      DO 30 I=1,NM
      DO 35 J=1,NG
      J1=KTAB(J)
      J1=NKTAB(J1)
      IF (J1-I) 35,36,35
36    DO 37 L=1,NG
      L1=KTAB(L)
      L1=NKTAB(L1)
37    PARA(J,L)=ARR(I,L1)
35    CONTINUE
30    CONTINUE
      DO 40 I=1,NK
      RS(I)=0.
      QS(I)=0.
      DO 41 J=1,NG
      RS(I)=RS(I)+NY(I,J)*R(J)
41    QS(I)=QS(I)+NY(I,J)*Q(J)
40    XL(I)=5.*(RS(I)-QS(I))-RS(I)+1.
      IF (NR)  34,34,39
34    WRITE (6,80)
80    FORMAT (//,'  GROUP CONSTANTS AND INTERACTION PARAMETERS   (R, Q, AND
     *ND A(I,J))  ',/)
      DO 81 I=1,NG
81    WRITE (6,82) R(I),Q(I),(PARA(I,J),J=1,NG)
82    FORMAT (2F10.4,8F10.2)
      WRITE (6,83)
83    FORMAT (///,3X,'MOLECULAR FUNCTIONAL GROUPS'//)
      DO 84 I=1,NK
84    WRITE (6,85) I,(NY(I,J),J=1,NG)
85    FORMAT (2I5,10I3)
39    CONTINUE
      RETURN
      END
```

```
C     ********************************************************************
C     *
C     *   SUBROUTINE GRES (P,X,GAM,NK,NG)
C     *
C     ********************************************************************
C
C
C     CALCULATION OF RESIDUAL ACTIVITY COEFFICIENTS LESS THE REFERENCE
C     PART STEMMING FROM  GROUP K IN PURE COMPONENT I   (THE LATTER IS
C     CALCULATED IN GREF).
C
      SUBROUTINE GRES (P,X,GAM,NK,NG)
C
      DIMENSION P(8,8),GAM(5),S1(8),XG(8),S4(8),TH(8),GAML(8),X(5)
      COMMON NY(5,8),R(8),Q(8),RS(5),QS(5),XL(5),NR
      S3=0.
      S2=0.
      DO 10 K=1,NG
      S1(K)=0.
      DO 11 I=1,NK
11    S1(K)=S1(K)+NY(I,K)*X(I)
10    S2=S2+S1(K)
      DO 13 K=1,NG
      XG(K)=S1(K)/S2
13    S3=S3+Q(K)*XG(K)
      DO 15 K=1,NG
      S4(K)=0.
15    TH(K)=Q(K)*XG(K)/S3
      DO 16 K=1,NG
      DO 16 I=1,NG
16    S4(K)=TH(I)*P(I,K)+S4(K)
      DO 20 K=1,NG
      G=1.-ALOG(S4(K))
      DO 21 I=1,NG
21    G=G-TH(I)*P(K,I)/S4(I)
20    GAML(K)=Q(K)*G
      DO 30 I=1,NK
      G=0.
      DO 31 J=1,NG
31    G=NY(I,J)*GAML(J)+G
30    GAM(I)=EXP(G)
      RETURN
      END
```

```
C     ******************************************************************
C     *
C     *  SUBROUTINE GREF(GAM,PARA,PARB,T,NK,NG)
C     *
C     ******************************************************************
C
C
C     CALCULATION OF RESIDUAL REFERENCE ACTIVITY COEFFICIENT
C
      SUBROUTINE GREF(GAM,PARA,PARB,T,NK,NG)
C
      DIMENSION PARA(8,8),PARB(8,8),X(5),GAMX(5),GAM(5)
      DO 10 I=1,NG
      DO 10 J=1,NG
   10 PARB(I,J)=EXP(-PARA(I,J)/T)
      DO 20 I=1,NK
      DO 21 J=1,NK
   21 X(J)=0.
      X(I)=1.
      CALL GRES(PARB,X,GAMX,NK,NG)
   20 GAM(I)=GAMX(I)
      RETURN
      END




C     ******************************************************************
C     *
C     *  SUBROUTINE GCOMB(X,GAMMA,NK,NG)
C     *
C     ******************************************************************
C
C
C     CALCULATION OF COMBINATORIAL PART
C
      SUBROUTINE GCOMB(X,GAMMA,NK,NG)
C
      DIMENSION X(5),GAMMA(5)
      COMMON NY(5,8),R(8),Q(8),RS(5),QS(5),XL(5),NR
      QSS=0.
      RSS=0.
      XLS=0.
      DO 10 I=1,NK
      QSS=QSS+QS(I)*X(I)
      RSS=RSS+RS(I)*X(I)
   10 XLS=XLS+XL(I)*X(I)
      DO 20 I=1,NK
      A=5.*QS(I)*ALOG(QS(I)/QSS*RSS/RS(I))+XL(I)-RS(I)/RSS*XLS
   20 GAMMA(I)=RS(I)/RSS*EXP(A)
      RETURN
      END
```

```
C     ***********************************************************************
C     *
C     *   SUBROUTINE HCON(PARA,PARB,T,HENRY,GAMC,NK,NG)
C     *
C     ***********************************************************************
C
C
C     THE RESIDUAL INFINITE DILUTION ACTIVITY COEFFICIENTS ARE RETURNED
C     IN HENRY.  THE COMBINATORIAL INFINITE DILUTION ACTIVITY
C     COEFFICIENTS ARE RETURNED IN GAMC
C     BOTH ARE CALCULATED ACCORDING TO UNIFAC
C
      SUBROUTINE HCON(PARA,PARB,T,HENRY,GAMC,NK,NG)
C
      DIMENSION PARA(8,8),PARB(8,8),HENRY(5,5),X(5),GAM(5),GAMX(5)
      DIMENSION GAMC(5,5),CGAM(5)
      DO 10 I=1,NG
      DO 10 J=1,NG
   10 PARB(I,J)=EXP(-PARA(I,J)/T)
      DO 20 I=1,NK
      DO 21 J=1,NK
   21 X(J)=0.
      X(I)=1.
      CALL GCOMB(X,CGAM,NK,NG)
      CALL GRES(PARB,X,GAMX,NK,NG)
      GAM(I)=GAMX(I)
      DO 20 J=1,NK
      GAMC(I,J)=CGAM(J)
   20 HENRY(I,J)=GAMX(J)
      DO 25 I=1,NK
      DO 25 J=1,NK
   25 HENRY(I,J)=HENRY(I,J)/GAM(J)
      RETURN
      END
```

```
C     ****************************************************************
C     *
C     *   CALCULATION OF ACTIVITY COEFFICIENTS
C     *
C     ****************************************************************
C
C
C     ****************************************************************
C     *
C     *   THIS PROGRAM CALCULATES MIXTURE ACTIVITY COEFFICIENTS FROM
C     *   SPECIFIED COMPOSITION AND TEMPERATURE
C     *
C     ****************************************************************
C
C
C     INPUT
C     1.  1 CARD, TEXT(80A1)
C           TEXT = IDENTFICATION OF THE SYSTEM
C     2.  1 CARD, NK,NG(40I2)
C           NK   = THE NUMBER OF COMPONENTS
C           NG   = THE NUMBER OF DIFFERENT GROUPS
C     3.  2*NK CARDS, ITAB(80I2)
C           ITAB = THE NUMBER OF GROUPS OF TYPE K IN MOLECULE I
C                  (SEE 'SUBGROUPS' IN SYSTM)
C     4.  1 CARD, T,X(8F10.4)
C           T    = TEMPERATURE IN K
C           X    = MOLE FRACTION
C
C     THE LAST CARD IS REPEATED FOR EACH NEW STARTING POINT
C     A BLANK CARD IS USED TO INDICATE END OF SYSTEM
C     THREE BLANK CARDS INDICATE END OF JOB
C
      DIMENSION X(5),GAMMA(5),TEXT(80)
      DIMENSION ITAB(5,56)
      COMMON NY(5,8),R(8),Q(8),RS(5),QS(5),XL(5),NR
C
C     THIS COMMON STATEMENT IS EXPLAINED FURTHER IN SUBROUTINE UNIFA
C
C     GAMMA= ACTIVITY COEFFICIENT
C     NR   = THE FIRST TIME THE UNIFA SUBROUTINE IS CALLED, NR MUST BE
C            ZERO. IN ALL SUBSEQUENT CALLS FOR THE SAME MIXTURE,
C            NR SHOULD BE GREATER THAN ZERO
C
  100 FORMAT (40I2)
  101 FORMAT (8F10.4)
  102 FORMAT (4X,F10.1,10F10.4)
  103 FORMAT (///,1X,'TEMPERATURE',1X,'*** MOLE FRACTIONS AND ACTIVITY C
     *OEFFICIENTS FOR EACH COMPONENT ***',/)
  104 FORMAT (///,80A1,/)
  105 FORMAT (80A1)
      NT=56
  199 READ (5,105) TEXT
      READ (5,100) NK,NG
```

```fortran
      IF (NK.EQ.0) GO TO 99
      WRITE (6,104) TEXT
      NR=0
      DO 1 I=1,NK
    1 READ (5,100) (ITAB(I,J),J=1,NT)
   98 READ (5,101) T, (X(I),I=1,NK)
      IF (T.EQ.0.) GO TO 199
      CALL UNIFA (NK,NG,ITAB,T,X,GAMMA)
      IF (NR.GT.0) GO TO 2
      WRITE (6,103)
    2 NR=NR+1
      WRITE (6,102) T, (X(I),GAMMA(I),I=1,NK)
      GO TO 98
   99 CONTINUE
      STOP
      END
```

```
***********************************************************************
*
*   SAMPLE INPUT FOR CALCULATION OF ACTIVITY COEFFICIENTS
*
***********************************************************************

DIETHYL AMINE - N-HEPTANE
 2  3
 2  1                                                              68
                                                                    1
 2  4
 2  5

308.15      .4       .6
328.15      .42      .58

WATER - ETHANOL
 2  2
                                            40
                                             1
                                    34
                                     1

350.        .5       .5
```

```
******************************************************************
*
*   SAMPLE OUTPUT, CALCULATION OF ACTIVITY COEFFICIENTS
*
******************************************************************


DIETHYL AMINE - N-HEPTANE


GROUP CONSTANTS AND INTERACTION PARAMETERS   (R, Q, AND A(I,J))

    0.9011    0.8480     0.0       0.0      255.70
    0.6744    0.5400     0.0       0.0      255.70
    1.4337    1.2440    65.33     65.33       0.0


MOLECULAR FUNCTIONAL GROUPS

    1    2    1    1
    2    2    5    0


TEMPERATURE *** MOLE FRACTIONS AND ACTIVITY COEFFICIENTS FOR EACH COMPONENT *

      308.1   0.4000   1.2267   0.6000   1.0875
      328.1   0.4200   1.1965   0.5800   1.0896


WATER - ETHANOL


GROUP CONSTANTS AND INTERACTION PARAMETERS   (R, Q, AND A(I,J))

    2.1055    1.9720     0.0      285.40
    0.9200    1.4000   -148.50      0.0


MOLECULAR FUNCTIONAL GROUPS

    1    0    1
    2    1    0


TEMPERATURE *** MOLE FRACTIONS AND ACTIVITY COEFFICIENTS FOR EACH COMPONENT

      350.0   0.5000   1.3035   0.5000   1.2147
```

```
C     ********************************************************************
C     *
C     *   PREDICTION OF LIQUID-LIQUID EQUILIBRIUM COMPOSITIONS
C     *
C     ********************************************************************
C
C
C
C
C     ********************************************************************
C     *
C     *   THIS PROGRAM CALCULATES THE LIQUID - LIQUID EQUILIBRIUM COMPO-
C     *   SITIONS AND THE AMOUNTS OF THE TWO LIQUID PHASES FOR A SYSTEM
C     *   CONTAINING UP TO 5 COMPONENTS. THE SAMPLE INPUT TO THE PROGRAM
C     *   IS A SPECIFICATION OF THE COMPONENTS AND THE TOTAL COMPOSITION
C     *   OF THE SYSTEM. THE UNIFAC METHOD IS USED TO CALCULATE THE
C     *   ACTIVITY COEFFICIENTS.
C     *
C     ********************************************************************
C
C
C     INPUT
C     1.  1 CARD,TEXT(80A1)
C           TEXT = IDENTFICATION OF THE SYSTEM
C     2.  1 CARD, NK,NG(40I2)
C           NK   = NUMBER OF COMPONENTS
C           NG   = NUMBER OF DIFFERENT GROUPS
C     3.  2*NK CARDS, ITAB(80I2)
C           ITAB = UNIFAC SPECIFICATION OF GROUPS
C                  COMPONENT 1 = SOLVENT NUMBER 1
C                  COMPONENT NK = SOLVENT NUMBER 2
C                  TWO CARDS PER COMPONENT
C     4.  1 CARD, T,(Z(I),I=1,NK)(6F10.4)
C           T    = TEMPERATURE IN K
C           Z(I) = TOTAL MOLES OF COMP. I IN THE MIXTURE, ALL I
C
C     THE LAST CARD IS REPEATED FOR EACH NEW STARTING POINT
C     A BLANK CARD IS USED TO INDICATE END OF SYSTEM
C     FOR EACH NEW SYSTEM START WITH TEXT
C     THREE BLANK CARDS INDICATE END OF JOB
C
      DIMENSION Z(5),FD(5),FDD(5),XD(5),XDD(5),X(5),GAMD(5),GAMDD(5),XDN
     *(5),XDDN(5)
      DIMENSION ITAB(5,56)
      DIMENSION TEXT(80)
      COMMON NY(5,8),R(8),Q(8),RS(5),QS(5),XL(5),NR
C
C
C     XD    = THE MOLEFRACTION IN PHASE D
C     XDD   = THE MOLEFRACTION IN PHASE DD
C     GAMD  = ACTIVITY COEFFICIENTS IN PHASE D
C     GAMDD = ACTIVITY COEFFICIENTS IN PHASE DD
C     FD    = TESTFUNCTION FOR THE EQUILIBRIUM CONDITION
C     FD    = XD*GAMD - XDD*GAMDD
```

```
C        NR    = THE FIRST TIME THE UNIFA-SUBROUTINE IS CALLED THE VARIABLE
C                NR MUST HAVE THE VALUE 0 (NR IS PLACED IN COMMON). THE
C                FOLLOWING TIMES NR MAY HAVE ANY VALUE DIFFERENT FROM 0. THE
C                PROGRAM WILL ALSO WORK IF NR=0 BUT IT WILL NOT BE AS FAST
C                AS POSSIBLE.
C
  100 FORMAT(40I2)
  101 FORMAT(6F10.4)
  102 FORMAT(////,80A1,//,4X,'MOLE FRACTIONS',26X,'ACTIVITY COEFFICIENTS',
     *',/)
  103 FORMAT(F10.4,4X,F10.4,16X,F10.2,4X,F10.2)
  104 FORMAT('THE UNIFAC METHOD IS UNABLE TO PREDICT THE PHASE SPLIT ',/
     */,' THE COMPOSITION IS',//,5F10.4)
  105 FORMAT(80A1)
  106 FORMAT(1H0,//,' THE TOTAL COMPOSITION OF THE SYSTEM IS',//,5F10.4)
  107 FORMAT(1H0,/,' THE AMOUNT OF PHASE 1 IS   ',3X,F10.4,//,' THE AMOUN
     *T OF PHASE 2 IS    ',3X,F10.4,///)
      NT=56
  199 READ(5,105)TEXT
      READ(5,100)NK,NG
      IF (NK.EQ.0) GO TO 99
      NK1=NK+1
      NR=0
      DELTA1=.001
      DELTA2=.0001
      DO 1 I=1,NK
    1 READ(5,100) (ITAB(I,J),J=1,NT)
   98 READ(5,101)T,(Z(I),I=1,NK)
      IF(T.EQ.0.) GO TO 199
      S=0.
      DO 2 I=1,NK
    2 S=S+Z(I)
      DO 3 I=1,NK
    3 FD(I)=.5*Z(I)
      FD(1)=.9*Z(1)
      FD(NK)=.1*Z(NK)
      SFD=0.
      SFDD=0.
      DO 4 I=1,NK
      FDD(I)=Z(I)-FD(I)
      SFD=SFD+FD(I)
    4 SFDD =SFDD+FDD(I)
      DO 5 I=1,NK
      XD(I)=FD(I)/SFD
    5 XDD(I)=FDD(I)/SFDD
    6 CONTINUE
      CALL UNIFA(NK,NG,ITAB,T,XD,GAMD)
      NR=NR+1
      CALL UNIFA(NK,NG,ITAB,T,XDD,GAMDD)
      TSFD=0.
      TSFDD=0.
      DO 7 I=1,NK
      FD(I)=Z(I)/(1.+SFDD*GAMD(I)/SFD/GAMDD(I))
      FDD(I)=Z(I)-FD(I)
      TSFD=TSFD+FD(I)
      TSFDD=TSFDD+FDD(I)
    7 CONTINUE
      DO 8 I=1,NK
```

```
      XDN(I)=FD(I)/TSFD
    8 XDDN(I)=FDD(I)/TSFDD
      DO 9 I=1,NK
      IF(ABS(XD(I)-XDN(I)).GT.DELTA1) GOTO 10
    9 CONTINUE
      IF(ABS(TSFD-SFD).GT.DELTA2) GOTO 10
      GOTO 12
   10 DO 11 I=1,NK
      XD(I)=XDN(I)
   11 XDD(I)=XDDN(I)
      SFD=TSFD
      SFDD=TSFDD
      GOTO 6
   12 WRITE(6,102)TEXT
      IF(ABS(XD(1)-XDD(1)).LE..01) WRITE(6,104)XD(1),XD(2),XD(3)
      DO 13 I=1,NK
      FD(I)=XD(I)*GAMD(I)-XDD(I)*GAMDD(I)
   13 WRITE(6,103)XD(I),XDD(I),GAMD(I),GAMDD(I),FD(I)
      WRITE(6,106)(Z(I),I=1,NK)
      WRITE(6,107)TSFD,TSFDD
      GO TO 98
   99 STOP
      END
```

```
***********************************************************************
*
*    SAMPLE INPUT, PREDICTION OF LIQUID-LIQUID EQUILIBRIUM COMPOSITIONS
*
***********************************************************************

     WATER (1) - N-PROPANOL (2) - N-HEXANE (3) AT 37.8 DEG. C
  3  4                                         40
                                                1

  1                                 28
                                     1
  2  4
  2  4

311.         50.          5.          50.
311.         50.         10.          25.
311.         50.         30.          25.
311.         50.         40.          25.
311.         50.         50.          25.
```

```
***********************************************************************
*
*   SAMPLE OUTPUT, PREDICTION OF LIQUID-LIQUID EQUILIBRIUM COMPOSITIONS
*
***********************************************************************


GROUP CONSTANTS AND INTERACTION PARAMETERS   (R, Q, AND A(I,J))

    0.9011    0.8480     0.0       0.0      737.50   1318.00
    0.6744    0.5400     0.0       0.0      737.50   1318.00
    1.8788    1.6640   -87.93    -87.93       0.0     285.40
    0.9200    1.4000   580.60    580.60    -148.50     0.0



MOLECULAR FUNCTIONAL GROUPS


    1    0    0    0    1
    2    1    0    1    0
    3    2    4    0    0



WATER (1) - N-PROPANOL (2) - N-HEXANE (3) AT 37.8 DEG. C

  MOLE FRACTIONS                        ACTIVITY COEFFICIENTS

    0.9554    0.0021                      1.02        454.59
    0.0446    0.0506                     10.94          9.63
    0.0000    0.9472                  24174.49          1.02


THE TOTAL COMPOSITION OF THE SYSTEM IS

   50.0000    5.0000   50.0000

THE AMOUNT OF PHASE 1 IS        52.2171

THE AMOUNT OF PHASE 2 IS        52.7829
```

WATER (1) - N-PROPANOL (2) - N-HEXANE (3) AT 37.8 DEG. C

MOLE FRACTIONS   ACTIVITY COEFFICIENTS

   0.8605   0.0031   1.12   313.87
   0.1388   0.0722   4.07   7.83
   0.0007   0.9248   1338.97   1.03

THE TOTAL COMPOSITION OF THE SYSTEM IS

  50.0000   10.0000   25.0000

THE AMOUNT OF PHASE 1 IS   58.0105

THE AMOUNT OF PHASE 2 IS   26.9895

WATER (1) - N-PROPANOL (2) - N-HEXANE (3) AT 37.8 DEG. C

MOLE FRACTIONS   ACTIVITY COEFFICIENTS

   0.6265   0.0028   1.56   351.68
   0.3557   0.0652   1.53   8.35
   0.0178   0.9320   53.69   1.03

THE TOTAL COMPOSITION OF THE SYSTEM IS

  50.0000   30.0000   25.0000

THE AMOUNT OF PHASE 1 IS   79.6999

THE AMOUNT OF PHASE 2 IS   25.3001

WATER (1) - N-PROPANOL (2) - N-HEXANE (3) AT 37.8 DEG. C

   MOLE FRACTIONS   ACTIVITY COEFFICIENTS

```
   0.5457      0.0028              1.79      351.11
   0.4204      0.0653              1.30        8.34
   0.0339      0.9319             28.17        1.03
```

THE TOTAL COMPOSITION OF THE SYSTEM IS

  50.0000   40.0000   25.0000

THE AMOUNT OF PHASE 1 IS   91.5068

THE AMOUNT OF PHASE 2 IS   23.4932

WATER (1) - N-PROPANOL (2) - N-HEXANE (3) AT 37.8 DEG. C

   MOLE FRACTIONS   ACTIVITY COEFFICIENTS

```
   0.4801      0.0028              2.02      346.18
   0.4673      0.0662              1.17        8.27
   0.0526      0.9310             18.16        1.03
```

THE TOTAL COMPOSITION OF THE SYSTEM IS

  50.0000   50.0000   25.0000

THE AMOUNT OF PHASE 1 IS   104.0303

THE AMOUNT OF PHASE 2 IS   20.9696

```
C     ******************************************************************
C     *
C     *   UNIQUAC PARAMETERS FROM UNIFAC
C     *
C     ******************************************************************
C
C
C     ******************************************************************
C     *
C     *   CALCULATION OF UNIQUAC OR WILSON PARAMETERS FROM UNIFAC USING THE
C     *   INFINITE DILUTION ACTIVITY COEFFICIENTS ONLY
C     *
C     ******************************************************************
C
C
C     INPUT
C     1.   1 CARD, TEXT(40A2)
C             TEXT = ANY TEXT TO IDENTIFY THE PROBLEM
C     2.   1 CARD, TEMP1,TEMP2(2F10.2)
C             TEMP = THE TWO TEMPERATURES IN K,
C             AT WHICH THE PARAMETERS ARE WANTED
C     3.   1 CARD, NK,NG,NMOD(40I2)
C             NK   = NUMBER OF COMPONENTS
C             NG   = NUMBER OF DIFFERENT GROUPS
C             NMOD = 1, WILSON PARAMETERS ARE CALCULATED
C             NMOD = 2, UNIQUAC PARAMETERS ARE CALCULATED
C     4.   2*NK CARDS, ITAB(80I2)
C             ITAB = UNIFAC SPECIFICATION OF GROUPS,
C                    I.E. THE NUMBER OF GROUPS OF KIND J IN MOLECULE I,
C                    TWO CARDS PER COMPONENT
C
C     OUTPUT   FOR UNIQUAC  TAU(J,I) IN EQUATION (3.9)
C                   J = ROW, I = COLUMN
C              FOR WILSON  LAMDA(I,J) IN EQUATION (3.5)
C                   I = ROW, J = COLUMN
C                   BOTH LAMDA AND TAU ARE LINEARIZED
C                   WITH RESPECT TO TEMPERATURE
C                   TAU(J,I) = A(J,I) + B(J,I)*T
C
      DIMENSION PARA(8,8),PARB(8,8),HENRY(5,5),TEMP(3)
      DIMENSION ITAB(5,56),NTEXT(40),CC(5,5),DD(5,5,2),GAMC(5,5)
      COMMON NY(5,8),R(8),Q(8),RS(5),QS(5),XL(5),NR
C
   86 FORMAT (40I2)
   87 FORMAT (//)
   88 FORMAT(40A2)
   89 FORMAT(3F10.2)
      NR=0
      NT=56
      READ(5,88) NTEXT
      READ(5,89) TEMP(1),TEMP(2)
      READ (5,86) NK,NG,NMOD
      DO 22 I=1,NK
```

```
   22 READ (5,86) (ITAB(I,J),J=1,NT)
      CALL SYSTM(PARA,NK,NG,ITAB)
      DO  50 KK=1,2
      TT=TEMP(KK)
      CALL HCON(PARA,PARB,TT,HENRY,GAMC,NK,NG)
      WRITE(6,87)
      WRITE(6,88) NTEXT
      WRITE(6,12) TT
   12 FORMAT (/,'     *****  T = ',F10.2,' K',/)
      WRITE(6,13)
   13 FORMAT (' INFINITE DILUTION ACTIVITY COEFFICIENTS '/)
      DO 1 I=1,NK
      DO 1 J=1,NK
    1 HENRY(I,J)=HENRY(I,J)*GAMC(I,J)
      DO 10 I=1,NK
   10 WRITE (6,3) I,(HENRY(I,J),J=1,NK)
    3 FORMAT (' COMPONENT',I2,' IS SOLVENT',5E12.4)
      IF (NMOD.NE.1) GO TO 2
      WRITE (6,23)
   23 FORMAT (/,' LAMBDA(I,J) =   (V(J)/V(I))*EXP(-DELTA(LAMBDA)/RT) ',/)
      DO 4 I=1,NK
      XL(I)=0.
      RS(I)=1.
    4 QS(I)=1.
      GO TO 21
    2 WRITE(6,15)
   15 FORMAT(/,'    TAU(J,I) = EXP(-DELTA(U)/RT) ',/)
      DO 5 I=1,NK
      DO 5 J=1,NK
    5 HENRY(I,J)=HENRY(I,J)/GAMC(I,J)
   21 NK1=NK-1
      DO 30 I=1,NK1
      I1=I+1
      DO 30 J=I1,NK
      B1=1.-ALOG(HENRY(I,J))/QS(J)
      B2=1.-ALOG(HENRY(J,I))/QS(I)
      A=.001
      IF(ABS(B1+B2-2.)-1.E-5)201,202,202
  201 A=1.
      GO TO 36
  202 IF(B1-1.)35,35,203
  203 IF(B2-1.)35,35,204
  204 IF(EXP(B2-1.).LT.B1) A=EXP(B1)
   35 S=EXP(B2-A)
      F=ALOG(A)+S-B1
      DF=S-1./A
      DA=F/DF
      A=A+DA
      IF (ABS(DA)-1.E-4) 36,36,35
   36 CC(I,J)=A
   30 CC(J,I)=B1-ALOG(A)
      DO 60 I=1,NK
      CC(I,I)=1.
   60 WRITE (6,85) (CC(I,J),J=1,NK)
   85 FORMAT (5E15.7)
      DO 70 I=1,NK
      DO 70 J=1,NK
   70 DD(I,J,KK)=CC(I,J)
```

```
   50 CONTINUE
      DO 80 I=1,NK
      DO 80 J=1,NK
      ALF=(DD(I,J,2)-DD(I,J,1))/(TEMP(2)-TEMP(1))
      DD(I,J,1)=DD(I,J,1)-TEMP(1)*ALF
   80 DD(I,J,2)=ALF
      WRITE(6,14)
   14 FORMAT(//,' THE PUNCHED OUTPUT CONSISTS OF')
      WRITE(6,16)
   16 FORMAT('      COMPONENT NUMBER, Q, R, AND L FOR EACH COMPONENT (UNI
     1QUAC ONLY) ')
      WRITE(6,17)
   17 FORMAT ('      COMPONENT NUMBER, KK, TEMP1, TEMP2, AND A(I,J) OR B(
     1I,J), J=1,2...N')
      WRITE(6,18)
   18 FORMAT ('         IT IS A(I,J) IF KK=1, AND IT IS B(I,J) IF KK=2.
     1   TAU  = A + B*T')
      IF (NMOD.EQ.1) GO TO 97
      WRITE(6,19)
   19 FORMAT (/,' COMP           Q              R              L    '/)
      DO 95 I=1,NK
      WRITE(7,91)I,QS(I),RS(I),XL(I)
      WRITE(6,91)I,QS(I),RS(I),XL(I)
   95 CONTINUE
   97 WRITE(6,20)
   20 FORMAT (/,' I KK  TEMP1   TEMP2   A(I,J) OR B(I,J) -------------'/)
   91 FORMAT(I5,3F15.5)
      DO 96 KK=1,2
      DO 96 I=1,NK
      WRITE(7,92)I,KK,TEMP(1),TEMP(2),(DD(I,J,KK),J=1,NK)
      WRITE(6,92)I,KK,TEMP(1),TEMP(2),(DD(I,J,KK),J=1,NK)
   92 FORMAT(2I2,2F8.1,5E12.5)
   96 CONTINUE
      STOP
      END
```

```
************************************************************************
*
*    SAMPLE INPUT, UNIQUAC PARAMETERS FROM UNIFAC
*
************************************************************************


1,2-DICHLOROETHANE (1) - N-PROPANOL (2) - TOLUENE (3) - ACETONE (4)
340.      360.
  4 6
           4
  2  2                          28
  1                              1
                 18   22
                  5    1
                                          44
  1                                        1
```

```
***********************************************************************
*
*   SAMPLE OUTPUT, UNIQUAC PARAMETERS FROM UNIFAC
*
***********************************************************************
```

GROUP CONSTANTS AND INTERACTION PARAMETERS   (R, Q, AND A(I,J))

0.9011	0.8480	0.0	61.13	76.50	737.50	476.40	35.9?
0.5313	0.4000	-11.12	0.0	167.00	477.00	25.77	-18.8?
1.2663	0.9680	-69.70	-146.80	0.0	469.00	-52.10	-114.1?
1.8788	1.6640	-87.93	-64.13	-99.38	0.0	48.16	-38.2?
1.6724	1.4880	26.76	140.10	365.80	129.20	0.0	-191.7?
1.4654	1.2640	91.46	4.68	122.90	368.60	286.30	0.0

MOLECULAR FUNCTIONAL GROUPS

```
     1   0  0  0  0  0  2
     2   1  0  0  1  0  0
     3   0  5  1  0  0  0
     4   1  0  0  0  1  0
```

1,2-DICHLOROETHANE (1) - N-PROPANOL (2) - TOLUENE (3) - ACETONE (4)

***** T =   340.00 K

INFINITE DILUTION ACTIVITY COEFFICIENTS

COMPONENT 1 IS SOLVENT	0.1000E 01	0.4443E 01	0.9949E 00	0.9309E 00
COMPONENT 2 IS SOLVENT	0.2932E 01	0.1000E 01	0.3554E 01	0.1833E 01
COMPONENT 3 IS SOLVENT	0.9935E 00	0.4997E 01	0.1000E 01	0.1411E 01
COMPONENT 4 IS SOLVENT	0.9155E 00	0.2112E 01	0.1825E 01	0.1000E 01

TAU(J,I) = EXP(-DELTA(U)/RT)

0.1000000E 01	0.5244893E 00	0.1528933E 01	0.5625153E 00
0.1056402E 01	0.1000000E 01	0.1190861E 01	0.9912975E 00
0.6062148E 00	0.4592944E 00	0.1000000E 01	0.1276591E 01
0.1608894E 01	0.7481847E 00	0.6533245E 00	0.1000000E 01

1,2-DICHLOROETHANE (1) - N-PROPANOL (2) - TOLUENE (3) - ACETONE (4)

***** T =   360.00 K

INFINITE DILUTION ACTIVITY COEFFICIENTS

```
COMPONENT 1 IS SOLVENT   0.1000E 01   0.4009E 01   0.9999E 00   0.9412E 00
COMPONENT 2 IS SOLVENT   0.2786E 01   0.1000E 01   0.3410E 01   0.1774E 01
COMPONENT 3 IS SOLVENT   0.9972E 00   0.4530E 01   0.1000E 01   0.1391E 01
COMPONENT 4 IS SOLVENT   0.9272E 00   0.2019E 01   0.1761E 01   0.1000E 01
```

TAU(J,I) = EXP(-DELTA(U)/RT)

```
0.1000000E 01   0.5486530E 00   0.1511588E 01   0.5835495E 00
0.1052296E 01   0.1000000E 01   0.1179046E 01   0.9926152E 00
0.6159329E 00   0.4832428E 00   0.1000000E 01   0.1262265E 01
0.1567512E 01   0.7607085E 00   0.6706786E 00   0.1000000E 01
```

THE PUNCHED OUTPUT CONSISTS OF
    COMPONENT NUMBER, Q, R, AND L FOR EACH COMPONENT (UNIQUAC ONLY)
    COMPONENT NUMBER, KK, TEMP1, TEMP2, AND A(I,J) OR B(I,J), J=1,2...N
    IT IS A(I,J) IF KK=1, AND IT IS B(I,J) IF KK=2.   TAU = A + B*T

```
COMP        Q              R              L

  1       2.52800        2.93080        0.08320
  2       2.51200        2.77990       -0.44040
  3       2.96800        3.92280        1.85120
  4       2.33600        2.57350       -0.38600

I KK  TEMP1    TEMP2    A(I,J) OR B(I,J) -------------

1 1   340.0    360.0 0.10000E 01 0.11371E 00 0.18238E 01 0.20493E 00
2 1   340.0    360.0 0.11262E 01 0.10000E 01 0.13917E 01 0.96890E 00
3 1   340.0    360.0 0.44101E 00 0.52172E-01 0.10000E 01 0.15201E 01
4 1   340.0    360.0 0.23124E 01 0.53528E 00 0.35831E 00 0.10000E 01
1 2   340.0    360.0 0.0         0.12082E-02-0.86722E-03 0.10517E-02
2 2   340.0    360.0-0.20533E-03 0.0        -0.59075E-03 0.65881E-04
3 2   340.0    360.0 0.48590E-03 0.11974E-02 0.0        -0.71630E-03
4 2   340.0    360.0-0.20691E-02 0.62619E-03 0.86770E-03 0.0
```

APPENDIX 3

ESTIMATION OF UNIFAC PARAMETERS

This appendix gives the program used for estimation of UNIFAC parameters. The program is based on Nelder-Mead's extended simplex minimization method as explained in Section 5.4. The program can only treat binary vapor-liquid equilibrium data.

This appendix contains: Page

    INPUT SPECIFICATIONS TO THE PROGRAM   256
    LISTING OF THE PROGRAM   259
    SAMPLE INPUT   270
    SAMPLE OUTPUT   271
    EXPLANATIONS TO THE OUTPUT   274

INPUT SPECIFICATIONS

a.   NCOMP, NG, NOBS, LAAF, NOIT   Format 5I3

    NCOMP - Number of different components in the binary mixtures to be fitted simultaneously (Max 10).

    NG   - Number of different groups (Max 10).

    NOBS   - Number of data points (Max 100).

    LAAF   - LAAF=1: The experimental activity coefficients are read.

            LAAF=2: The logarithms to the experimental activity coefficients are read.

    NOIT   - NOIT=1: Parameter estimation is performed.

            NOIT=2: No parameter estimation is performed. The activity coefficients are calculated based on the given R, Q, and A matrix.

b.   FREE TEXT (16 cards)   Format 40A2

c.   NUM(I), I=1,NCOMP   Format 10I3

NUM(I) — Number attached to component I. The components should be arranged in the same order as the cards in the NNY matrix (see next items).

d. NNY MATRIX   Format 10I2

NNY(I,K), K=1,NG — One card is included for each component. The card for component I presents the number of groups of kind K in molecule I.

e. KG(K), K=1,NG   Format 10I3

KG(K) — Number attached to group K. (These numbers are given in Table 4.3).

f. T(N), (NU(N,I), XXX(N,I), GME(N,I), I=1,2), N=1,NOBS
                                           Format F7.2,3(I2,2F7.4))

One card is included for each data point N.

T(N) — Temperature in K.

NU(N,I) — Number attached to component I. This number should be the same as given in NUM(I).

XXX(N,I) — Liquid mole fraction.

GME(N,I) — LAAF=1: Experimental activity coefficient.

LAAF=2: Logarithm to the experimental activity coefficient.

g. R(K), Q(K), A(K,J), J=1,NG   Format 8F10.4

For each group K a card is given.

R(K) — Group volume of group K.

Q(K) — Group area of group K.

A(K,J) — Interaction parameter between group K and group J. (An unknown parameter is normally given the value of unity).

h. NPAR, KRIT, IDEN   Format 3I3

This card is only included, if NOIT=1.

NPAR — Number of parameters to be estimated (Max 4).

KRIT — The objective function to be minimized is defined as: KRIT=1: The sum of the squared differences between the experimental and calculated activity coefficients.

KRIT=2: The sum of the squared differences between the logarithms to the experimental and calculated activity coefficients.

IDEN: Number of identical pairs of interaction parameters to be estimated. IDEN is equal to zero if no pairs are identical. IDEN is equal to at least two if some pairs are identical.

EXAMPLE: In a mixture with 5 different groups we want to estimate the interaction between groups 3 and 5 i.e. A(3,5) and A(5,3). If A(4,5) = A(3,5) and A(5,4) = A(5,3) we have IDEN=2.

i. IPAR(I), I=1,NPAR   Format 4I3

  This card is only included, if NOIT=1.

  IPAR is a vector indicating which parameters are to be estimated.

  In the above example IPAR is 3 and 5.

j. JENS(J), J=1,2*IDEN   Format 20I3

  This card is only included, if NOIT=1 and IDEN>0.

  JENS is a vector indicating the identical pairs of parameters to be estimated. (In the above example JENS is 3,5, 4 and 5).

k. X(1,I), I=1,NPAR   Format 4F10.3

  This card is only included, if NOIT=1

  X(1,I) is an initial guess on parameter I. No initial parameters must be zero.

```
C     ***************************************************************
C     *
C     *   UNIFAC PARAMETER ESTIMATION BY MEANS OF
C     *   NELDER-MEAD'S EXTENDED SIMPLEX MINIMIZATION METHOD
C     *
C     ***************************************************************
C
C
C
C
      DIMENSION R(10),RS(10),QS(10),XL(10),KG(10),NTEXT(40),NUM(10),IPAR
     1(4),PAR(4),DEV(100,2),JENS(10),NU(100,2)
      DIMENSION X(5,4),F(5),XB(4),XS(4),XM(4),XE(4),XX(4),XR(4),XK(4)
      COMMON T(100),NM(100,2),XXX(100,2),GME(100,2),GMC(100,2),GMR(100,2
     1),NNY(10,10),Q(10),A(10,10)
      DATA NC,NP/5,6/
      READ(NC,16)NCOMP,NG,NOBS,LAAF,NOIT
   16 FORMAT(2013)
C
C         NCOMP= NUMBER OF COMPONENTS
C         NG   = NUMBER OF DIFFERENT GROUPS
C         NOBS = NUMBER OF DATA POINTS
C         LAAF,  LAAF=1 THE EXPERIMENTAL ACTIVITY COEFFICIENTS
C                       ARE READ
C                LAAF=2 THE LOGARITHMS TO THE EXPERIMENTAL
C                       ACTIVITY COEFFICIENTS ARE READ
C         NOIT,  NOIT=1 PARAMETER ESTIMATION IS PERFORMED
C                NOIT=2 NO PARAMETER ESTIMATION. THE ACTIVITY
C                       COEFFICIENTS ARE CALCULATED BASED ON
C                       THE GIVEN R,Q AND A MATRIX
C
      IF(NOIT.EQ.2) GO TO 1301
      WRITE(NP,15)
   15 FORMAT(1H1,'    PARAMETER ESTIMATION')
      GO TO 1302
 1301 WRITE(NP,1303)
 1303 FORMAT(1H1,'     CALCULATION OF THE ACTIVITY COEFFICIENTS BASED ON
     1THE GIVEN R,Q AND A MATRIX')
 1302 CONTINUE
      DO 14 I=1,16
      READ(NC,13)NTEXT
   14 WRITE(NP,13)NTEXT
   13 FORMAT(40A2)
      READ(NC,16)(NUM(I),I=1,NCOMP)
C
C         NUM GIVES THE NUMBERS ATTACHED TO
C         THE DIFFERENT COMPONENTS
C
      DO 1304 I=1,NCOMP
 1304 READ(NC,2)(NNY(I,K),K=1,NG)
    2 FORMAT(40I2)
C
C         NNY(I,K) IS THE MATRIX GIVING THE NUMBER OF GROUPS OF
```

```
C              KIND K IN MOLECULE I
C
       READ(NC,16)(KG(K),K=1,NG)
C
C         KG(K) IS THE NUMBER ATTACHED TO GROUP K
C
       DO 100 N=1,NOBS
  100  READ(NC,101)T(N),(NU(N,I),XXX(N,I),GME(N,I),I=1,2)
  101  FORMAT (F7.2,3(I2,2F7.4))
C
C          N        = DATA POINT NUMBER
C          T(N)     = TEMPERATURE IN K
C          NU(N,I)  = NUMBER ATTACHED TO COMPONENT I
C                     (NU(N,I)=NUM(I))
C          XXX(N,I) = LIQUID MOLE FRACTION
C          GME(N,I),  LAAF=1 EXPERIMENTAL ACTIVITY COEFFICIENT
C                     LAAF=2 LOGARITHM TO THE EXPERIMENTAL ACTIVITY
C                            COEFFICIENT
C
       IF(LAAF-1)820,820,821
  821  CONTINUE
       DO 822 N=1,NOBS
       DO 822 I=1,2
       GGG=GME(N,I)
       GME(N,I)=EXP(GGG)
  822  CONTINUE
  820  CONTINUE
       DO 705 I=1,NG
       READ(NC,704)R(I),Q(I),(A(I,J),J=1,NG)
  704  FORMAT(8F10.4)
C
C          R(I) IS THE GROUP VOLUME OF GROUP I
C          Q(I) IS THE GROUP AREA OF GROUP I
C          A(I,J) IS THE GROUP INTERACTION PARAMETER
C          BETWEEN GROUPS I AND J
C
  705  CONTINUE
       DO 7 I=1,NCOMP
       RS(I)=0.
       QS(I)=0.
       DO 8 J=1,NG
       RS(I)=RS(I)+NNY(I,J)*R(J)
       QS(I)=QS(I)+NNY(I,J)*Q(J)
    8  CONTINUE
    7  XL(I)=5.*(RS(I)-QS(I))-RS(I)+1.
       WRITE(NP,5)
    5  FORMAT(1H0,'    COMPONENT /   GROUPS',/)
       WRITE(NP,12)(KG(I),I=1,NG)
   12  FORMAT(17X,10I3)
       DO 10 I=1,NCOMP
   10  WRITE(NP,6)NUM(I),(NNY(I,J),J=1,NG)
    6  FORMAT(10X,I3,4X,10I3)
       WRITE(NP,3)
    3  FORMAT(1H0,'  GROUP NO     GROUP R       GROUP Q',/)
       DO 11 I=1,NG
   11  WRITE(NP,4)KG(I),R(I),Q(I)
    4  FORMAT(4X,I3,5X,2E12.4)
       WRITE(NP,1400)
```

```
 1400 FORMAT(1H0,'  GROUP INTERACTION PARAMETERS',/)
      DO 1401 I=1,NG
 1401 WRITE(NP,1402)(A(I,J),J=1,NG)
 1402 FORMAT(10E12.4)
      DO 102 N=1,NOBS
      DO 102 I=1,2
      DO 104 J=1,NCOMP
      IF(NU(N,I)-NUM(J))104,103,104
  103 NM(N,I)=J
      GO TO 102
  104 CONTINUE
  102 CONTINUE
      CALL PFAC3(RS,QS,XL,NOBS)
      IF(NOIT-1)830,830,831
  831 CONTINUE
      CALL PFAC4(NG,NOBS)
      DO 832 NR=1,NOBS
      DO 832 I=1,2
      GMR(NR,I)=GMC(NR,I)+GMR(NR,I)
      GMR(NR,I)=EXP(GMR(NR,I))
  832 CONTINUE
      GO TO 833
  830 CONTINUE
      READ(NC,16)NPAR,KRIT,IDEN
C
C        NPAR IS THE NUMBER OF PARAMETERS TO
C        BE ESTIMATED
C
C        KRIT DETERMINES THE OBJECTIVE FUNCTION
C        IF KRIT=1, THE SUM OF THE SQUARED DIFFERENCES BETWEEN
C        THE EXPERIMENTAL AND CALCULATED ACTIVITY COEFFICIENTS
C        IS MINIMIZED
C        IF KRIT=2, THE LOGARITHMS OF THE ACTIVITY COEFFICIENTS
C        ARE USED
C
C        IDEN IS THE NUMBER OF IDENTICAL PAIRS
C        OF INTERACTION PARAMETERS
C
      READ(NC,16)(IPAR(I),I=1,NPAR)
C
C        IPAR IS THE VECTOR INDICATING THE
C        PARAMETERS TO BE ESTIMATED IN THE
C        A MATRIX
C
      IF(IDEN)950,950,951
  951 IDEN=2*IDEN
      READ(NC,16)(JENS(J),J=1,IDEN)
  950 CONTINUE
C
C        JENS IS THE VECTOR INDICATING THE IDENTICAL
C        PAIRS OF PARAMETERS
C
      NN=NPAR+1
      N=NPAR
      READ(NC,400)(X(1,I),I=1,NPAR)
C
C        X(1,I)= ROW OF INITIAL PARAMETERS
C        NO INITIAL PARAMETER MUST BE ZERO
```

```
C
      SA=1.E-6
C
C         SA IS THE STANDARD ERROR AS DEFINED BY
C         NELDER-MEAD
C
      DO 201 J=2,NN
      DO 201 I=1,N
      IF(J-I-1)202,203,202
  203 X(J,I)=1.1*X(1,I)
      GO TO 201
  202 X(J,I)=X(1,I)
  201 CONTINUE
      WRITE(NP,300)(IPAR(I),I=1,N)
  300 FORMAT(1H0,'   INITIAL PARAMETERS',4I3/)
      DO 204 J=1,NN
  204 WRITE(NP,400)(X(J,I),I=1,N)
      DO 1 J=1,NN
      DO 21 I=1,N
   21 XX(I)=X(J,I)
      CALL FMIN(NPAR,IPAR,PAR,NOBS,NG,XX,FF,KRIT,JENS,IDEN)
    1 F(J)=FF
      NF=NN
C
C         NF IS THE NUMBER OF CALCULATIONS OF F
C
      ALFA=1.
      BETA=0.5
      GAMMA=2.
      ITER=0
      JPR=0
  400 FORMAT(8F10.3)
C
C         ESTIMATION OF THE LOWEST VALUE OF F=FB
C
   25 FB=F(1)
      DO 98 I=1,N
   98 XB(I)=X(1,I)
      JB=1
      DO 31 J=2,NN
      IF(FB-F(J))31,31,108
  108 FB=F(J)
      JB=J
      DO 41 I=1,N
   41 XB(I)=X(J,I)
   31 CONTINUE
C
C         ESTIMATION OF THE HIGHEST VALUE OF F=FS
C
      FS=F(1)
      DO 51 I=1,N
   51 XS(I)=X(1,I)
      JS=1
      DO 61 J=2,NN
      IF(FS-F(J))111,61,61
  111 FS=F(J)
      JS=J
      DO 71 I=1,N
```

```
   71 XS(I)=X(J,I)
   61 CONTINUE
C
C         CALCULATION OF THE CENTROID XM(I) OF POINTS
C         EXCLUDING XS(I)
C
      DO 81 I=1,N
   81 XM(I)=-XS(I)
      DO 9 J=1,NN
      DO 122 I=1,N
  122 XM(I)=XM(I)+X(J,I)
    9 CONTINUE
      DO 121 I=1,N
  121 XM(I)=XM(I)/FLOAT(N)
C
C         REFLECTION
C
      DO 131 I=1,N
  131 XR(I)=XM(I)+ALFA*(XM(I)-XS(I))
      CALL FMIN(NPAR,IPAR,PAR,NOBS,NG,XR,FR,KRIT,JENS,IDEN)
      NF=NF+1
C
C         EXPANSION
C
      IF(FR-FB)141,151,151
  141 DO 161 I=1,N
  161 XE(I)=XM(I)+GAMMA*(XR(I)-XM(I))
      CALL FMIN(NPAR,IPAR,PAR,NOBS,NG,XE,FE,KRIT,JENS,IDEN)
      NF=NF+1
      IF(FE-FB)17,18,18
   17 DO 19 I=1,N
      X(JS,I)=XE(I)
   19 XS(I)=XE(I)
      F(JS)=FE
C
C         CALCULATION OF THE HALTING CRITERION
C
   27 FM=0.
      DO 20 J=1,NN
   20 FM=FM+F(J)
      FM=FM/FLOAT(NN)
      FRMS=0.
      DO 22 J=1,NN
   22 FRMS=(F(J)-FM)**2+FRMS
      RMS=SQRT(FRMS/FLOAT(N))
      ITER=ITER+1
      JPR=JPR+1
      IF(ITER-200)500,500,23
  500 CONTINUE
      IF(JPR-1)902,902,903
  903 CONTINUE
      IF(JPR-6)901,904,904
  904 JPR=1
  902 CONTINUE
      WRITE(NP,107)ITER,NF
  107 FORMAT(1H0,'    ITERATION',I4,'    NUMBER OF CALLS FOR THE SUBROUTIN
     1E',I5)
      WRITE(NP,109)
```

```
      109 FORMAT('       PARAMETERS')
          WRITE(NP,400)(X(JS,I),I=1,N)
          WRITE(NP,106)F(JS),RMS
      106 FORMAT(1H ,'  FMIN=',E14.5,'    SD=',E14.5)
      901 CONTINUE
          IF(RMS-SA)23,23,25
C
C         NEW SIMPLEX
C         FE GREATER THAN FB
C
       18 DO 26 I=1,N
          X(JS,I)=XR(I)
       26 XS(I)=XR(I)
          F(JS)=FR
          FS=FR
          GO TO 27
C
C         NEW SIMPLEX
C         FR GREATER THAN FB
C
      151 DO 30 J=1,NN
          IF(J-JS)28,30,28
       28 IF(FR-F(J))18,18,30
       30 CONTINUE
          IF(FR-FS)91,91,32
       91 DO 33 I=1,N
          X(JS,I)=XR(I)
       33 XS(I)=XR(I)
          F(JS)=FR
          FS=FR
       32 DO 34 I=1,N
       34 XK(I)=XM(I)+BETA*(XS(I)-XM(I))
          CALL FMIN(NPAR,IPAR,PAR,NOBS,NG,XK,FK,KRIT,JENS,IDEN)
          NF=NF+1
C
C         NEW SIMPLEX
C         AFTER CONTRACTION
C
          IF(FK-FS)35,35,36
       35 DO 37 I=1,N
          X(JS,I)=XK(I)
       37 XS(I)=XK(I)
          F(JS)=FK
          FS=FK
          GO TO 27
       36 DO 38 J=1,NN
          DO 39 I=1,N
       39 X(J,I)=(X(J,I)+XB(I))/2.
       38 CONTINUE
          GO TO 27
       23 WRITE(NP,905)
      905 FORMAT(1H0,'      FINAL PARAMETERS')
          WRITE(NP,400)(X(JS,I),I=1,N)
          WRITE(NP,106)F(JS),RMS
      833 CONTINUE
          DO 906 N=1,NOBS
          DO 906 I=1,2
          DEV(N,I)=(GMR(N,I)-GME(N,I))*100./GME(N,I)
```

```
906 CONTINUE
    WRITE(NP,207)
207 FORMAT(1H0,'    TEMP  NUMBER     X              GAMEXP         GAMCAL    DEV
   1',/)
    DO 222 N=1,NOBS
    WRITE(NP,16)N
    DO 223 I=1,2
223 WRITE(NP,96)T(N),NU(N,I),XXX(N,I),GME(N,I),GMR(N,I),DEV(N,I)
 96 FORMAT(F8.2,I5,F8.4,2F14.4,F6.1)
222 CONTINUE
    IF(NOIT-1)835,835,836
835 CONTINUE
    WRITE(NP,1400)
    DO 112 I=1,NG
112 WRITE(NP,1402)(A(I,J),J=1,NG)
    WRITE(NP,996)
996 FORMAT(1H0,'     THE SUM OF THE SQUARED DIFFERENCES BETWEEN THE')
    IF(KRIT-1)997,997,998
998 WRITE(NP,995)
995 FORMAT('       LOGARITHMS TO THE')
997 WRITE(NP,999)
999 FORMAT('     EXPERIMENTAL AND CALCULATED ACTIVITY COEFFICIENTS IS
   1 MINIMIZED')
836 CONTINUE
    STOP
    END
```

```
C     ************************************************************************
C     *
C     *   SUBROUTINE PFAC3
C     *
C     ************************************************************************
C
C
C     ******************
C         SUBROUTINE PFAC3 CALCULATES THE COMBINATORIAL PART
C         OF THE ACTIVITY COEFFICIENTS
C     ******************
C
      SUBROUTINE PFAC3(RS,QS,XL,NOBS)
C
      DIMENSION THETA(2),PHI(2),RS(10),QS(10),XL(10)
      COMMON T(100),NM(100,2),XXX(100,2),GME(100,2),GMC(100,2),GMR(100,2
     1),NNY(10,10),Q(10),A(10,10)
      DO 3 N=1,NOBS
      SQ=0.
      SR=0.
      SXL=0.
      DO 2 I=1,2
      J=NM(N,I)
      SXL=SXL+XL(J)*XXX(N,I)
      SQ=SQ+QS(J)*XXX(N,I)
    2 SR=SR+RS(J)*XXX(N,I)
      DO 3 I=1,2
      J=NM(N,I)
      THETA(I)=QS(J)/SQ
      PHI(I)=RS(J)/SR
      GMC(N,I)=ALOG(PHI(I))+5.*QS(J)*ALOG(THETA(I)/PHI(I))+XL(J)-PHI(I)*
     1SXL
    3 CONTINUE
      RETURN
      END
```

```
C     ********************************************************************
C     *
C     *   SUBROUTINE PFAC4
C     *
C     ********************************************************************
C
C
C     ********************
C           SUBROUTINE PFAC4 CALCULATES THE RESIDUAL PART
C           OF THE ACTIVITY COEFFICIENTS
C     ********************
C
      SUBROUTINE PFAC4(NG,NOBS)
C
      DIMENSION GMOL(10),ATET(10),ANYK(10),BNYK(10),GK(10,3),P(10,10)
      COMMON T(100),NM(100,2),XXX(100,2),GME(100,2),GMC(100,2),GMR(100,2
     1),NNY(10,10),Q(10),A(10,10)
C
C           CALCULATION OF THE PSI MATRIX
C
      DO 250 NR=1,NOBS
      DO 7 I=1,NG
      DO 7 J=1,NG
    7 P(I,J)=EXP(-A(I,J)/T(NR))
C
C           CALCULATION OF GROUP MOLE FRACTIONS
C
      DO 105 II=1,3
      IF(II-2)100,100,101
  100 SNYK=0.
C
C           PURE COMPONENT
C
      J=NM(NR,II)
      DO 12 K=1,NG
   12 SNYK=SNYK+FLOAT(NNY(J,K))
      DO 13 K=1,NG
   13 GMOL(K)=FLOAT(NNY(J,K))/SNYK
      GO TO 102
  101 SNYK=0.
C
C           MIXTURE
C
      DO 2 I=1,2
      J=NM(NR,I)
      DO 2 K=1,NG
    2 SNYK=SNYK+FLOAT(NNY(J,K))*XXX(NR,I)
      DO 3 K=1,NG
      GNYK=0.
      DO 4 I=1,2
      J=NM(NR,I)
    4 GNYK=GNYK+FLOAT(NNY(J,K))*XXX(NR,I)
    3 GMOL(K)=GNYK/SNYK
```

```
C
C         CALCULATION OF GROUP AREA FRACTIONS
C
  102 SNYK=0.
      DO 5 K=1,NG
    5 SNYK=SNYK+Q(K)*GMOL(K)
      DO 6 K=1,NG
    6 ATET(K)=Q(K)*GMOL(K)/SNYK
C
C         CALCULATION OF GAMMA K
C
      DO 9 K=1,NG
      ANYK(K)=0.
      DO 10 M=1,NG
      SNYK=0.
      DO 8 N=1,NG
    8 SNYK=SNYK+ATET(N)*P(N,M)
   10 ANYK(K)=ATET(M)*P(K,M)/SNYK +ANYK(K)
      BNYK(K)=0.
      DO 11 M=1,NG
   11 BNYK(K)=BNYK(K)+ATET(M)*P(M,K)
      BNYK(K)=ALOG(BNYK(K))
    9 GK(K,II)=Q(K)*(1.-BNYK(K)-ANYK(K))
  105 CONTINUE
      DO 201 I=1,2
      J=NM(NR,I)
      SNYK=0.
      DO 200 K=1,NG
  200 SNYK=SNYK+FLOAT(NNY(J,K))*(GK(K,3)-GK(K,I))
  201 GMR(NR,I)=SNYK
  250 CONTINUE
      RETURN
      END
```

```
C     **********************************************************************
C     *
C     *   SUBROUTINE FMIN
C     *
C     **********************************************************************
C
C
C     *******************
C           SUBROUTINE FMIN CALCULATES F(THE OBJECTIVE FUNCTION)
C           AS A FUNCTION OF A SET OF PARAMETERS
C     *******************
C
      SUBROUTINE FMIN(NPAR,IPAR,PAR,NOBS,NG,XX,FF,KRIT,JENS,IDEN)
C
      DIMENSION IPAR(4),PAR(4),XX(4),JENS(10)
      COMMON T(100),NM(100,2),XXX(100,2),GME(100,2),GMC(100,2),GMR(100,2
     1),NNY(10,10),Q(10),A(10,10)
      DO 2 I=1,NPAR,2
      KI=IPAR(I)
      KJ=IPAR(I+1)
      A(KI,KJ)=XX(I)
    2 A(KJ,KI)=XX(I+1)
      IF(IDEN)9,9,8
    8 KKI=JENS(1)
      KKJ=JENS(2)
      DO 7 J=3,IDEN,2
      IKI=JENS(J)
      IKJ=JENS(J+1)
      A(IKI,IKJ)=A(KKI,KKJ)
      A(IKJ,IKI)=A(KKJ,KKI)
    7 CONTINUE
    9 CONTINUE
      CALL PFAC4(NG,NOBS)
      DO 200 NR=1,NOBS
      DO 200 I=1,2
      GMR(NR,I)=GMC(NR,I)+GMR(NR,I)
      GMR(NR,I)=EXP(GMR(NR,I))
  200 CONTINUE
      FF=0.
      DO 3 N=1,NOBS
      DO 3 I=1,2
      IF(KRIT-1)10,10,20
   10 FF=FF+(GMR(N,I)-GME(N,I))**2
      GO TO 3
   20 GCAL=GMR(N,I)
      GEXP=GME(N,I)
      FF=FF+(ALOG(GCAL)-ALOG(GEXP))**2
    3 CONTINUE
      RETURN
      END
```

```
******************************************************************
*
*   SAMPLE INPUT (UNIFAC PARAMETER ESTIMATION)
*
******************************************************************

    ³   ⁶   ⁹  ¹²  ¹⁵
    5|  5| 13|  1|  1|


    **************
    *            *
    * CH2CO/CCOH *
    *            *
    **************


ACETONE(11)-ETHANOL(20) AT 305.15, CAN.J.RES. 24B,254,(1946)
ETHANOL(20)-2.BUTANONE(12) AT 1 ATM, ZH.PRIKL.KHIM. 41,589,(1968)
ISOPROPANOL(1)-2.PENTANONE(16) AT 1 ATM, IEC 45,1803,(1953)


    ³   ⁶   ⁹  ¹²  ¹⁵
   11| 20| 12| 16|  1|
    1       1
    ²  ⁴  ¹    ¹⁰
    1| 1|      1|
    1  2       1
    1       1
       1  2 17 15 22
305.1511 0.0750 2.051620 0.9250 1.0055
305.1511 0.5000 1.218720 0.5000 1.2304
305.1511 0.8000 1.035620 0.2000 1.6686
305.1511 1.0000 1.000020 0.0000 2.3166
348.1520 0.2380 1.454612 0.7620 1.0506
347.5520 0.3970 1.249712 0.6030 1.1154
347.5520 0.6220 1.096512 0.3780 1.2825
348.4520 0.7860 1.023412 0.2140 1.5091
350.1520 0.9340 1.004612 0.0660 1.7549
371.37 1 0.0755 1.522116 0.9245 0.9678
361.71 1 0.3720 1.239516 0.6280 1.0060
358.59 1 0.5845 1.097416 0.4155 1.1203
356.76 1 0.7815 1.025716 0.2185 1.2816
0.9011      0.848       0.          0.          737.5   737.5   476.4
0.6744      0.54        0.          0.          737.5   737.5   476.4
2.1055      1.972      -87.93      -87.93       0.      0.      1.
1.878       1.66       -87.93      -87.93       0.      0.      1.
1.6724      1.488       26.76       26.76       1.      1.      0.
    2  2  2
    3  5
    3  5  4  5
100.     100.
```

```
***************************************************************************
*                                                                         *
*    SAMPLE OUTPUT (UNIFAC PARAMETER ESTIMATION)                          *
*                                                                         *
***************************************************************************


    PARAMETER ESTIMATION



    ***************
    *             *
    * CH2CO/CCOH  *
    *             *
    ***************


ACETONE(11)-ETHANOL(20) AT 305.15, CAN.J.RES. 24B,254,(1946)
ETHANOL(20)-2.BUTANONE(12) AT 1 ATM, ZH.PRIKL.KHIM. 41,589,(1968)
ISOPROPANOL(1)-2.PENTANONE(16) AT 1 ATM, IEC 45,1803,(1953)


        COMPONENT /   GROUPS

                      1   2  17  15  22
              11      1   0   0   0   1
              20      0   0   1   0   0
              12      1   1   0   0   1
              16      1   2   0   0   1
               1      1   0   0   1   0

    GROUP NO      GROUP R         GROUP Q

         1       0.9011E 00      0.8480E 00
         2       0.6744E 00      0.5400E 00
        17       0.2105E 01      0.1972E 01
        15       0.1878E 01      0.1660E 01
        22       0.1672E 01      0.1488E 01

GROUP INTERACTION PARAMETERS

 0.0           0.0           0.7375E 03    0.7375E 03    0.4764E 03
 0.0           0.0           0.7375E 03    0.7375E 03    0.4764E 03
-0.8793E 02   -0.8793E 02    0.0           0.0           0.1000E 01
-0.8793E 02   -0.8793E 02    0.0           0.0           0.1000E 01
 0.2676E 02    0.2676E 02    0.1000E 01    0.1000E 01    0.0

    INITIAL PARAMETERS    3    5
    100.000     100.000
    110.000     100.000
    100.000     110.000
```

```
ITERATION    1    NUMBER OF CALLS FOR THE SUBROUTINE      5
   PARAMETERS
115.000      80.000
FMIN=    0.24502E 00    SD=    0.81657E-01

ITERATION    6    NUMBER OF CALLS FOR THE SUBROUTINE     15
   PARAMETERS
116.426      34.141
FMIN=    0.55602E-01    SD=    0.28932E-02

ITERATION   11    NUMBER OF CALLS FOR THE SUBROUTINE     25
   PARAMETERS
112.457      40.795
FMIN=    0.54494E-01    SD=    0.29760E-04

ITERATION   16    NUMBER OF CALLS FOR THE SUBROUTINE     35
   PARAMETERS
116.547      37.169
FMIN=    0.54425E-01    SD=    0.43163E-05

     FINAL PARAMETERS
116.871      36.987
FMIN=    0.54424E-01    SD=    0.44564E-06
```

TEMP	NUMBER	X	GAMEXP	GAMCAL	DEV
1					
305.15	11	0.0750	2.0516	1.8338	-10.6
305.15	20	0.9250	1.0055	1.0037	-0.2
2					
305.15	11	0.5000	1.2187	1.2090	-0.8
305.15	20	0.5000	1.2304	1.1902	-3.3
3					
305.15	11	0.8000	1.0356	1.0331	-0.2
305.15	20	0.2000	1.6686	1.5987	-4.2
4					
305.15	11	1.0000	1.0000	1.0000	0.0
305.15	20	0.0	2.3166	2.1522	-7.1
5					
348.15	20	0.2380	1.4546	1.5374	5.7
348.15	12	0.7620	1.0506	1.0463	-0.4
6					
347.55	20	0.3970	1.2497	1.3013	4.1
347.55	12	0.6030	1.1154	1.1307	1.4
7					
347.55	20	0.6220	1.0965	1.1057	0.8
347.55	12	0.3780	1.2825	1.3384	4.4
8					
348.45	20	0.7860	1.0234	1.0320	0.8
348.45	12	0.2140	1.5091	1.5755	4.4
9					
350.15	20	0.9340	1.0046	1.0029	-0.2
350.15	12	0.0660	1.7549	1.8755	6.9
10					
371.37	1	0.0755	1.5221	1.5662	2.9
371.37	16	0.9245	0.9678	1.0028	3.6
11					
361.71	1	0.3720	1.2395	1.2482	0.7
361.71	16	0.6280	1.0060	1.0743	6.8
12					
358.59	1	0.5845	1.0974	1.1069	0.9
358.59	16	0.4155	1.1203	1.2020	7.3
13					
356.76	1	0.7815	1.0257	1.0298	0.4
356.76	16	0.2185	1.2816	1.4075	9.8

GROUP INTERACTION PARAMETERS

0.0	0.0	0.7375E 03	0.7375E 03	0.4764E 03
0.0	0.0	0.7375E 03	0.7375E 03	0.4764E 03
-0.8793E 02	-0.8793E 02	0.0	0.0	0.1169E 03
-0.8793E 02	-0.8793E 02	0.0	0.0	0.1169E 03
0.2676E 02	0.2676E 02	0.3699E 02	0.3699E 02	0.0

THE SUM OF THE SQUARED DIFFERENCES BETWEEN THE
LOGARITHMS TO THE
EXPERIMENTAL AND CALCULATED ACTIVITY COEFFICIENTS IS MINIMIZED

EXPLANATIONS TO THE OUTPUT

The first matrix is the NNY matrix showing the number and type of groups in the different components. The first row gives the numbers of the groups as presented in the input by KG. The first column gives the numbers of the components as identified by NUM: In the sample output we see that component 12 (2-Butanone) is built from one group of type 1 (CH_3), one group of type 2 (CH_2) and one group of type 22 (CH_3CO).

The following two matrices give the group volumes R and areas Q and the group interaction parameters A.

The matrix of INITIAL PARAMETERS represents the vertices of the initial simplex.

The results of the parameter estimation are printed every fifth time the standard deviation SD (Equation 5.22) is checked against the preset value of 10^{-6}. The current "best" parameters and the corresponding value of the objective function FMIN are printed.

The experimental and calculated activity coefficients (GAMEXP and GAMCAL) are given in the table as functions of temperature (TEMP) and liquid mole fractions (X) for each data point. The values GAMCAL are calculated by means of UNIFAC with the parameters as presented in the final group interaction parameter matrix. The components are identified by integers (NUMBER), and DEV are the relative deviations in per cent between GAMCAL and GAMEXP.

Since only a few data points are included for this illustration, this example does not lead to parameters presented in Table 4.4.

APPENDIX 4

DISTILLATION PROGRAMS

This appendix lists the "full" and "simplified" versions of the distillation program referred to in Chapter 9. The appendix contains:

		Page
1.	Input specifications to distillation programs	
	Full version	276
	Simplified version	279
2.	List of symbols used in the distillation programs	281
3.	Full version: program listing	285
	sample input	307
	sample output	309
4.	Simplified version: program listing	327
	sample input	335
	sample output	336

For the full version, the in- and output shown here correspond to the complex column problem shown in Figure 9.8. In addition to this, calculations are in the same job carried out for a simple distillation column operating on the same mixture.

For the simplified version, the sample in- and output correspond to the four-component extractive distillation problem with phenol as solvent, see Table 9.3. Extractive distillation problems have more difficult convergence characteristica than the ordinary distillation problems. Usually the simplified version requires 6-8 iterations to converge as opposed to 11 iterations as found here. Note that in the extractive distillation calculations, a temperature profile similar to the final profile is obtained after a few iterations. The first several iterations, which diverge rather than converge, are used to establish reasonable concentration profiles from the initial, very "wild" guess.

Subroutine GAUSL, which is called by both of the main distillation programs, is listed in Appendix 1. Otherwise all of the needed subprograms are listed in this appendix.

INPUT TO DISTILLATION PROGRAM (full version)

<u>a</u>. NK Format I5
NK - number of components (integer)

<u>b</u>. NOAC, NOFUG, NODIM, NOEX and NDIM. Format 5I5

NDIM is the number of the dimerizing component. If a control variable is zero or negative, the corresponding nonideality is NOT included.

NOAC - Activity coefficients
NOFUG - Fugacity coefficients
NODIM - Vapor phase dimerization
NOEX - Excess enthalpy (by means of the UNIQUAC-parameters in item <u>d</u>)

<u>c</u>. Text card. Format 40A2 Any text

<u>d</u>. Input for the calculation of nonidealities (PARIN)

1: UNIQUAC parameters. (Only if NOAC is greater than zero). The cards are obtained as output from the program which calculates the UNIQUAC parameters from UNIFAC. According to Appendix 2 the input is:

q_k, r_k, and ℓ_k Format 5X,3F15.5
One card per component.

The matrix A(J,I) (same as $A_{ji}^{(0)}$) I=1,NK;J=1,NK
Format 20X,5E12.5. One card per component.

The matrix B(J,I) (same as $A_{ji}^{(1)}$) I=1,NK;J=1,NK
Format 20X,5E12.5. One card per component.

$\tau_{ji} = A_{ji}^{(0)} + A_{ji}^{(1)} T$ (see Equation (3.9))

2: Second virial coefficients (only if NOFUG is greater than zero). For i = 1,NK:

$B_{i,j}(T_1)$, j = 1,NK   Format 10X, 5F10.3

$B_{i,j}(T_2)$, j = 1,NK   Format 10X, 5F10.3

The second virial coefficients may be based on experimental data, or they may be estimated from the Hayden-O'Connell correlation as shown in Appendix 1, program BIJ.

3: T_1 (the temperature at which the $B_{i,j}(T_1)$'s are calculated) and T_2 (similarly for $B_{i,j}(T_2)$), Format 2F10.2. T_1 and T_2 are in K.

4: A, B in dimerization equation (only if NODIM is greater than zero). Format 10X, 5F10.2 (see Table 2.3).

e. ANT(K,J), K = 1,3   Format 3F10.3

ANT(K,J) are the three Antoine constants for component J. One card for each component.

f. ENP(J,K), K = 1,4   Format 4F10.3

ENP(J,K) gives the enthalpy for pure component J. The enthalpy is defined by a linearized equation such that the enthalpy for a liquid is defined as:

$h = h_o + c_\ell \cdot t$, and for a vapor

$H = H_o + c_v \cdot t$

K = 1 corresponds to h_o, K = 2 to c_ℓ, K = 3 to H_o, and K = 4 corresponds to c_v. The temperature is here denoted by t, which is referred to any standard temperature. In reading these parameters, one must refer to the same standard temperature. If c_ℓ, c_v, and H^o (the heat of evaporation at the standard temperature) are known, the parameters are defined as follows:

$h_o = -c_\ell \cdot t_o$

$H_o = H^o - c_v \cdot t_o$

where t_o is the standard temperature
the unit must be energy unit/mole $°C$
One card for each component

g. NST, NFEED, NSL, NSV, NCON   Format 5I5
where

   NST — Number of stages, including reboiler and condenser

        NFEED   — Number of feed plates
        NSL   — Number of liquid sidestreams
        NSV   — Number of vapor sidestreams
        NCON: If NCON is zero, the calculation stops

h. DEST, RFLX, P, TT, TB   Format 5F10.5

    DEST — Amount of distillate, moles/hour
    RFLX — Reflux ratio
    P   — Pressure, atm
    TT   — Guessed temperature at top of column, $^\circ$C
    TB   — Guessed temperature at bottom of column, $^\circ$C

i. KMU   Format I5

    $KMU \leq 0$: A Murphree stage efficiency is to be read for each stage

    $KMU > 0$: One constant efficiency is to be read

j. Murphree stage efficiencies, MUS or MU(I)   Format F10.5

    1: If KMU is greater than zero, only one card is read (the constant stage efficiency)

    2: If KMU is less than one, one card must be read for each stage (including reboiler and condenser)

k. NF — feed plate number   Format I5

l. TF(NF), FKV(NF), (FSTR(NF,J), J = 1,NK)   Format 8F10.5
Only one card

TF(NF) is the temperature of the feed at plate NF
FKV(NF) is the fraction of feed at plate NF which appears as vapor (the most important specification of these two specifications)

FSTR(NF,J), J = 1,NK is the feed of each component on plate NF

Items **k** and **l** are repeated NFEED times

m. NF — liquid sidestream plate number   Format I5

n. SL(NF) total liquid sidestream (moles/hour) at plate number NF,   Format F10.5

    Items **m** and **n** are repeated NSL times

o. NF – vapor sidestream plate number   Format I5

p. SV(NF) – total vapor sidestream (moles/hour) at plate
    number NF,   Format F10.5

   Items o and p are repeated NSV times

   The program now, after calculation of the column, steps
   back to read item g: NST, NFEED ....

INPUT TO DISTILLATION PROGRAM (simplified version)

a. NK   Format I5
    NK – Number of components

b. NOAC   Format I5

    If NOAC is less than unity, the activity coefficients are
    equated to one

c. Text card (any text)   Format 40A2

d. UNIQUAC parameters (Only if NOAC is greater than zero).
    See full version, item d.

e. ANT(K,J), K = 1,3   Format 3F10.3
    See full version, item e.

f. NST, NFEED, NSL, NSV, NCON   Format 5I5
    See full version, item g.

g. DEST, RFLX, P, TT, TB   Format 5F10.5
    See full version, item h.

h. NF – feed plate number   Format I5

i. FKV(NF), (FSTR(NF,J), J = 1,NK)   Format 8F10.5
    Only one card

    FKV(NF) is the fraction of feed at plate NF which appears
    as vapor

    FSTR(NF,J), J = 1,NK is the feed of each component on
    plate NF

    Items h and i are repeated NFEED times

j. NF – liquid sidestream plate number   Format I5

k. SL(NF) total molar liquid sidestream at plate number NF

   Items j and k are repeated NSL times   Format F10.5

l. NF - vapor sidestream plate number   Format I5

m. SV(NF) - total molar vapor sidestream at plate number NF

   Items l and m are repeated NSV times   Format F10.5

   The program will after calculations step back to read item f: NST, NFEED ....

IN BOTH DISTILLATION PROGRAMS:

All streams have the unit of moles/hour; the temperature unit is $^{\circ}$C; the units for Antoine parameters are the usual mm Hg and $^{\circ}$C.

For a column one may have up to:

   5 components
   30 stages
   30 sidestreams
   30 feed streams

without changing the program. If more components or stages are needed, change the appropriate DIMENSION statements.

LIST OF SYMBOLS, MAIN DISTILLATION PROGRAMS

TMAX	=	Limit for temperature corrections
NK	=	Number of components
NOAC:		If NOAC is less than unity, activity coefficients are equated to one
NOFUG:		If NOFUG is less than unity, fugacity coefficients are equated to one
NODIM:		If NODIM is less than unity, no vapor phase dimerization is included
NOEX:		If NOEX is less than unity, the excess enthalpy is equated to zero
NKO	=	NK
NAVN	=	Any text
ANT(K,J)	=	Antoine constants for component J
ENP(J,K)	=	Enthalpy parameters for component J
NST	=	Number of stages
NFEED	=	Number of feed stages
NSL	=	Number of liquid phase sidestreams
NSV	=	Number of vapor phase sidestreams
DEST	=	Amount of distillate
RFLX	=	Reflux ratio
P	=	Total pressure
TT	=	Guessed temperature at top of column
TB	=	Guessed temperature at bottom of column
KMU:		If KMU is greater than zero, constant stage efficiencies are assumed
MU(I)	=	Stage efficiency on stage I
MUS	=	Constant stage efficiency
SV(I)	=	Vapor sidestream on stage I
SL(I)	=	Liquid sidestream on stage I
FKV(I)	=	Fraction of feed on stage I as vapor
FSTR(I,NK1)	=	Total feed on stage I
HF(I)	=	Feed enthalpy
NF	=	Feed stage number (is also used for other purposes)
TF(NF)	=	Temperature of feed
FSTR(I,J)	=	Feed of component J on stage I

FV(I)	=	Total vapor flow from stage I
FL(I)	=	Total liquid flow from stage I
FEED(J)	=	Total feed of component J
FEED(NK1)	=	Total feed
BU	=	Bottoms product
XDEST	=	Total amount of product in sidestreams
FVMAX	=	Limit for vapor flow correction
FLMAX	=	Limit for liquid flow correction
T(I)	=	Temperature on stage I
FVV(I,J)	=	Vapor flow of component J from stage I
FLL(I,J)	=	Liquid flow of component J from stage I
XFK(I,J)	=	Separation factor for component J on stage I
DFK(I,J,K)	=	DFAC(I,J,K) from subroutine KFAC
D(I,J)	=	Discrepancy function for component J on stage I or corrections to these
BMAT(I,J,K)	=	B - matrices (See Reference [1], Chapter 9)
AMAT(J,K)	=	A - matrices
CM(J,K)	=	A matrix used in GAUSL
QQ	=	Limitation on variables
RES	=	Sum of squares of discrepancy functions
PROD(1,I,J)	=	Product streams as liquid of component J on stage I
PROD(2,I,J)	=	Product streams as vapor of component J on stage I
FINM(1,I,J)	=	Liquid mole fraction of component J on stage I
FINM(2,I,J)	=	Vapor mole fraction of component J on stage I

Subroutines

Subroutine: EXCES

HE(I)	=	Contribution to h^E from component I
TE(I)	=	Temperature on stage I
X(J)	=	Mole fraction of component J
Q(I)	=	UNIQUAC pure-component parameter

Subroutine: EXCOR

DEX(I,J)	=	Derivatives with respect to component liquid flow or temperature

Subroutine: MQUAC

R(I),Q(I),XL(I),THETA(I),PHI(I),PARAM(J,I,K),PAR(J),THS(I) are UNIQUAC parameters

GAM(J)   =   Activity coefficient of component J

Subroutine: FUG

BI1(I,J)	=	Second virial coefficient at temperature T1
BI2(I,J)	=	Second virial coefficient at temperature T2
BI(I,J)	=	Second virial coefficient at temperature T
BMIX	=	Mixture second virial coefficient
FI(J)	=	Fugacity coefficient of component J
ZNOD	=	Vapor phase correction for non-dimerizing components
ZDIM	=	Vapor phase correction for dimerizing component

Subroutine: ONEDI

ZDI	=	Correction factor for the dimerizing component
ZNODT	=	Correction factor for non-dimerizing components

Subroutine: KFAC

PAR(J)	=	Partial pressure of component J
YY(J)=Y(J)	=	Mole fraction of component J in vapor phase

XX(J)	=	Mole fraction of component J in liquid phase
XKF(I,J)	=	Equilibrium ratio for component J on stage I
DFAC(I,J,K)	=	Derivatives of the K-factors with respect to component liquid and vapor flow rates and temperature

Subroutine: ENT

HL(I)	=	Liquid phase enthalpy on stage I
HV(I)	=	Vapor phase enthalpy on stage I
HVV(I,J)	=	Vapor phase enthalpy of component J on stage I
HLL(I,J)	=	Liquid phase enthalpy of component J on stage I
CV(I),CL(I)	=	Derivatives of enthalpy with respect to temperature

Subroutine: PARIN

A	=	A in the expression $-\log_{10} K_A = A - B/T$
BX	=	B in the above expression

```
C     ***********************************************************************
C     *
C     *   MAIN DISTILLATION PROGRAM, FULL VERSION
C     *
C     ***********************************************************************
C
C
      IMPLICIT REAL(M)
      DIMENSION FLMAX(5),FVMAX(5)
      DIMENSION PROD(2,30,5)
      DIMENSION FEED(6),FL(30),FV(30),XFK(30,5),FLL(30,5),T(30)
      DIMENSION FVV(30,5),HV(30),HL(30),CV(30),CL(30),HVV(30,5),HLL(30,5
     *),HF(30),TF(30)
      DIMENSION AMAT(11,11),BMAT(30,11,11),D(30,11)
      REAL*8 CM(11,22)
      DIMENSION SL(30),SV(30),FKV(30),FSTR(30,6),DFK(30,5,11)
      DIMENSION ANT(3,5),NAVN(40)
      DIMENSION MU(30)
      DIMENSION HHE(30),DEXH(30,6)
      DIMENSION FINM(2,30,5)
      DIMENSION ANTM(2,5)
      COMMON NDIM,NOAC,NOFUG,NODIM,NOEX,NK,Q(5),R(5),XL(5),PARAM(5,5,2),
     *ENP(5,4),BI1(5,5),BI2(5,5),T1,T2,A,BX
C
C     TMAX IS THE MAXIMUM TEMPERATURE VARIATION BETWEEN ITERATIONS
      TMAX=10.
C
C     NK IS THE NUMBER OF COMPONENTS
C
      READ(5,8) NK
    8 FORMAT(16I5)
C
C     IF THE CONSTANTS BELOW ARE GREATER THAN ZERO, RESPECTIVELY
C     ACTIVITY COEFFICIENTS, FUGACITY COEFFICIENTS, VAPOR-PHASE DIMERIZA-
C     TION, AND EXCESS ENTHALPIES WILL BE CALCULATED.  NDIM IS THE NUMBER
C     OF THE DIMERIZING COMPONENT
C
      READ(5,8) NOAC,NOFUG,NODIM,NOEX,NDIM
      NKO=NK
      NK1=NK+1
      NK2=2*NK+1
      NK11=NK1+1
      NK21=NK2-1
C     READ UNIQUAC PARAMETERS
      CALL PARIN(NAVN)
      WRITE(6,12) NAVN
   12 FORMAT(1H1,40A2)
      IF (NOAC.GT.0) WRITE(6,801)
  801 FORMAT(/,'     INCLUDING ACTIVITY-COEFFICIENTS    ')
      IF (NOFUG.GT.0) WRITE(6,802)
  802 FORMAT(/,'     INCLUDING FUGACITY-COEFFICIENTS    ')
      IF (NODIM.GT.0) WRITE(6,803)
  803 FORMAT(/,'     INCLUDING DIMERIZATION    ')
```

```
      IF (NOEX.GT.0) WRITE(6,804)
  804 FORMAT(/,'          INCLUDING EXCESS ENTHALPY        ')
      DO 17 J=1,NK
C
C     READ ANTOINE CONSTANTS
C
   17 READ(5,13)(ANT(K,J),K=1,3)
      DO 19 J=1,NK
      ANTM(1,J)=ANT(1,J)
      ANT(1,J)=2.302585*(ANT(1,J)-2.880814)
      ANTM(2,J)=ANT(2,J)
   19 ANT(2,J)=2.302585*ANT(2,J)
      DO 1 J=1,NK
C
C     READ ENTHALPY CONSTANTS
C
    1 READ(5,2) (ENP(J,K),K=1,4)
    2 FORMAT(5F10.3)
C
C     READ COLUMN SPECIFICATIONS
C
  333 READ(5,8) NST,NFEED,NSL,NSV,NCON
      IF (NCON .EQ. 0) GO TO 555
      READ(5,13) DEST,RFLX,P,TT,TB
   13 FORMAT(8F10.5)
C
C     READ MURPHREE STAGE EFFICIENCIES.  IF KMU IS POSITIVE, ONE
C      CONSTANT EFFICIENCY IS READ INTO MUS
C
      READ(5,8)KMU
      IF (KMU) 1001,1001,1002
 1001 CONTINUE
      DO 1000 I=2,NST
 1000 READ(5,1003)MU(I)
      GO TO 1004
 1002 READ(5,1003) MUS
      DO 1005 I=2,NST
 1005 MU(I)=MUS
 1003 FORMAT(F10.5)
 1004 CONTINUE
      MU(1)=1.
      DO 200 I=1,NST
      SL(I)=0.
      SV(I)=0.
      FKV(I)=0.
      FSTR(I,NK1)=0.
      DO 200 J=1,NK
  200 FSTR(I,J)=0.
      DO 1010 I=1,NST
 1010 HF(I)=0.
      DO 201 I=1,NFEED
      READ(5,8) NF
C
C     READ FEED SPECIFICATIONS
C
      READ(5,13)TF(NF),FKV(NF),(FSTR(NF,J),J=1,NK)
      DO 201 J=1,NK
      HF(NF)=HF(NF)+FSTR(NF,J)*(FKV(NF)*(ENP(J,3)+ENP(J,4)*TF(NF))+(1.-F
```

```
         *KV(NF))*(ENP(J,1)+ENP(J,2)*TF(NF)))
  201 FSTR(NF,NK1)=FSTR(NF,NK1)+FSTR(NF,J)
      IF (NSL.EQ.0) GO TO 203
      DO 202 I=1,NSL
      READ(5,8) NF
  202 READ(5,13) SL(NF)
  203 IF (NSV.EQ.0) GO TO 204
      DO 205 I=1,NSV
      READ(5,8) NF
  205 READ(5,13) SV(NF)
C
C     FIRST GUESS OF THE COMPONENT FLOWS AND THE TOTAL FLOWS.
C
C
C     THE TOTAL FLOWS ARE ESTIMATED BY ASSUMING CONSTANT FLOWS FROM
C     STAGE TO STAGE IN EACH SECTION OF THE COLUMN.
C     AS FOR THE COMPONENT FLOWS,THE FIRST GUESSES ARE MADE BY THE FOL-
C     LOWING EQUATIONS
C                       FVV(I,J)=FEED(J)/FEED(NK1)*FV(I), AND
C                       FLL(I,J)=FEED(J)/FEED(NK1)*FL(I), WHERE
C                       NK1=NUMBER OF COMPONENTS +ONE
C                       FEED(NK1)=TOTAL FEED
C                       FV=TOTAL VAPOR FLOW
C                       FL=TOTAL LIQUID FLOW
C
  204 FV(NST)=DEST
      FL(NST)=DEST*RFLX
      FV(NST-1)=FL(NST)-FSTR(NST,NK1)+SV(NST)+SL(NST)+DEST
      DO 210 II=3,NST
      IF (NST .LE. 2) GO TO 210
      I=NST+2-II
      FL(I)=FL(I+1)-SL(I)+(1.-FKV(I))*FSTR(I,NK1)
      FV(I-1)=FV(I)+SV(I)-FKV(I)*FSTR(I,NK1)
  210 CONTINUE
      FL(1)=FL(2)-SL(1)+(1.-FKV(1))*FSTR(1,NK1)
      FL(1)=FL(1)-FV(1)
      DO 211 J=1,NK1
      FEED(J)=0.
      DO 211 I=1,NST
  211 FEED(J)=FEED(J)+FSTR(I,J)
      BU=0.
      DO 212 J=1,NKO
  212 BU=BU+FEED(J)
      XDEST=0.
      DO 213 I=1,NST
  213 XDEST=XDEST+SV(I)+SL(I)
      BU=BU-DEST-XDEST
      WRITE(6,18)
   18 FORMAT(/,' FEED COMPOSITION AND ANTOINE PARAMETERS ',//)
      WRITE(6,23)(J,FEED(J),ANTM(1,J),ANTM(2,J),ANT(3,J),J=1,NK)
   23 FORMAT(I3,F12.3,5X,3F12.3)
C
C     MAXIMUM DEVIATIONS IN COMPONENT FLOWS BETWEEN ITERATIONS
C
      DO 24 J=1,NK
      FVMAX(J)=FEED(J)*RFLX
   24 FLMAX(J)=FEED(J)*RFLX
      WRITE(6,16)NST,DEST,RFLX,P
```

```
   16 FORMAT(//,' NUMBER OF STAGES         ',I9/,' DISTILLATE
     1',F12.3/,' REFLUX RATIO             ',F15.6/,' TOTAL PRESSURE
     1  ',F12.3//)
      WRITE(6,215)
  215 FORMAT(//,' FLOW CONFIGURATION ',//,' I       FL       FV       SL
     1       SV      FKV     FEED STREAMS',/)
      DO 216 I=1,NST
  216 WRITE(6,217) I,FL(I),FV(I),SL(I),SV(I),FKV(I),(FSTR(I,J),J=1,NK)
  217 FORMAT(I3,3X,10F8.1)
C
C     GUESS OF THE TEMPERATURE PROFILE IN THE COLUMN.
C
      DO 26 I=1,NST
   26 T(I)=TB+(I-1)*(TT-TB)/NST
      DO 3 I=1,NST
      DO 3 J=1,NK
      FVV(I,J)=FEED(J)/FEED(NK1)*FV(I)
    3 FLL(I,J)=FEED(J)/FEED(NK1)*FL(I)
  500 CALL KFAC(NST,NKO,T,ANT,P,FLL,FVV,XFK,DFK)
      CALL ENT(NST,T,FLL,FVV,HLL,HVV,HL,HV,CL,CV)
      CALL EXCES(NST,T,FLL,HHE)
      CALL EXCOR(NST,T,FLL,DEXH)
C
C     D(I,J) IS THE MATRIX CONSISTING OF THE CALCULATED DISCREPANCY FUNC-
C     TIONS ON PLATE I FOR COMPONENT J.
C     AS FOR PLATE I, D(I,J) CONSISTS OF
C
C                                      COMPONENT MATERIAL BALANCES(J=1,NK)
C                                      EQUILIBRIUM RELATIONSHIPS (J=NK+1,
C                                      2*NK-1)
C                                      ENTHALPY BALANCES (J=2*NK+1)
C     NOTE THAT FOR I=1 AND I=NST (NST=NUMBER OF PLATES)THERE ARE SPE-
C     CIAL RELATIONS IN DEFINING THE ENTHALPY BALANCES.
C
      DO 30 I=1,NST
      DO 31 J=1,NKO
      D(I,J)=FVV(I,J)*(1.+SV(I)/FV(I))+FLL(I,J)*(1.+SL(I)/FL(I))-FSTR(I,
     *J)
      IF (I.GT.1) D(I,J)=D(I,J)-FVV(I-1,J)
      IF (I.LT.NST) D(I,J)=D(I,J)-FLL(I+1,J)
   31 CONTINUE
      DO 32 J=NK1,NK21
      K=J-NKO
      D(I,J)=MU(I)*XFK(I,K)*FV(I)*FLL(I,K)/FL(I)-FVV(I,K)
      IF (I.GT.1) D(I,J)=D(I,J)+(1.-MU(I))*FVV(I-1,K)*FV(I)/FV(I-1)
   32 CONTINUE
      IF (I.EQ.1) GO TO 27
      IF (I.EQ.NST) GO TO 28
      D(I,NK2)=(1.+SV(I)/FV(I))*HV(I)+(1.+SL(I)/FL(I))*HL(I)-HF(I)-HV(I-
     *1)-HL(I+1)+(FL(I)+SL(I))*HHE(I)-FL(I+1)*HHE(I+1)
      GO TO 29
   27 D(I,NK2)=BU
      DO 101 J=1,NKO
  101 D(I,NK2)=D(I,NK2)-FLL(1,J)
      GO TO 29
   28 D(I,NK2)=RFLX*DEST
      DO 102 J=1,NKO
  102 D(I,NK2)=D(I,NK2)-FLL(NST,J)
```

```
      GO TO 29
   29 CONTINUE
   30 CONTINUE
C
C     DEFINITION OF THE JACOBIAN MATRIX. IT CONSISTS OF PARTIAL DERIVATI
C     VES OF THE PLATE CORRECTION FUNCTIONS WITH RESPECT TO THE
C     VARIABLES ON PLATE I.
C
C
C     IN THE DESCRIPTION OF THE JACOBIAN MATRIX (B-MATRIX) THE CORREC-
C     TION FUNCTIONS WILL BE STATED AS F AND THE INDEPENDENT VARIABLES
C     WILL BE STATED AS V.
C
      DO 40 I=1,NST
      DO 41 J=1,NKO
      DO 42 K=1,NKO
C
C                F=MATERIAL BALANCES
C                V=COMPONENT FLOW IN LIQUID PHASE
C
   42 BMAT(I,J,K)=-FLL(I,J)*SL(I)/(FL(I)**2)
      BMAT(I,J,J)=BMAT(I,J,J)+(FL(I)+SL(I))/FL(I)
C
C                F=MATERIAL BALANCES
C                V=TEMPERATURE
C
      BMAT(I,J,NK1)=0.
      DO 43 KK=NK11,NK2
C
C                F=MATERIAL BALANCES
C                V=COMPONENT FLOW IN VAPOR PHASE
C
      K=KK-NK1
   43 BMAT(I,J,KK)=-FVV(I,J)*SV(I)/(FV(I)**2)
      BMAT(I,J,J+NK1)=BMAT(I,J,J+NK1)+(1.+SV(I)/FV(I))
   41 CONTINUE
      DO 44 JJ=NK1,NK21
      J=JJ-NKO
      DO 45 K=1,NKO
C
C                F=EQUILIBRIUM RELATIONSHIPS
C                V=COMPONENT FLOW IN LIQUID PHASE
C
   45 BMAT(I,JJ,K)=MU(I)*FV(I)*FLL(I,J)/FL(I)*(DFK(I,J,K)-XFK(I,J)/FL(I)
     *)
      BMAT(I,J+NKO,J)=BMAT(I,J+NKO,J)+MU(I)*FV(I)*XFK(I,J)/FL(I)
C
C                F=EQUILIBRIUM RELATIONSHIPS
C                V=TEMPERATURE
C
      BMAT(I,JJ,NK1)=MU(I)*FV(I)*FLL(I,J)*DFK(I,J,NK1)/FL(I)
      DO 46 KK=NK11,NK2
C
C                F=EQUILIBRIUM RELATIONSHIPS
C                V=COMPONENT FLOW IN VAPOR PHASE
C
      K=KK-NK1
      BMAT(I,JJ,KK)=MU(I)*FLL(I,J)*(XFK(I,J)+DFK(I,J,KK)*FV(I))/FL(I)
```

```
   46 IF (I.GT.1) BMAT(I,JJ,KK)=BMAT(I,JJ,KK)+(1.-MU(I))*FVV(I-1,J)/FV(I
     *-1)
      BMAT(I,J+NKO,J+NK1)=BMAT(I,J+NKO,J+NK1)-1.
   44 CONTINUE
C
C              NOTE THE SPECIAL EQUATION FOR I=1 AND I=NST
C
      IF (I.EQ.1) GO TO 103
      IF (I.EQ.NST) GO TO 103
      DO 47 K=1,NKO
C
C                  F=ENTHALPY BALANCES
C                  V=COMPONENT FLOW IN LIQUID PHASE
C
   47 BMAT(I,NK2,K)=-HL(I)*SL(I)/(FL(I)**2)+(1.+SL(I)/FL(I))*HLL(I,K)+HH
     *E(I)+(FL(I)+SL(I))*DEXH(I,K)
C
C                  F=ENTHALPY BALANCES
C                  V=TEMPERATURE
C
      BMAT(I,NK2,NK1)=CV(I)*(1.+SV(I)/FV(I))+CL(I)*(1.+SL(I)/FL(I))+(FL(
     *I)+SL(I))*DEXH(I,NK1)
      DO 48 KK=NK11,NK2
C
C                  F=ENTHALPY BALANCES
C                  V=COMPONENT FLOW IN VAPOR PHASE
C
      K=KK-NK1
   48 BMAT(I,NK2,KK)=-HV(I)*SV(I)/(FV(I)**2)+(1.+SV(I)/FV(I))*HVV(I,K)
      GO TO 104
  103 DO 105 K=1,NK2
  105 BMAT(I,NK2,K)=0.
      DO 106 K=1,NKO
  106 BMAT(I,NK2,K)=-1.
  104 CONTINUE
   40 CONTINUE
C
C     DEFINITION OF THE FIRST JACOBIAN MATRIX CONSISTING OF THE PARTIAL
C     DERIVATIVES OF THE I'TH PLATE CORRECTION FUNCTION WITH RESPECT TO
C     THE VARIABLES ON PLATE I-1 (A-MATRIX,I=1)
C     NOTE THAT THE NEXT JACOBIAN MATRIX (I=2,3....) IS DEFINED IN THE
C     DO LOOP NUMBERED 50
C
      DO 74 J=1,NK2
      DO 74 K=1,NK2
   74 AMAT(J,K)=0.
      DO 70 J=1,NKO
C
C                  F=MATERIAL BALANCES
C                  V=COMPONENT FLOW IN VAPOR PHASE
C
   70 AMAT(J,J+NK1)=-1.
C
C                  F=ENTHALPY BALANCES
C                  V=TEMPERATURE
C
      AMAT(NK2,NK1)=-CV(1)
```

```
      YX=(1-MU(2))*FV(2)/(FV(1)**2)
C
C                 F=EQUILIBRIUM RELATIONSHIPS
C                 V=COMPONENT FLOW IN VAPOR PHASE
C
      DO 71 JJ=NK1,NK21
      J=JJ-NKO
      DO 72 KK=NK11,NK2
      K=KK-NK1
   72 AMAT(JJ,KK)=-YX*FVV(1,J)
   71 AMAT(J+NKO,J+NK1)=AMAT(J+NKO,J+NK1)+YX*FV(1)
      DO 73 KK=NK11,NK2
C
C                 F=ENTHALPY BALANCES
C                 V=COMPONENT FLOW IN VAPOR PHASE
C
      K=KK-NK1
   73 AMAT(NK2,KK)=-HVV(1,K)
C
C          NOTE THAT THE SIGN BEFORE D(I,J) IS CHANGED DUE TO THE NEW-
C     TON-RAPHSON EQUATION.
C
      DO 57 I=1,NST
      DO 57 J=1,NK2
   57 D(I,J)=-D(I,J)
C
C          SOLUTION OF THE BLOCK TRIDIAGONAL EQUATION.
C
      DO 50 I=2,NST
      DO 51 J=1,NK2
C
C     THE B-MATRIX AND THE A-MATRIX ARE STORED IN THE MATRIX CM. NOTE
C     THAT BOTH A AND B ARE STORED IN THE TRANSPOSED FORM.
C
      DO 51 K=1,NK2
   51 CM(J,K)=BMAT(I-1,K,J)
      DO 52 K=1,NK2
      KK=K+NK2
      DO 59 J=1,NK2
   59 CM(J,KK)=AMAT(K,J)
   52 CONTINUE
C
C     FORMATION OF THE NEXT A-MATRIX.
C
      DO 84 J=1,NK2
      DO 84 K=1,NK2
   84 AMAT(J,K)=0.
      IF (I.EQ.NST) GO TO 85
      DO 80 J=1,NKO
C
C                 F=MATERIAL BALANCES
C                 V=COMPONENT FLOW IN VAPOR PHASE
C
   80 AMAT(J,J+NK1)=-1.
      DO 81 JJ=NK1,NK21
C
C                 F=EQUILIBRIUM RELATIONSHIPS
C                 V=COMPONENT FLOW IN VAPOR PHASE
```

```
C
      J=JJ-NKO
      DO 82 KK=NK11,NK2
      K=KK-NK1
      X=(1.-MU(I+1))*FV(I+1)/(FV(I)**2)
   82 AMAT(JJ,KK)=-X*FVV(I,J)
   81 AMAT(J+NKO,J+NK1)=AMAT(J+NKO,J+NK1)+X*FV(I)
C
C              NOTE THAT FOR PLATE NST (TOP OF COLUMN),WE HAVE SPECIAL
C     RELATIONS (GO TO 85).
C
      IF (I.EQ.NST-1) GO TO 85
C
C                   F=ENTHALPY BALANCES
C                   V=TEMPERATURE
C
      AMAT(NK2,NK1)=-CV(I)
      DO 83 KK=NK11,NK2
C
C                   F=ENTHALPY BALANCES
C                   V=COMPONENT FLOW IN VAPOR PHASE
C
      K=KK-NK1
   83 AMAT(NK2,KK)=-HVV(I,K)
   85 CONTINUE
C
C     SOLUTION TO    Q=A  *B   (INVERSE).
C                      N-1  N-1
C
C     NOTE THAT THE SOLUTION IS TRANSPOSED.
C
      CALL GAUSL(11,22,NK2,NK2,CM)
      DO 53 J=1,NK2
      DO 53 K=1,NK2
      KK=NK2+K
      CM(K,J)=CM(J,KK)
   53 CONTINUE
C
C     ALLIGNMENT OF THE MATRICES FOR BLOCK ELIMINATION
C
C     NOTE THAT THE JACOBIAN MATRIX, CONSISTING OF THE PARTIAL DERIVATI-
C     VES OF THE I'TH PLATE CORRECTION FUNCTIONS WITH RESPECT TO THE VA-
C     RIABLES ON PLATE I+1, IS NOT STORED, BECAUSE OF ITS SIMPLE STRUC-
C     TURE
C
      DO 54 J=1,NK2
      DO 54 K=1,NKO
      IF (I.EQ.2) BMAT(I,J,K)=BMAT(I,J,K)+CM(J,K)
      IF (I.EQ.2) GO TO 58
      BMAT(I,J,K)=BMAT(I,J,K)+CM(J,K)+CM(J,NK2)*(HLL(I,K)+HHE(I)+FL(I)*D
     *EXH(I,K))
   58 CONTINUE
   54 CONTINUE
      IF (I.EQ.2) GO TO 150
      DO 55 J=1,NK2
   55 BMAT(I,J,NK1)=BMAT(I,J,NK1)+CM(J,NK2)*(CL(I)+FL(I)*DEXH(I,NK1))
  150 CONTINUE
      DO 56 J=1,NK2
```

```
      DO 56 K=1,NK2
   56 D(I,J)=D(I,J)-CM(J,K)*D(I-1,K)
   50 CONTINUE
C
C     CORRECTION OF THE VARIABLES ACCORDING TO THE NEWTON-RAPHSON METHOD
C     THE SIMPLE STRUCTURE OF THE COMPLETE JACOBIAN MATRIX IS UTILIZED.
C
      DO 60 KK=1,NST
      I=NST+1-KK
      DO 61 J=1,NK2
      CM(J,NK2+1)=D(I,J)
      DO 61 K=1,NK2
   61 CM(J,K)=BMAT(I,J,K)
      IF (I-NST) 62,63,63
   62 DO 64 J=1,NKO
   64 CM(J,NK2+1)=CM(J,NK2+1)+D(I+1,J)
      IF (I.EQ.1) GO TO 63
      DO 65 J=1,NKO
   65 CM(NK2,NK2+1)=CM(NK2,NK2+1)+D(I+1,J)*(HLL(I+1,J)+HHE(I+1)+FL(I+1)*
     *DEXH(I+1,J))
      CM(NK2,NK2+1)=CM(NK2,NK2+1)+D(I+1,NK1)*(CL(I+1)+FL(I+1)*DEXH(I+1,N
     *K1))
   63 CALL GAUSL(11,22,NK2,1,CM)
      DO 66 J=1,NK2
   66 D(I,J)=CM(J,NK2+1)
      DO 67 J=1,NKO
C
C     EVALUATION OF THE CORRECTIONS. IF THEY ARE TOO GREAT, THEY ARE
C     SIMPLY THE SPECIFIED LIMITS FOR THE FLOW- AND TEMPERATURE CORREC-
C     TIONS. (FLMAX,FVMAX,TMAX)
C
      QQ=1.
      IF (ABS(D(I,J))-FLMAX(J)) 222,222,223
  223 QQ=FLMAX(J)/ABS(D(I,J))
  222 CONTINUE
   67 FLL(I,J)=FLL(I,J)+QQ*D(I,J)
      QQ=1.
      IF (ABS(D(I,NK1))-TMAX) 233,233,234
  234 QQ=TMAX/ABS(D(I,NK1))
  233 CONTINUE
      T(I)=T(I)+D(I,NK1)*QQ
      DO 68 JJ=NK11,NK2
      J=JJ-NK1
      QQ=1.
      IF (ABS(D(I,JJ))-FVMAX(J)) 243,243,244
  244 QQ=FVMAX(J)/ABS(D(I,JJ))
  243 CONTINUE
   68 FVV(I,J)=FVV(I,J)+D(I,JJ)*QQ
      DO 1121 J=1,NKO
C
C     THE SIGNS OF THE VARIABLES ARE CHECKED.
C
      IF (FLL(I,J)) 121,121,122
  121 FLL(I,J)=0.0001
  122 CONTINUE
      IF (FVV(I,J)) 131,131,132
  131 FVV(I,J)=0.0001
  132 CONTINUE
```

```
 1121 CONTINUE
   60 CONTINUE
      DO 90 I=1,NST
      FL(I)=0.
      FV(I)=0.
      DO 90 J=1,NKO
      FL(I)=FL(I)+FLL(I,J)
   90 FV(I)=FV(I)+FVV(I,J)
C
C     THE COMPONENT FLOWS ARE MODIFIED SO THAT THE TOTAL FLOWS ARE THE
C     SAME AS THOSE KNOWN BEFOREHAND (AROUND TOP AND BOTTOM)
C
      WRITE(6,21)
   21 FORMAT(//,' STAGE    TEMP.    TOTAL FLOW    COMPONENT FLOW    (ALL F
     *LOWS ARE LIQUID FLOWS)',/)
      DO 1021 I=1,NST
      WRITE(6,22) I,T(I),FL(I),(FLL(I,J),J=1,NKO)
 1021 CONTINUE
   22 FORMAT(I4,F9.2,F12.2,5X,5F12.3)
      WRITE(6,323)
  323 FORMAT(//,' STAGE    TEMP.    TOTAL FLOW    COMPONENT FLOW (ALL FLOW
     *S ARE VAPOR FLOWS)',/)
      DO 1022 I=1,NST
      WRITE(6,324)I,T(I),FV(I),(FVV(I,J),J=1,NKO)
 1022 CONTINUE
  324 FORMAT(I4,F9.2,F12.2,5X,5F12.3)
      RES=0.
      DO 501 I=1,NST
C
C     SUM OF SQUARES OF CORRECTIONS TO INDEPENDENT VARIABLES
C
      DO 501 J=1,NK2
  501 RES=RES+D(I,J)**2
      WRITE(6,502) RES
  502 FORMAT(/,'   RESIDUE = ',E12.3)
      IF (ABS(RES)-.1) 600,600,500
  600 DO 612 I=1,NST
      DO 612 J=1,NK
      PROD(1,I,J)=FLL(I,J)*SL(I)/FL(I)
  612 PROD(2,I,J)=FVV(I,J)*SV(I)/FV(I)
      DO 611 J=1,NKO
      PROD(1,1,J)=FLL(1,J)
  611 PROD(2,NST,J)=FVV(NST,J)
      DO 613 K=1,2
      IF (K.EQ.1) WRITE(6,614)
      IF (K.EQ.2) WRITE(6,615)
  614 FORMAT(//,' PRODUCT FLOW, LIQUID PHASE ',/)
  615 FORMAT(//,' PRODUCT FLOW, VAPOR PHASE ',/)
      DO 613 I=1,NST
  613 WRITE(6,605) I,(PROD(K,I,J),J=1,NKO)
  605 FORMAT(I4,7X,8F10.3)
      WRITE(6,603)
  603 FORMAT(////,'   K-FACTOR PROFILE IN COLUMN ',//)
      DO 604 I=1,NST
  604 WRITE(6,605) I,(XFK(I,J),J=1,NKO)
      DO 720 I=1,NST
      XT=0.
      YT=0.
```

```
      DO 721 J=1,NK
      XT=XT+FLL(I,J)
  721 YT=YT+FVV(I,J)
      DO 720 J=1,NK
      FINM(1,I,J)=FLL(I,J)/XT
  720 FINM(2,I,J)=FVV(I,J)/YT
      DO 722 K=1,2
      IF (K.EQ.1) WRITE(6,723)
      IF (K.EQ.2) WRITE(6,724)
  723 FORMAT(//,' MOLEFRACTION PROFILE IN COLUMN (LIQUID PHASE)',/)
  724 FORMAT(//,' MOLEFRACTION PROFILE IN COLUMN (VAPOR PHASE)',/)
      DO 722 I=1,NST
  722 WRITE(6,605) I,(FINM(K,I,J),J=1,NK)
C
C     THE HEAT DUTIES IN REBOILER AND CONDENSER ARE CALCULATED.
C
      CALL ENT(NST,T,FLL,FVV,HLL,HVV,HL,HV,CL,CV)
      QC=HV(NST-1)-HL(NST)-HV(NST)
      QR=HV(1)+HL(1)-HL(2)
      WRITE(6,725) QR
      WRITE(6,726) QC
  725 FORMAT(//,' THE HEAT TO BE SUPPLIED IN THE REBOILER IS   ',E12.4,'
     * (CAL/HR)',/)
  726 FORMAT(//,' THE HEAT TO BE REMOWED BY THE CONDENSER IS   ',E12.4,'
     * (CAL/HR)',/)
      GO TO 333
  555 STOP
      END
```

```
C     ***********************************************************************
C     *
C     *   SUBROUTINES FOR DISTILLATION PROGRAM, FULL VERSION.
C     *
C     ***********************************************************************
C
C
C     ***********************************************************************
C     *
C     *   SUBROUTINE MQUAC(TEMP,X,GAM)
C     *
C     ***********************************************************************
C
C
C     THIS SUBROUTINE CALCULATES THE ACTIVITY COEFFICIENTS IN THE LIQUID
C     PHASE ACCORDING TO THE UNIQUAC MODEL.
C
      SUBROUTINE MQUAC(TEMP,X,GAM)
C
      DIMENSION X(5),GAM(5),THETA(5),PHI(5),THS(5),PAR(5,5)
      COMMON NDIM,NOAC,NOFUG,NODIM,NOEX,NK,Q(5),R(5),XL(5),PARAM(5,5,2),
     *ENP(5,4),BI1(5,5),BI2(5,5),T1,T2,A,BX
      IF (NOAC.LT.1) GO TO 5
      THETS=0.
      PHS=0.
      XLS=0.
      DO 6 I=1,NK
      THETS = THETS + Q(I)*X(I)
      PHS = PHS + R(I)*X(I)
    6 XLS = XLS + XL(I)*X(I)
      DO 7 I=1,NK
      THETA(I) = Q(I)*X(I)/THETS
    7 PHI(I) = X(I)*R(I)/PHS
      DO 8 I=1,NK
      THS(I)=0.
      DO 8 J=1,NK
      PAR(J,I)=PARAM(J,I,1)+TEMP*PARAM(J,I,2)
    8 THS(I)= THS(I) + THETA(J)*PAR(J,I)
      DO 10 I=1,NK
      GA = ALOG(R(I)/PHS) + 5.*Q(I)*ALOG(Q(I)/R(I)*PHS/THETS)
      GA=GA+XL(I)-R(I)/PHS*XLS
      GB = 1. - ALOG(THS(I))
      DO 11 J=1,NK
   11 GB = GB - THETA(J)*PAR(I,J)/THS(J)
   10 GAM(I) = EXP(GA + Q(I)*GB)
      GO TO 4
    5 DO 3 J=1,NK
    3 GAM(J)=1.
    4 CONTINUE
      RETURN
      END
```

```
C     ******************************************************************
C     *
C     *   SUBROUTINE PARIN(NAME)
C     *
C     ******************************************************************
C
C
C     INPUT FOR THE SUBROUTINES MQUAC, FUG, AND ONEDI.
C     FOR FURTHER EXPLANATION, SEE THESE SUBROUTINES.
C
      SUBROUTINE PARIN(NAME)
C
      DIMENSION NAME(40)
      COMMON NDIM,NOAC,NOFUG,NODIM,NOEX,NK,Q(5),R(5),XL(5),PARAM(5,5,2),
     *ENP(5,4),BI1(5,5),BI2(5,5),T1,T2,A,BX
      READ(5,1) NAME
    1 FORMAT(40A2)
      IF (NOAC.LT.1) GO TO 4
      DO 5 I=1,NK
    5 READ(5,2) Q(I),R(I),XL(I)
    2 FORMAT(5X,3F15.5)
      DO 10 K=1,2
      DO 10 I=1,NK
   10 READ(5,3)(PARAM(I,J,K),J=1,NK)
    3 FORMAT(20X,5E12.5)
    4 CONTINUE
      IF (NOFUG.LT.1) GO TO 6
      DO 7 I=1,NK
      READ(5,11) (BI1(I,J),J=1,NK)
    7 READ(5,11)(BI2(I,J),J=1,NK)
   11 FORMAT(10X,5F10.3)
      READ(5,8) T1,T2
    8 FORMAT(2F10.3)
    6 CONTINUE
      IF (NODIM.LT.1) GO TO 9
      READ(5,11) A,BX
    9 CONTINUE
      RETURN
      END
```

```
C     ********************************************************************
C     *
C     *    SUBROUTINE EXCOR(NST,TEM,FLL,DEX)
C     *
C     ********************************************************************
C
C
C     TEMPERATURE IN DEGREES CELSIUS
C
C     THIS SUBROUTINE CALCULATES THE PARTIAL DERIVATIVES OF THE EXCES
C     ENTHALPY WITH RESPECT TO COMPONENT FLOW IN LIQUID PHASE AND TEMPE-
C     RATURE. BOTH ARE EVALUATED NUMERICALLY.
C
      SUBROUTINE EXCOR(NST,TEM,FLL,DEX)
C
      DIMENSION FLM(30,5),FLL(30,5),TEM(30),HE(30),HEX(30),TEMX(30),DEX(
     *30,6)
      COMMON NDIM,NOAC,NOFUG,NODIM,NOEX,NK,Q(5),R(5),XL(5),PARAM(5,5,2),
     *ENP(5,4),BI1(5,5),BI2(5,5),T1,T2,A,BX
      NK1=NK+1
      IF (NOEX.LT.1) GO TO 5
      CALL EXCES(NST,TEM,FLL,HE)
      DO 1 J=1,NK
      DO 2 I=1,NST
      DO 8 K=1,NK
    8 FLM(I,K)=FLL(I,K)
    2 FLM(I,J)=FLL(I,J)+1.
      CALL EXCES(NST,TEM,FLM,HEX)
      DO 1 I=1,NST
    1 DEX(I,J)=HEX(I)-HE(I)
      DO 3 I=1,NST
    3 TEMX(I)=TEM(I)+20.
      CALL EXCES(NST,TEMX,FLL,HEX)
      DO 4 I=1,NST
    4 DEX(I,NK+1)=(HEX(I)-HE(I))/20.
      GO TO 7
    5 DO 6 I=1,NST
      DO 6 J=1,NK1
    6 DEX(I,J)=0.
    7 CONTINUE
      RETURN
      END
```

```
C     ***********************************************************************
C     *
C     *   SUBROUTINE ENT(NST,T,FLL,FVV,HLL,HVV,HL,HV,CL,CV)
C     *
C     ***********************************************************************
C
C
C     THIS SUBROUTINE CALCULATES
C                                 THE PURE COMPONENT ENTHALPY.
C                                 THE TOTAL ENTHALPY OF A PHASE.
C                                 THE PARTIAL DERIVATIVES OF THE TOTAL
C                                 ENTHALPY WITH RESPECT TO TEMPERATURE.
C     THE ENTHALPY IS CALCULATED AFTER
C                                 H=HO+C*T.
C     NOTE THAT IN THE INPUT, THE LIQUID PARAMETERS ARE READ FIRST,
C     THEN THE VAPOR PARAMETERS.
C     THE ENTHALPIES OF EVAPORATION HAVE TO BE REFERRED TO THE SAME
C     TEMPERATURE.
C
      SUBROUTINE ENT(NST,T,FLL,FVV,HLL,HVV,HL,HV,CL,CV)
C
      DIMENSION T(30),FLL(30,5),FVV(30,5),HLL(30,5),HVV(30,5),HL(30),HV(
     *30),CL(30),CV(30)
      COMMON NDIM,NOAC,NOFUG,NODIM,NOEX,NK,Q(5),R(5),XL(5),PARAM(5,5,2),
     *ENP(5,4),BI1(5,5),BI2(5,5),T1,T2,A,BX
      DO 3 I=1,NST
      HL(I)=0.
      HV(I)=0.
      CL(I)=0.
      CV(I)=0.
      DO 3 J=1,NK
      HLL(I,J)=ENP(J,1)+ENP(J,2)*T(I)
      HVV(I,J)=ENP(J,3)+ENP(J,4)*T(I)
      HL(I)=HL(I)+HLL(I,J)*FLL(I,J)
      HV(I)=HV(I)+HVV(I,J)*FVV(I,J)
      CL(I)=CL(I)+ENP(J,2)*FLL(I,J)
    3 CV(I)=CV(I)+ENP(J,4)*FVV(I,J)
      RETURN
      END
```

```
C     *****************************************************************
C     *
C     *    SUBROUTINE EXCES(NST,ATE,FLL,HE)
C     *
C     *****************************************************************
C
C
C     TEMPERATURE IN DEGREES CELSIUS
C
C     THIS SUBROUTINE CALCULATES THE EXCESS ENTHALPY OF A LIQUID MIXTURE
C     THE ENTHALPY OF A MIXTURE IS DEFINED AS THE IDEAL ENTHALPY +THE EX
C     CES ENTHALPY. THE CALCULATION IS BASED ON THE UNIQUAC PARAMETERS.
C
C     NOTE THAT THIS METHOD IS NOT BELIEVED TO ACCURATELY PREDICT EXCESS
C     ENTHALPIES.
C
      SUBROUTINE EXCES(NST,ATE,FLL,HE)
C
      DIMENSION TE(30),X(5),THETA(5),HFAC(5),HHFA(5),HE(30),FLL(30,5),HM
     *(30),ATE(30)
      COMMON NDIM,NOAC,NOFUG,NODIM,NOEX,NK,Q(5),R(5),XL(5),PARAM(5,5,2),
     *ENP(5,4),BI1(5,5),BI2(5,5),T1,T2,A,BX
      IF (NOEX.LT.1) GO TO 7
      DO 11 I=1,NST
      HE(I)=0.
   11 TE(I)=ATE(I)+273.16
      DO 10 K=1,NST
      XT=0.
      DO 5 J=1,NK
    5 XT=XT+FLL(K,J)
      DO 6 J=1,NK
    6 X(J)=FLL(K,J)/XT
      THETS=0.
      DO 1 I=1,NK
    1 THETS=THETS+Q(I)*X(I)
      DO 2 J=1,NK
    2 THETA(J)=Q(J)*X(J)/THETS
      DO 3 I=1,NK
      HFAC(I)=0.
      HHFA(I)=0.
      DO 3 J=1,NK
      HFAC(I)=HFAC(I)+THETA(J)*(PARAM(J,I,1)+PARAM(J,I,2)*TE(K))*ALOG(PA
     *RAM(J,I,1)+PARAM(J,I,2)*TE(K))
    3 HHFA(I)=HHFA(I)+THETA(J)*(PARAM(J,I,1)+PARAM(J,I,2)*TE(K))
      DO 4 I=1,NK
    4 HM(I)=Q(I)*X(I)*HFAC(I)/HHFA(I)
      DO 12 I=1,NK
   12 HE(K)=HE(K)-1.9872*TE(K)*HM(I)
   10 CONTINUE
      GO TO 9
    7 DO 8 I=1,NST
    8 HE(I)=0.
    9 CONTINUE
```

```
RETURN
END
```

```
C     ********************************************************************
C     *
C     *   SUBROUTINE ONEDI(P,ATEM,Y,AKA,ZDI,ZNODI)
C     *
C     ********************************************************************
C
C
C     TEMPERATURE IN KELVIN
C
C     THIS SUBROUTINE CALCULATES THE VAPORPHASE DIMERIZATION CORRECTIONS
C     WHEN NO MORE THAN ONE COMPONENT DIMERIZES.
C     FOR THE DIMERIZING COMPONENT, THE FUGACITY COEFFICIENT IS MODI-
C     FIED BY THE FACTOR ZDI. FOR ALL THE OTHER COMPONENTS, THE FUGACITY
C     COEFFICIENTS ARE MODIFIED BY THE FACTOR ZNODI.
C     FOR FURTHER EXPLANATION, SEE CHAPTER 2
C
C     SUBROUTINE ONEDI(P,ATEM,Y,AKA,ZDI,ZNODI)
C
      COMMON NDIM,NOAC,NOFUG,NODIM,NOEX,NK,Q(5),R(5),XL(5),PARAM(5,5,2),
     *ENP(5,4),BI1(5,5),BI2(5,5),T1,T2,A,BX
      IF (Y-0.0001) 1,1,2
    2 AKA=EXP((2.880814-A+BX/ATEM)*2.302585)
      AKOR=4.*AKA*P*Y*(2.-Y)
      ZDI=(SQRT(1.+AKOR)-1.)*2./AKOR
      ZNODI=2.*(1.-Y+SQRT(1.+AKOR))/(2.-Y)/(1.+SQRT(1.+AKOR))
      GO TO 3
    1 ZDI=1.
      ZNODI=1.
    3 CONTINUE
      RETURN
      END
```

```
C     ****************************************************************
C     *
C     *   SUBROUTINE FUG(TT,P,Y,FI,PAR)
C     *
C     ****************************************************************
C
C
C     TEMPERATURE IN KELVIN
C
C     BI1 AND BI2 ARE SECOND VIRIAL COEFFICIENTS  AT TEMPERATURES T1 AND
C     T2 (KELVIN)
C
C     THIS SUBROUTINE CALCULATES THE FUGACITY COEFFICIENTS IN THE VAPOR
C     PHASE. THE DIMERIZATION EFFECT IS INCLUDED IN THE CALCULATION.
C     THE FUGACITY COEFFICIENTS ARE EVALUATED BY THE MEANS OF THE VIRI-
C     AL EQUATION.
C     FOR FURTHER EXPLANATION, SEE APPENDIX 1.
C
      SUBROUTINE FUG(TT,P,Y,FI,PAR)
C
      DIMENSION Y(5),FIFAC(5),FI(5),B(5,5),PAR(5)
      COMMON NDIM,NOAC,NOFUG,NODIM,NOEX,NK,Q(5),R(5),XL(5),PARAM(5,5,2),
     *ENP(5,4),BI1(5,5),BI2(5,5),T1,T2,A,BX
      IF (NOFUG.LT.1) GO TO 7
      DO 4 I=1,NK
      DO 4 J=1,NK
    4 B(I,J)=BI1(I,J)+(BI2(I,J)-BI1(I,J))/(T2-T1)*(TT-T1)
      BMIX=0.
      DO 1 I=1,NK
      DO 1 J=1,NK
    1 BMIX=BMIX+Y(I)*Y(J)*B(I,J)
      DO 2 J=1,NK
    2 FIFAC(J)=0.
      DO 3 J=1,NK
      DO 3 K=1,NK
    3 FIFAC(J)=FIFAC(J)+Y(K)*B(J,K)
      DO 10 J=1,NK
   10 FI(J)=EXP((2.*FIFAC(J)-BMIX)*P/82.05/TT-B(J,J)*PAR(J)*P/82.05/TT)
      IF (NODIM.LT.1) GO TO 6
      GO TO 8
    7 DO 9 J=1,NK
    9 FI(J)=1.
      IF (NODIM.LT.1) GO TO 6
    8 CONTINUE
      IF (NODIM.GT.0) CALL ONEDI(P,TT,Y(NDIM),AK,ZDIM,ZNOD)
      DO 5 J=1,NK
    5 FI(J)=FI(J)*ZNOD
      FI(NDIM)=ZDIM
    6 CONTINUE
      RETURN
      END
```

```
C     ********************************************************************
C     *
C     *   SUBROUTINE KFAC(NST,NKO,T,ANT,P,FLL,FVV,XKF,DFAC)
C     *
C     ********************************************************************
C
C
C     TEMPERATURE IN DEGREES CELSIUS
C
C     THIS SUBROUTINE CALCULATES THE SEPARATION FACTOR AND THE PARTIAL
C     DERIVATIVES OF THE SEPARATION FACTOR WITH RESPECT TO COMPONENT
C     FLOW IN THE VAPOR PHASE, COMPONENT FLOW IN THE LIQUID PHASE, AND
C     TEMPERATURE.
C     EXCEPT FOR THE TEMPERATURE DEPENDENCE, ALL THE DERIVATIVES ARE
C     EVALUATED NUMERICALLY.
C     IN THE CALCULATION OF DERIVATIVES, IT IS ASSUMED THAT THE TEMPERA-
C     TURE DEPENDENCE OF THE ACTIVITY (LIQUID PHASE) AND FUGACITY (VA-
C     POR PHASE) CAN BE NEGLECTED EXCEPT IN THE CASE OF DIMERIZATION,
C     WHERE THE DERIVATIVE OF THE VAPOR PHASE FUGACITY WITH RESPECT TO
C     TEMPERATURE IS EVALUATED NUMERICALLY.
C
      SUBROUTINE KFAC(NST,NKO,T,ANT,P,FLL,FVV,XKF,DFAC)
C
      DIMENSION XKF(30,5),DFAC(30,5,11)
      DIMENSION T(30),ANT(3,5),FLL(30,5),PAR(5),XX(5),GAM(5),GAMX(5)
      DIMENSION FVV(30,5),YY(5),Y(5),FI(5),FIX(5),FIXM(5)
      COMMON NDIM,NOAC,NOFUG,NODIM,NOEX,NK,Q(5),R(5),XL(5),PARAM(5,5,2),
     *ENP(5,4),BI1(5,5),BI2(5,5),T1,T2,A,BX
      NK1=NKO+1
C
C     CALCULATION OF THE SEPARATION FACTOR.
C
      DO 10 I=1,NST
      XT=0.
      YT=0.
      DO 11 J=1,NKO
      YT=YT+FVV(I,J)
      XT=XT+FLL(I,J)
   11 PAR(J)=EXP(ANT(1,J)-ANT(2,J)/(T(I)+ANT(3,J)))/P
      DO 12 J=1,NKO
      YY(J)=FVV(I,J)/YT
      Y(J)=YY(J)
   12 XX(J)=FLL(I,J)/XT
      CALL MQUAC(T(I)+273.16,XX,GAM)
      IF (NODIM.GT.0) CALL ONEDI(P,T(I)+273.16,Y(NDIM),AKA,ZDIM,ZNODIM)
      IF (NODIM.GT.0) PAR(NDIM)=(SQRT(1.+4.*AKA*PAR(NDIM)*P)-1.)/2./AKA/
     *P
      CALL FUG(T(I)+273.16,P,YY,FI,PAR)
      DO 13 J=1,NKO
   13 XKF(I,J)=PAR(J)*GAM(J)/FI(J)
C
C     IF THE INTEGER PARAMETER NOAC IS LESS THAN ONE ,WE ASSUME, THAT
```

```
C       THE LIQUID PHASE FORMS AN IDEAL MIXTURE. IN THAT CASE THE PARTIAL
C       DERIVATIVES OF THE SEPARATION FACTOR WITH RESPECT TO COMPONENT
C       FLOW IN THE LIQUID PHASE ARE SIMPLY ZERO (GO TO 1).
C
        IF (NOAC.LT.1) GO TO 1
C
C       CALCULATION OF THE DERIVATIVE OF THE SEPARATION FACTOR WITH RE-
C       SPECT TO COMPONENT FLOW IN THE LIQUID PHASE.
C
        DO 14 K=1,NKO
        DO 15 J=1,NKO
   15   XX(J)=FLL(I,J)/(XT+1.)
        XX(K)=(FLL(I,K)+1.)/(XT+1.)
        CALL MQUAC(T(I)+273.16,XX,GAMX)
        DO 14 J=1,NKO
   14   DFAC(I,J,K)=PAR(J)*GAMX(J)/FI(J)-XKF(I,J)
        GO TO 40
    1   DO 3 J=1,NKO
        DO 3 K=1,NK
    3   DFAC(I,J,K)=0.
   40   CONTINUE
C
C       IF THE INTEGER PARAMETER NOFUG IS LESS THAN ONE WE ASSUME, THAT
C       THE VAPOR PHASE FORMS AN IDEAL MIXTURE. IN THAT CASE THE PARTIAL
C       DERIVATIVES OF THE SEPERATION FACTOR WITH RESPECT TO COMPONENT
C       FLOW IN THE VAPOR PHASE ARE SIMPLY ZERO (GO TO 41).
C
        IF (NOFUG.LT.1) GO TO 41
C
C       CALCULATION OF THE DERIVATIVE OF THE SEPARATION FACTOR WITH RE-
C       SPECT TO COMPONENT FLOW IN THE VAPOR PHASE.
C
        DO 16 K=1,NK
        DO 17 J=1,NK
   17   YY(J)=FVV(I,J)/(YT+1.)
        YY(K)=(FVV(I,K)+1.)/(YT+1.)
        CALL FUG(T(I)+273.16,P,YY,FIX,PAR)
        DO 16 J=1,NK
   16   DFAC(I,J,K+1+NK)=PAR(J)*GAM(J)/FIX(J)-XKF(I,J)
        GO TO 42
   41   DO 43 J=1,NKO
        DO 43 K=1,NKO
   43   DFAC(I,J,K+NKO+1)=0.
   42   CONTINUE
C
C       CALCULATION OF THE DERIVATIVE OF THE SEPARATION FACTOR WITH RE-
C       SPECT TO TEMPERATURE.
C
        DO 18 J=1,NKO
   18   DFAC(I,J,NKO+1)=XKF(I,J)*ANT(2,J)/(T(I)+ANT(3,J))**2
C
C       IF THE INTEGER PARAMETER NODIM IS LESS THAN ONE WE ASSUME ,THAT WE
C       HAVE NO DIMERIZATION.IN THAT CASE THERE IS NO CORRECTION MADE FOR
C       THE DERIVATIVE.
C
        IF (NODIM.GT.0) CALL ONEDI(P,T(I)+278.16,Y(NDIM),AKAS,ZDIMM,ZNO)
        IF (NODIM.GT.0) PAR(NDIM)=EXP(ANT(1,NDIM)-ANT(2,NDIM)/(T(I)+5.+ANT
       *(3,NDIM)))
```

```
      IF (NODIM.GT.0) PAR(NDIM)=(SQRT(1.+4.*AKAS*PAR(NDIM))-1.)/2./AKAS/
     *P
      IF (NODIM.GT.0) CALL FUG(T(I)+278.16,P,Y,FIXM,PAR)
C
C     CALCULATION OF THE DERIVATIVE OF THE SEPARATION FACTOR WITH RE-
C     SPECT TO TEMPERATURE IN CASE, THERE IS A DIMERIZATING COMPONENT.
C
      IF (NODIM.GT.0) DFAC(I,NDIM,NK1)=(PAR(NDIM)*GAM(NDIM)/FIXM(NDIM)-X
     *KF(I,NDIM))/5.
   10 CONTINUE
      RETURN
      END
```

```
*************************************************************************
*
*    SAMPLE INPUT (FULL VERSION)
*
*************************************************************************

     5|
     4|   10|    15|    20|    25|
     2|    2|     2|     2|     4|
ETHANOL (1) - N-PROPANOL (2) - WATER (3) - ACETIC ACID (4)
     1        1.97199        2.10550         -0.43799
     2        2.51200        2.77990         -0.44039
     3        1.40000        0.92000         -2.32000
  2  4 4      2.07200        2.20240         -0.55039
  1  1 353.   393.       0.10000E 01 0.12273E 01 0.19283E 00-0.21691E 00
  2  1 353.   393.       0.77161E 00 0.10000E 01 0.87732E 00 0.33463E-01
  3  1 353.   393.       0.18128E 01 0.20593E 00 0.10000E 01-0.67430E-01
  4  1 353.   393.       0.41135E 01 0.27263E 01 0.29816E 01 0.10000E 01
  1  2 353.   393.       0.00000E 00-0.20330E-03 0.90927E-03 0.13950E-02
  2  2 353.   393.       0.13147E-03 0.00000E 00-0.24844E-03 0.11243E-02
  3  2 353.   393.      -0.11735E-02 0.96916E-03 0.00000E 00 0.13032E-02
  4  2 353.   393.      -0.49680E-02-0.25139E-02-0.37084E-02 0.00000E 00
 353.   1     -920.083    -477.755   -223.862   -616.148
 373.   1     -727.402    -418.917   -194.041   -539.903
 353.   2     -477.755   -1160.935   -252.530   -763.736
 373.   2     -418.917    -931.554   -219.091   -668.275
 353.   3     -223.862    -252.530   -652.368   -323.959
 373.   3     -194.041    -219.091   -479.629   -604.094
 353.   4     -326.105    -394.054   -183.889  -1160.955
 373.   4     -285.070    -344.738   -159.758   -931.554
       353.       373.
         8      10.108      3018.
   8.04494     1554.3       222.65
   7.99733     1569.7       209.50
   7.96681     1668.2       228.0
   7.18807     1416.7       211.
  -3032.32       37.904     7903.72      18.676
  -2812.80       35.160     7722.40      26.770
    -1440.       18.        9230.62       8.705
   -2246.4       28.09      4374.75      17.920
```

```
         5    10   15   20   25
    → 30|   2|   1|   0|   2|
         40.       2.        1.       100.      80.
      2
      7      .7
         85.        .         .       12.5     12.5      25.
     23
        110.       1.       25.       12.5     12.5
     10
         20.
    →15    1    0    0    1
         50.       2.        1.        85.     105.
      2
 1.
      7
85.       0.       25.       25.       25.       25.
      0
```

NOTE THAT TWO COLUMN CONFIGURATIONS ARE SPECIFIED
(INDICATED BY ARROWS)

```
****************************************************************
*
*   SAMPLE OUTPUT (FULL VERSION)
*
****************************************************************
```

1 ETHANOL (1) - N-PROPANOL (2) - WATER (3) - ACETIC ACID (4)

    INCLUDING ACTIVITY-COEFFICIENTS

    INCLUDING FUGACITY-COEFFICIENTS

    INCLUDING DIMERIZATION

    INCLUDING EXCES-ENTHALPY

FEED COMPOSITION AND ANTOINE PARAMETERS

1	25.000	8.045	1554.300	222.650
2	25.000	7.997	1569.700	209.500
3	25.000	7.967	1668.210	228.000
4	25.000	7.188	1416.700	211.000

    NUMBER OF STAGES   30
    DISTILLATE   40.000
    REFLUX RATIO   2.000000
    TOTAL PRESSURE   1.000

FLOW CONFIGURATION

I	FL	FV	SL	SV	FKV	FEED STREAMS		
1	40.0	70.0	0.0	0.0	0.0	0.0	0.0	0.0	0.0
2	110.0	70.0	0.0	0.0	0.0	0.0	0.0	0.0	0.0
3	110.0	70.0	0.0	0.0	0.0	0.0	0.0	0.0	0.0
4	110.0	70.0	0.0	0.0	0.0	0.0	0.0	0.0	0.0
5	110.0	70.0	0.0	0.0	0.0	0.0	0.0	0.0	0.0
6	110.0	70.0	0.0	0.0	0.0	0.0	0.0	0.0	0.0
7	110.0	70.0	0.0	0.0	0.0	0.0	12.5	12.5	25.0
8	60.0	70.0	0.0	0.0	0.0	0.0	0.0	0.0	0.0
9	60.0	70.0	0.0	0.0	0.0	0.0	0.0	0.0	0.0
10	60.0	70.0	20.0	0.0	0.0	0.0	0.0	0.0	0.0
11	80.0	70.0	0.0	0.0	0.0	0.0	0.0	0.0	0.0
12	80.0	70.0	0.0	0.0	0.0	0.0	0.0	0.0	0.0
13	80.0	70.0	0.0	0.0	0.0	0.0	0.0	0.0	0.0
14	80.0	70.0	0.0	0.0	0.0	0.0	0.0	0.0	0.0
15	80.0	70.0	0.0	0.0	0.0	0.0	0.0	0.0	0.0

16	80.0	70.0	0.0	0.0	0.0	0.0	0.0	0.0	0.0
17	80.0	70.0	0.0	0.0	0.0	0.0	0.0	0.0	0.0
18	80.0	70.0	0.0	0.0	0.0	0.0	0.0	0.0	0.0
19	80.0	70.0	0.0	0.0	0.0	0.0	0.0	0.0	0.0
20	80.0	70.0	0.0	0.0	0.0	0.0	0.0	0.0	0.0
21	80.0	70.0	0.0	0.0	0.0	0.0	0.0	0.0	0.0
22	80.0	70.0	0.0	0.0	0.0	0.0	0.0	0.0	0.0
23	80.0	120.0	0.0	0.0	1.0	25.0	12.5	12.5	0.0
24	80.0	120.0	0.0	0.0	0.0	0.0	0.0	0.0	0.0
25	80.0	120.0	0.0	0.0	0.0	0.0	0.0	0.0	0.0
26	80.0	120.0	0.0	0.0	0.0	0.0	0.0	0.0	0.0
27	80.0	120.0	0.0	0.0	0.0	0.0	0.0	0.0	0.0
28	80.0	120.0	0.0	0.0	0.0	0.0	0.0	0.0	0.0
29	80.0	120.0	0.0	0.0	0.0	0.0	0.0	0.0	0.0
30	80.0	40.0	0.0	0.0	0.0	0.0	0.0	0.0	0.0

STAGE	TEMP.	TOTAL FLOW	COMPONENT FLOW	(ALL FLOWS ARE LIQUID FLOWS)		
1	90.00	47.14	0.000	13.469	3.258	30.415
2	90.67	124.67	2.678	39.530	30.994	51.465
3	91.33	122.78	6.846	37.675	33.196	45.062
4	92.00	121.50	9.128	36.479	33.968	41.924
5	92.67	120.85	10.399	35.795	34.239	40.417
6	93.33	120.50	11.044	35.427	34.340	39.689
7	94.00	120.22	11.259	35.245	34.392	39.329
8	92.51	68.22	10.594	22.280	21.420	13.922
9	90.75	67.33	12.184	22.623	21.589	10.939
10	89.43	66.59	13.555	22.692	21.456	8.890
11	88.57	86.06	18.831	29.626	27.735	9.871
12	87.92	85.70	19.902	29.493	27.360	8.949
13	87.40	85.46	20.814	29.349	26.973	8.319
14	86.96	85.27	21.620	29.209	26.589	7.853
15	86.58	85.12	22.358	29.078	26.209	7.478
16	86.23	85.00	23.056	28.955	25.832	7.153
17	85.92	84.88	23.742	28.837	25.448	6.853
18	85.62	84.77	24.442	28.718	25.046	6.564
19	85.33	84.66	25.185	28.590	24.612	6.277
20	85.05	84.56	26.006	28.443	24.125	5.987
21	84.77	84.46	26.945	28.263	23.561	5.689
22	84.48	84.36	28.058	28.033	22.886	5.380
23	84.67	84.27	29.425	27.729	22.056	5.057
24	85.33	85.39	30.093	28.219	24.855	2.227
25	86.00	84.65	30.583	27.840	26.224	0.000
26	86.67	86.07	31.214	27.197	27.660	0.000
27	87.33	87.53	32.171	26.173	29.187	0.000
28	88.00	89.07	33.804	24.530	30.738	0.000
29	88.67	90.75	36.866	21.822	32.060	0.000
30	89.33	92.65	43.027	17.181	32.439	0.000

STAGE	TEMP.	TOTAL FLOW	COMPONENT FLOW	(ALL FLOWS	ARE VAPOR	FLOWS)
1	90.00	84.67	9.821	26.061	27.736	21.049
2	90.67	82.78	13.989	24.206	29.938	14.647
3	91.33	81.50	16.271	23.010	30.710	11.508
4	92.00	80.85	17.542	22.326	30.981	10.001
5	92.67	80.50	18.187	21.958	31.082	9.273
6	93.33	80.22	18.402	21.776	31.134	8.913
7	94.00	78.22	17.737	21.311	30.662	8.507
8	92.51	77.33	19.326	21.654	30.830	5.524
9	90.75	76.59	20.697	21.723	30.698	3.475
10	89.43	76.06	22.005	21.642	30.374	2.041
11	88.57	75.70	23.075	21.509	29.999	1.119
12	87.92	75.46	23.988	21.365	29.612	0.490
13	87.40	75.27	24.794	21.225	29.228	0.024
14	86.96	75.47	25.531	21.094	28.849	0.000
15	86.58	75.67	26.229	20.971	28.471	0.000
16	86.23	75.86	26.915	20.853	28.087	0.000
17	85.92	76.04	27.616	20.734	27.686	0.000
18	85.62	76.22	28.359	20.606	27.251	0.000
19	85.33	76.40	29.179	20.459	26.764	0.000
20	85.05	76.60	30.118	20.279	26.201	0.000
21	84.77	76.81	31.232	20.049	25.525	0.000
22	84.48	77.04	32.599	19.745	24.695	0.000
23	84.67	131.00	58.267	32.736	39.994	0.000
24	85.33	132.48	58.756	32.357	41.363	0.000
25	86.00	133.90	59.387	31.713	42.800	0.000
26	86.67	135.36	60.344	30.690	44.326	0.000
27	87.33	136.90	61.978	29.046	45.877	0.000
28	88.00	138.58	65.039	26.338	47.199	0.000
29	88.67	140.48	71.200	21.697	47.578	0.000
30	89.33	47.83	28.173	4.516	15.139	0.000

RESIDUE = 0.544E 05

STAGE	TEMP.	TOTAL FLOW	COMPONENT FLOW	(ALL FLOWS	ARE LIQUID	FLOWS)
1	100.00	40.00	0.388	11.757	3.221	24.634
2	100.67	122.56	1.216	41.993	17.267	62.084
3	98.90	117.69	1.786	41.115	25.291	49.499
4	96.65	114.25	2.608	39.858	30.467	41.313
5	95.34	112.19	3.254	38.823	33.735	36.377
6	94.59	110.97	3.668	38.050	35.809	33.444
7	94.18	110.24	3.878	37.506	37.165	31.696
8	89.04	58.13	3.813	24.046	24.783	5.492
9	87.24	57.19	4.548	24.716	26.236	1.694
10	86.32	56.79	5.213	24.755	26.820	0.000
11	85.99	76.95	8.176	32.897	35.879	0.000
12	85.82	77.07	9.052	32.439	35.577	0.000
13	85.71	77.06	10.086	31.900	35.070	0.000
14	85.59	77.01	11.308	31.296	34.405	0.000
15	85.46	76.99	12.682	30.659	33.654	0.000

16	85.30	77.00	14.221	29.975	32.804	0.000
17	85.12	77.02	15.922	29.241	31.855	0.000
18	84.93	77.05	17.777	28.460	30.809	0.000
19	84.72	77.08	19.776	27.631	29.671	0.000
20	84.50	77.11	21.909	26.758	28.447	0.000
21	84.26	77.16	24.164	25.848	27.145	0.000
22	84.01	77.20	26.521	24.906	25.774	0.000
23	83.85	77.26	28.965	23.946	24.352	0.000
24	83.87	79.79	29.026	22.456	28.306	0.000
25	83.97	79.82	29.197	20.867	29.753	0.000
26	84.06	79.80	29.538	19.405	30.852	0.000
27	84.15	79.80	30.277	18.000	31.524	0.000
28	84.22	79.83	31.720	16.520	31.585	0.000
29	84.22	79.88	34.331	14.766	30.783	0.000
30	84.16	80.00	38.675	12.483	28.842	0.000

STAGE	TEMP.	TOTAL FLOW	COMPONENT FLOW (ALL FLOWS ARE VAPOR FLOWS)			
1	100.00	82.56	0.828	30.237	14.046	37.450
2	100.67	77.69	1.397	29.358	22.070	24.866
3	98.90	74.25	2.220	28.102	27.245	16.680
4	96.65	72.19	2.866	27.066	30.514	11.743
5	95.34	70.97	3.279	26.294	32.587	8.810
6	94.59	70.24	3.489	25.749	33.943	7.062
7	94.18	68.13	3.425	24.789	34.061	5.859
8	89.04	67.19	4.160	25.460	35.514	2.060
9	87.24	66.69	4.825	25.499	36.099	0.264
10	86.32	66.95	5.617	25.192	36.144	0.000
11	85.99	67.07	6.492	24.734	35.842	0.000
12	85.82	67.06	7.527	24.195	35.335	0.000
13	85.71	67.01	8.749	23.591	34.670	0.000
14	85.59	67.00	10.122	22.954	33.919	0.000
15	85.46	67.00	11.662	22.270	33.069	0.000
16	85.30	67.02	13.363	21.536	32.120	0.000
17	85.12	67.05	15.217	20.755	31.074	0.000
18	84.93	67.08	17.216	19.926	29.936	0.000
19	84.72	67.12	19.350	19.053	28.712	0.000
20	84.50	67.16	21.604	18.143	27.410	0.000
21	84.26	67.20	23.962	17.201	26.039	0.000
22	84.01	67.26	26.405	16.241	24.617	0.000
23	83.85	119.79	51.467	27.251	41.071	0.000
24	83.87	119.82	51.638	25.662	42.518	0.000
25	83.97	119.80	51.978	24.200	43.617	0.000
26	84.06	119.80	52.717	22.795	44.289	0.000
27	84.15	119.83	54.161	21.315	44.350	0.000
28	84.22	119.88	56.771	19.561	43.548	0.000
29	84.22	120.00	61.116	17.278	41.607	0.000
30	84.16	40.00	22.441	4.795	12.765	0.000

RESIDUE = 0.108E 05

STAGE	TEMP.	TOTAL FLOW	COMPONENT FLOW	(ALL FLOWS	ARE LIQUID	FLOWS)
1	108.46	40.00	0.234	11.438	3.583	24.745
2	102.52	122.20	1.360	41.067	19.735	60.043
3	99.41	117.86	1.669	41.046	25.631	49.512
4	97.17	114.52	1.941	40.161	30.630	41.786
5	95.70	112.40	2.175	39.140	34.326	36.756
6	94.81	111.13	2.375	38.267	36.795	33.691
7	94.29	110.40	2.550	37.617	38.329	31.900
8	88.62	58.42	2.654	24.088	25.937	5.740
9	87.00	57.68	3.700	24.676	27.008	2.301
10	86.24	57.26	4.899	24.626	27.010	0.725
11	85.94	77.10	8.052	32.803	35.901	0.348
12	85.78	77.06	9.499	32.243	35.191	0.124
13	85.64	77.05	10.981	31.615	34.411	0.044
14	85.52	77.06	12.513	30.944	33.588	0.016
15	85.38	77.08	14.109	30.238	32.725	0.006
16	85.24	77.10	15.777	29.499	31.821	0.002
17	85.08	77.12	17.521	28.728	30.873	0.001
18	84.90	77.15	19.344	27.927	29.879	0.000
19	84.71	77.18	21.245	27.098	28.839	0.000
20	84.51	77.21	23.218	26.243	27.753	0.000
21	84.29	77.25	25.254	25.371	26.627	0.000
22	84.06	77.29	27.337	24.487	25.466	0.000
23	83.83	77.33	29.448	23.603	24.282	0.000
24	83.76	79.79	29.607	22.045	28.138	0.000
25	83.66	79.78	29.939	20.590	29.255	0.000
26	83.54	79.79	30.587	19.294	29.908	0.000
27	83.37	79.81	31.712	18.024	30.070	0.000
28	83.13	79.84	33.529	16.622	29.689	0.000
29	82.77	79.90	36.328	14.914	28.658	0.000
30	82.26	80.00	40.450	12.728	26.822	0.000

STAGE	TEMP.	TOTAL FLOW	COMPONENT FLOW	(ALL FLOWS	ARE VAPOR	FLOWS)
1	108.46	82.20	1.126	29.629	16.152	35.298
2	102.52	77.86	1.435	29.608	22.048	24.767
3	99.41	74.52	1.707	28.723	27.047	17.041
4	97.17	72.40	1.941	27.702	30.742	12.011
5	95.70	71.13	2.141	26.829	33.211	8.947
6	94.81	70.40	2.316	26.179	34.746	7.155
7	94.29	68.42	2.420	25.150	34.854	5.995
8	88.62	67.68	3.466	25.738	35.925	2.556
9	87.00	67.26	4.665	25.688	35.927	0.980
10	86.24	67.10	6.108	25.265	35.384	0.348
11	85.94	67.06	7.555	24.704	34.674	0.124
12	85.78	67.05	9.037	24.076	33.894	0.044
13	85.64	67.06	10.569	23.405	33.071	0.016
14	85.52	67.08	12.165	22.699	32.208	0.006
15	85.38	67.10	13.833	21.961	31.304	0.002

16	85.24	67.12	15.577	21.190	30.356	0.001
17	85.08	67.15	17.400	20.389	29.362	0.000
18	84.90	67.18	19.301	19.559	28.321	0.000
19	84.71	67.21	21.274	18.705	27.236	0.000
20	84.51	67.25	23.310	17.832	26.109	0.000
21	84.29	67.29	25.392	16.949	24.949	0.000
22	84.06	67.33	27.503	16.064	23.764	0.000
23	83.83	119.79	52.663	27.007	40.120	0.000
24	83.76	119.78	52.994	25.551	41.238	0.000
25	83.66	119.79	53.643	24.256	41.891	0.000
26	83.54	119.81	54.767	22.985	42.053	0.000
27	83.37	119.84	56.585	21.584	41.671	0.000
28	83.13	119.90	59.383	19.876	40.640	0.000
29	82.77	120.00	63.505	17.690	38.805	0.000
30	82.26	40.00	23.056	4.962	11.983	0.000

RESIDUE = 0.317E 03

STAGE	TEMP.	TOTAL FLOW	COMPONENT FLOW	(ALL FLOWS ARE LIQUID FLOWS)		
1	107.50	40.00	0.248	11.434	3.571	24.746
2	102.12	121.88	1.300	41.073	20.372	59.136
3	99.15	117.51	1.614	40.888	26.324	48.683
4	96.99	114.28	1.884	39.987	31.197	41.215
5	95.59	112.26	2.115	38.997	34.746	36.400
6	94.75	111.05	2.314	38.169	37.087	33.482
7	94.26	110.36	2.492	37.560	38.524	31.782
8	88.59	58.41	2.603	24.064	26.062	5.678
9	86.98	57.69	3.662	24.681	27.068	2.275
10	86.23	57.27	4.868	24.639	27.035	0.727
11	85.94	77.12	8.014	32.824	35.927	0.353
12	85.77	77.07	9.463	32.264	35.215	0.126
13	85.64	77.06	10.947	31.637	34.432	0.044
14	85.51	77.07	12.484	30.966	33.605	0.016
15	85.38	77.09	14.085	30.260	32.738	0.005
16	85.23	77.11	15.758	29.520	31.829	0.002
17	85.07	77.13	17.509	28.748	30.877	0.001
18	84.89	77.16	19.339	27.944	29.879	0.000
19	84.70	77.19	21.247	27.112	28.834	0.000
20	84.50	77.23	23.227	26.254	27.746	0.000
21	84.29	77.26	25.269	25.377	26.617	0.000
22	84.06	77.30	27.358	24.490	25.455	0.000
23	83.83	77.34	29.474	23.602	24.268	0.000
24	83.75	79.80	29.653	22.038	28.111	0.000
25	83.64	79.80	30.007	20.574	29.215	0.000
26	83.51	79.80	30.681	19.268	29.853	0.000
27	83.33	79.82	31.829	17.987	30.001	0.000
28	83.08	79.85	33.659	16.581	29.609	0.000
29	82.70	79.91	36.444	14.881	28.581	0.000
30	82.15	80.00	40.513	12.713	26.774	0.000

STAGE	TEMP.	TOTAL FLOW	COMPONENT FLOW	(ALL FLOWS	ARE VAPOR	FLOWS)
1	107.50	81.88	1.052	29.639	16.800	34.389
2	102.12	77.51	1.365	29.454	22.753	23.936
3	99.15	74.28	1.636	28.553	27.625	16.469
4	96.99	72.26	1.867	27.563	31.174	11.654
5	95.59	71.05	2.066	26.735	33.516	8.736
6	94.75	70.36	2.244	26.126	34.953	7.036
7	94.26	68.41	2.355	25.130	34.991	5.932
8	88.59	67.69	3.414	25.747	35.997	2.529
9	86.98	67.27	4.620	25.705	35.964	0.980
10	86.23	67.12	6.066	25.285	35.414	0.353
11	85.94	67.07	7.515	24.725	34.702	0.126
12	85.77	67.06	8.999	24.098	33.920	0.044
13	85.64	67.07	10.535	23.428	33.092	0.016
14	85.51	67.09	12.137	22.721	32.225	0.006
15	85.38	67.11	13.810	21.981	31.317	0.002
16	85.23	67.13	15.561	21.209	30.364	0.001
17	85.07	67.16	17.391	20.406	29.366	0.000
18	84.89	67.19	19.299	19.573	28.322	0.000
19	84.70	67.23	21.279	18.715	27.233	0.000
20	84.50	67.26	23.321	17.838	26.104	0.000
21	84.29	67.30	25.410	16.951	24.942	0.000
22	84.06	67.34	27.526	16.063	23.755	0.000
23	83.83	119.80	52.705	26.999	40.098	0.000
24	83.75	119.80	53.059	25.535	41.202	0.000
25	83.64	119.80	53.732	24.229	41.840	0.000
26	83.51	119.82	54.881	22.949	41.988	0.000
27	83.33	119.85	56.711	21.543	41.596	0.000
28	83.08	119.91	59.496	19.842	40.568	0.000
29	82.70	120.00	63.564	17.674	38.761	0.000
30	82.15	40.00	23.052	4.961	11.987	0.000

RESIDUE =   0.887E 01

STAGE	TEMP.	TOTAL FLOW	COMPONENT FLOW	(ALL FLOWS	ARE LIQUID	FLOWS)
1	107.50	40.00	0.248	11.433	3.572	24.746
2	102.14	121.92	1.298	41.071	20.346	59.202
3	99.17	117.54	1.612	40.896	26.287	48.742
4	97.00	114.30	1.883	40.002	31.160	41.257
5	95.60	112.27	2.114	39.013	34.717	36.426
6	94.76	111.06	2.314	38.182	37.067	33.497
7	94.26	110.36	2.492	37.570	38.511	31.789
8	88.59	58.41	2.603	24.070	26.055	5.682
9	86.98	57.69	3.662	24.685	27.064	2.277
10	86.23	57.27	4.868	24.641	27.034	0.727
11	85.94	77.12	8.014	32.825	35.926	0.353
12	85.77	77.07	9.463	32.265	35.215	0.126
13	85.64	77.06	10.947	31.637	34.433	0.045
14	85.51	77.07	12.484	30.967	33.606	0.016
15	85.38	77.09	14.085	30.260	32.739	0.006

16	85.23	77.11	15.758	29.520	31.830	0.002
17	85.07	77.14	17.509	28.748	30.878	0.001
18	84.89	77.16	19.339	27.944	29.879	0.000
19	84.70	77.19	21.247	27.112	28.835	0.000
20	84.50	77.23	23.227	26.254	27.747	0.000
21	84.29	77.26	25.269	25.377	26.618	0.000
22	84.06	77.30	27.358	24.490	25.455	0.000
23	83.83	77.34	29.474	23.602	24.268	0.000
24	83.75	79.80	29.653	22.038	28.111	0.000
25	83.64	79.80	30.007	20.574	29.216	0.000
26	83.51	79.80	30.680	19.268	29.854	0.000
27	83.33	79.82	31.828	17.987	30.002	0.000
28	83.07	79.85	33.659	16.582	29.609	0.000
29	82.70	79.91	36.443	14.881	28.581	0.000
30	82.15	80.00	40.512	12.714	26.775	0.000

STAGE	TEMP.	TOTAL FLOW	COMPONENT FLOW (ALL FLOWS ARE VAPOR FLOWS)			
1	107.50	81.92	1.050	29.637	16.774	34.456
2	102.14	77.54	1.363	29.462	22.715	23.996
3	99.17	74.30	1.634	28.568	27.588	16.511
4	97.00	72.27	1.866	27.579	31.145	11.680
5	95.60	71.06	2.065	26.749	33.495	8.751
6	94.76	70.36	2.244	26.136	34.939	7.043
7	94.26	68.41	2.355	25.137	34.983	5.935
8	88.59	67.69	3.414	25.752	35.992	2.531
9	86.98	67.27	4.620	25.708	35.962	0.981
10	86.23	67.12	6.065	25.286	35.414	0.353
11	85.94	67.07	7.515	24.726	34.702	0.126
12	85.77	67.06	8.999	24.099	33.920	0.045
13	85.64	67.07	10.535	23.428	33.093	0.016
14	85.51	67.09	12.136	22.721	32.226	0.006
15	85.38	67.11	13.810	21.982	31.317	0.002
16	85.23	67.14	15.561	21.209	30.365	0.001
17	85.07	67.16	17.391	20.406	29.367	0.000
18	84.89	67.19	19.298	19.573	28.322	0.000
19	84.70	67.23	21.278	18.715	27.234	0.000
20	84.50	67.26	23.320	17.839	26.105	0.000
21	84.29	67.30	25.410	16.951	24.942	0.000
22	84.06	67.35	27.526	16.064	23.755	0.000
23	83.83	119.80	52.705	26.999	40.099	0.000
24	83.75	119.80	53.058	25.535	41.203	0.000
25	83.64	119.80	53.732	24.229	41.841	0.000
26	83.51	119.82	54.880	22.949	41.989	0.000
27	83.33	119.85	56.710	21.543	41.596	0.000
28	83.07	119.91	59.495	19.843	40.569	0.000
29	82.70	120.00	63.563	17.675	38.762	0.000
30	82.15	40.00	23.052	4.961	11.987	0.000

RESIDUE =   0.346E-01

PRODUCT FLOW, LIQIUD PHASE

1	0.248	11.433	3.572	24.746
2	0.0	0.0	0.0	0.0
3	0.0	0.0	0.0	0.0
4	0.0	0.0	0.0	0.0
5	0.0	0.0	0.0	0.0
6	0.0	0.0	0.0	0.0
7	0.0	0.0	0.0	0.0
8	0.0	0.0	0.0	0.0
9	0.0	0.0	0.0	0.0
10	1.700	8.605	9.441	0.254
11	0.0	0.0	0.0	0.0
12	0.0	0.0	0.0	0.0
13	0.0	0.0	0.0	0.0
14	0.0	0.0	0.0	0.0
15	0.0	0.0	0.0	0.0
16	0.0	0.0	0.0	0.0
17	0.0	0.0	0.0	0.0
18	0.0	0.0	0.0	0.0
19	0.0	0.0	0.0	0.0
20	0.0	0.0	0.0	0.0
21	0.0	0.0	0.0	0.0
22	0.0	0.0	0.0	0.0
23	0.0	0.0	0.0	0.0
24	0.0	0.0	0.0	0.0
25	0.0	0.0	0.0	0.0
26	0.0	0.0	0.0	0.0
27	0.0	0.0	0.0	0.0
28	0.0	0.0	0.0	0.0
29	0.0	0.0	0.0	0.0
30	0.0	0.0	0.0	0.0

PRODUCT FLOW, VAPOR PHASE

1	0.0	0.0	0.0	0.0
2	0.0	0.0	0.0	0.0
3	0.0	0.0	0.0	0.0
4	0.0	0.0	0.0	0.0
5	0.0	0.0	0.0	0.0
6	0.0	0.0	0.0	0.0
7	0.0	0.0	0.0	0.0
8	0.0	0.0	0.0	0.0
9	0.0	0.0	0.0	0.0
10	0.0	0.0	0.0	0.0
11	0.0	0.0	0.0	0.0
12	0.0	0.0	0.0	0.0
13	0.0	0.0	0.0	0.0
14	0.0	0.0	0.0	0.0
15	0.0	0.0	0.0	0.0

16	0.0	0.0	0.0	0.0
17	0.0	0.0	0.0	0.0
18	0.0	0.0	0.0	0.0
19	0.0	0.0	0.0	0.0
20	0.0	0.0	0.0	0.0
21	0.0	0.0	0.0	0.0
22	0.0	0.0	0.0	0.0
23	0.0	0.0	0.0	0.0
24	0.0	0.0	0.0	0.0
25	0.0	0.0	0.0	0.0
26	0.0	0.0	0.0	0.0
27	0.0	0.0	0.0	0.0
28	0.0	0.0	0.0	0.0
29	0.0	0.0	0.0	0.0
30	23.052	4.961	11.987	0.000

K-FACTOR PROFILE IN COLUMN

1	2.064	1.266	2.294	0.680
2	1.843	1.151	1.981	0.539
3	1.741	1.110	1.809	0.445
4	1.666	1.087	1.674	0.375
5	1.617	1.077	1.580	0.328
6	1.589	1.074	1.520	0.299
7	1.572	1.074	1.483	0.281
8	1.286	0.937	1.211	0.167
9	1.205	0.895	1.142	0.122
10	1.173	0.870	1.111	0.100
11	1.168	0.858	1.101	0.092
12	1.170	0.849	1.096	0.089
13	1.175	0.840	1.092	0.088
14	1.180	0.832	1.089	0.087
15	1.185	0.822	1.085	0.088
16	1.189	0.812	1.081	0.088
17	1.192	0.801	1.076	0.088
18	1.194	0.790	1.071	0.088
19	1.196	0.777	1.066	0.089
20	1.196	0.763	1.060	0.089
21	1.195	0.749	1.054	0.089
22	1.193	0.735	1.048	0.088
23	1.190	0.720	1.042	0.088
24	1.195	0.753	0.988	0.089
25	1.199	0.766	0.960	0.090
26	1.202	0.774	0.938	0.090
27	1.203	0.775	0.920	0.090
28	1.201	0.768	0.902	0.090
29	1.193	0.749	0.885	0.090
30	1.178	0.718	0.866	0.089

MOLEFRACTION PROFILE IN COLUMN (LIQUID PHASE)

1	0.006	0.286	0.089	0.619
2	0.011	0.337	0.167	0.486
3	0.014	0.348	0.224	0.415
4	0.016	0.350	0.273	0.361
5	0.019	0.347	0.309	0.324
6	0.021	0.344	0.334	0.302
7	0.023	0.340	0.349	0.288
8	0.045	0.412	0.446	0.097
9	0.063	0.428	0.469	0.039
10	0.085	0.430	0.472	0.013
11	0.104	0.426	0.466	0.005
12	0.123	0.419	0.457	0.002
13	0.142	0.411	0.447	0.001
14	0.162	0.402	0.436	0.000
15	0.183	0.393	0.425	0.000
16	0.204	0.383	0.413	0.000
17	0.227	0.373	0.400	0.000
18	0.251	0.362	0.387	0.000
19	0.275	0.351	0.374	0.000
20	0.301	0.340	0.359	0.000
21	0.327	0.328	0.345	0.000
22	0.354	0.317	0.329	0.000
23	0.381	0.305	0.314	0.000
24	0.372	0.276	0.352	0.000
25	0.376	0.258	0.366	0.000
26	0.384	0.241	0.374	0.000
27	0.399	0.225	0.376	0.000
28	0.422	0.208	0.371	0.000
29	0.456	0.186	0.358	0.000
30	0.506	0.159	0.335	0.000

MOLEFRACTION PROFILE IN COLUMN (VAPOR PHASE)

1	0.013	0.362	0.205	0.421
2	0.018	0.380	0.293	0.309
3	0.022	0.384	0.371	0.222
4	0.026	0.382	0.431	0.162
5	0.029	0.376	0.471	0.123
6	0.032	0.371	0.497	0.100
7	0.034	0.367	0.511	0.087
8	0.050	0.380	0.532	0.037
9	0.069	0.382	0.535	0.015
10	0.090	0.377	0.528	0.005
11	0.112	0.369	0.517	0.002
12	0.134	0.359	0.506	0.001
13	0.157	0.349	0.493	0.000
14	0.181	0.339	0.480	0.000
15	0.206	0.328	0.467	0.000

16	0.232	0.316	0.452	0.000
17	0.259	0.304	0.437	0.000
18	0.287	0.291	0.422	0.000
19	0.317	0.278	0.405	0.000
20	0.347	0.265	0.388	0.000
21	0.378	0.252	0.371	0.000
22	0.409	0.239	0.353	0.000
23	0.440	0.225	0.335	0.000
24	0.443	0.213	0.344	0.000
25	0.449	0.202	0.349	0.000
26	0.458	0.192	0.350	0.000
27	0.473	0.180	0.347	0.000
28	0.496	0.165	0.338	0.000
29	0.530	0.147	0.323	0.000
30	0.576	0.124	0.300	0.000

THE HEAT TO BE SUPPLIED IN THE REBOILER IS   0.6664E 06   (CAL/HR)

THE HEAT TO BE REMOWED BY THE CONDENSER IS   0.7706E 06   (CAL/HR)

NEW COLUMN CONFIGURATION

FEED COMPOSITION AND ANTOINE PARAMETERS

1	25.000	8.045	1554.300	222.650
2	25.000	7.997	1569.700	209.500
3	25.000	7.967	1668.210	228.000
4	25.000	7.188	1416.700	211.000

NUMBER OF STAGES	15
DISTILLATE	50.000
REFLUX RATIO	2.000000
TOTAL PRESSURE	1.000

FLOW CONFIGURATION

I	FL	FV	SL	SV	FKV	FEED STREAMS			
1	50.0	150.0	0.0	0.0	0.0	0.0	0.0	0.0	0.0
2	200.0	150.0	0.0	0.0	0.0	0.0	0.0	0.0	0.0
3	200.0	150.0	0.0	0.0	0.0	0.0	0.0	0.0	0.0
4	200.0	150.0	0.0	0.0	0.0	0.0	0.0	0.0	0.0
5	200.0	150.0	0.0	0.0	0.0	0.0	0.0	0.0	0.0
6	200.0	150.0	0.0	0.0	0.0	0.0	0.0	0.0	0.0
7	200.0	150.0	0.0	0.0	0.0	25.0	25.0	25.0	25.0
8	100.0	150.0	0.0	0.0	0.0	0.0	0.0	0.0	0.0
9	100.0	150.0	0.0	0.0	0.0	0.0	0.0	0.0	0.0
10	100.0	150.0	0.0	0.0	0.0	0.0	0.0	0.0	0.0
11	100.0	150.0	0.0	0.0	0.0	0.0	0.0	0.0	0.0
12	100.0	150.0	0.0	0.0	0.0	0.0	0.0	0.0	0.0
13	100.0	150.0	0.0	0.0	0.0	0.0	0.0	0.0	0.0
14	100.0	150.0	0.0	0.0	0.0	0.0	0.0	0.0	0.0
15	100.0	50.0	0.0	0.0	0.0	0.0	0.0	0.0	0.0

STAGE	TEMP.	TOTAL FLOW	COMPONENT FLOW	(ALL FLOWS ARE LIQUID FLOWS)		
1	105.11	50.00	3.421	16.227	0.030	30.322
2	97.47	227.17	33.192	83.024	28.099	82.856
3	93.17	216.44	37.270	83.541	39.911	55.720
4	91.00	210.96	39.656	80.739	48.759	41.803
5	89.67	208.36	41.643	76.273	55.284	35.159
6	88.53	207.28	44.103	71.195	59.651	32.331
7	88.03	207.00	47.603	66.121	61.876	31.396
8	85.67	106.13	27.041	36.121	36.694	6.273
9	84.33	105.51	28.222	36.458	40.832	0.000
10	83.00	109.69	29.046	35.700	44.946	0.000
11	81.67	112.42	29.621	33.940	48.856	0.000
12	80.33	113.84	30.183	31.395	52.264	0.000
13	79.06	114.20	31.139	28.344	54.719	0.000
14	79.33	113.78	33.098	25.022	55.659	0.000
15	79.74	112.85	36.894	21.535	54.420	0.000

STAGE	TEMP.	TOTAL FLOW	COMPONENT FLOW	(ALL FLOWS	ARE VAPOR	FLOWS)
1	105.11	177.17	29.771	66.797	28.069	52.533
2	97.47	166.44	33.849	67.314	39.881	25.398
3	93.17	160.96	36.235	64.512	48.729	11.481
4	91.00	158.36	38.222	60.046	55.254	4.837
5	89.67	157.28	40.682	54.968	59.621	2.009
6	88.53	157.00	44.182	49.894	61.846	1.074
7	88.03	156.13	48.620	44.894	61.665	0.951
8	85.67	160.83	49.801	45.230	65.803	0.000
9	84.33	165.01	50.625	44.473	69.916	0.000
10	83.00	167.74	51.200	42.713	73.827	0.000
11	81.67	169.16	51.762	40.168	77.234	0.000
12	80.33	169.52	52.718	37.117	79.689	0.000
13	79.06	169.10	54.677	33.795	80.629	0.000
14	79.33	168.17	58.473	30.308	79.390	0.000
15	79.74	55.32	21.579	8.773	24.970	0.000

RESIDUE =   0.713E 05

STAGE	TEMP.	TOTAL FLOW	COMPONENT FLOW	(ALL FLOWS	ARE LIQUID	FLOWS)
1	101.27	50.00	3.725	15.590	5.685	25.000
2	93.20	210.78	26.410	71.136	53.823	59.407
3	88.90	201.36	31.881	63.484	72.310	33.683
4	87.83	199.94	35.012	56.109	81.426	27.397
5	87.87	200.36	37.739	52.641	82.588	27.396
6	87.97	200.70	41.275	51.921	79.480	28.029
7	87.97	200.86	45.681	51.827	74.992	28.361
8	85.20	100.28	25.388	25.979	45.480	3.430
9	84.56	99.90	25.808	25.373	48.152	0.566
10	84.39	99.87	25.902	25.787	48.097	0.082
11	84.30	99.88	26.296	26.512	47.060	0.011
12	84.21	99.91	27.138	26.878	45.891	0.001
13	84.12	99.96	28.555	26.482	44.920	0.000
14	83.90	99.94	30.904	25.138	43.902	0.000
15	83.52	100.00	35.119	22.751	42.130	0.000

STAGE	TEMP.	TOTAL FLOW	COMPONENT FLOW	(ALL FLOWS	ARE VAPOR	FLOWS)
1	101.27	160.78	22.685	55.546	48.137	34.40
2	93.20	151.36	28.156	47.894	66.625	8.68
3	88.90	149.94	31.287	40.519	75.740	2.39
4	87.83	150.36	34.015	37.051	76.903	2.39
5	87.87	150.70	37.550	36.331	73.795	3.02
6	87.97	150.86	41.956	36.237	69.306	3.36
7	87.97	150.28	46.664	35.389	64.795	3.43
8	85.20	149.90	47.083	34.783	67.467	0.56
9	84.56	149.87	47.177	35.197	67.412	0.08
10	84.39	149.88	47.571	35.922	66.374	0.01
11	84.30	149.91	48.414	36.288	65.206	0.00
12	84.21	149.96	49.830	35.892	64.234	0.00
13	84.12	149.94	52.180	34.548	63.217	0.00
14	83.90	150.00	56.394	32.161	61.445	0.00
15	83.52	50.00	21.275	9.410	19.315	0.00

RESIDUE =   0.162E 05

STAGE	TEMP.	TOTAL FLOW	COMPONENT FLOW	(ALL FLOWS ARE LIQUID FLOWS)		
1	102.38	50.00	2.946	15.736	6.318	25.000
2	95.04	214.73	21.715	76.064	49.205	67.744
3	90.53	203.84	28.553	70.764	65.296	39.231
4	89.03	201.21	33.573	64.405	72.803	30.433
5	88.59	200.83	37.802	60.068	74.296	28.666
6	88.36	200.82	42.011	57.158	73.272	28.380
7	88.16	200.86	46.403	54.846	71.277	28.337
8	85.34	100.24	25.719	27.583	43.629	3.304
9	84.67	99.86	26.353	27.198	45.798	0.507
10	84.55	99.86	26.653	27.088	46.046	0.070
11	84.50	99.86	27.058	26.918	45.878	0.009
12	84.44	99.87	27.797	26.577	45.499	0.001
13	84.31	99.89	29.152	25.921	44.820	0.000
14	84.08	99.93	31.635	24.696	43.599	0.000
15	83.64	100.00	36.149	22.468	41.383	0.000

STAGE	TEMP.	TOTAL FLOW	COMPONENT FLOW (ALL FLOWS ARE VAPOR FLOWS)			
1	102.38	164.73	18.769	60.328	42.887	42.744
2	95.04	153.84	25.608	55.028	58.977	14.231
3	90.53	151.21	30.627	48.669	66.484	5.433
4	89.03	150.83	34.856	44.333	67.977	3.665
5	88.59	150.82	39.065	41.422	66.953	3.380
6	88.36	150.86	43.457	39.111	64.958	3.337
7	88.16	150.24	47.773	36.847	62.311	3.304
8	85.34	149.86	48.407	36.462	64.480	0.507
9	84.67	149.86	48.707	36.353	64.727	0.070
10	84.55	149.86	49.112	36.183	64.560	0.009
11	84.50	149.87	49.851	35.842	64.180	0.001
12	84.44	149.89	51.206	35.186	63.502	0.000
13	84.31	149.93	53.689	33.960	62.281	0.000
14	84.08	150.00	58.203	31.732	60.064	0.000
15	83.64	50.00	22.054	9.264	18.682	0.000

RESIDUE = 0.143E 04

STAGE	TEMP.	TOTAL FLOW	COMPONENT FLOW	(ALL FLOWS ARE LIQUID FLOWS)		
1	102.38	50.00	2.947	15.733	6.320	25.000
2	95.07	214.70	21.546	76.135	49.450	67.564
3	90.66	203.98	28.159	71.031	65.317	39.471
4	89.15	201.28	33.139	64.909	72.594	30.640
5	88.67	200.86	37.505	60.598	73.980	28.775
6	88.41	200.83	41.874	57.557	72.969	28.429
7	88.19	200.87	46.355	55.092	71.063	28.355
8	85.36	100.23	25.700	27.712	43.514	3.308
9	84.68	99.85	26.327	27.289	45.732	0.508
10	84.56	99.86	26.639	27.151	45.995	0.070
11	84.51	99.86	27.053	26.967	45.835	0.009
12	84.44	99.87	27.788	26.611	45.474	0.001
13	84.32	99.89	29.141	25.941	44.811	0.000
14	84.08	99.93	31.626	24.706	43.598	0.000
15	83.64	100.00	36.144	22.473	41.383	0.000

STAGE	TEMP.	TOTAL FLOW	COMPONENT FLOW (ALL FLOWS ARE VAPOR FLOWS)			
1	102.38	164.70	18.599	60.402	43.131	42.564
2	95.07	153.98	25.212	55.298	58.997	14.471
3	90.66	151.28	30.192	49.176	66.274	5.640
4	89.15	150.86	34.558	44.865	67.661	3.775
5	88.67	150.83	38.927	41.824	66.649	3.428
6	88.41	150.87	43.408	39.359	64.744	3.355
7	88.19	150.23	47.753	36.979	62.194	3.308
8	85.36	149.85	48.380	36.555	64.412	0.508
9	84.68	149.86	48.692	36.418	64.676	0.070
10	84.56	149.86	49.106	36.234	64.515	0.009
11	84.51	149.87	49.842	35.878	64.154	0.001
12	84.44	149.89	51.195	35.208	63.491	0.000
13	84.32	149.93	53.679	33.973	62.278	0.000
14	84.08	150.00	58.197	31.739	60.064	0.000
15	83.64	50.00	22.053	9.267	18.680	0.000

RESIDUE = 0.377E 01

STAGE	TEMP.	TOTAL FLOW	COMPONENT FLOW (ALL FLOWS ARE LIQUID FLOWS)			
1	102.38	50.00	2.947	15.733	6.320	25.000
2	95.07	214.70	21.548	76.136	49.449	67.563
3	90.66	203.98	28.161	71.030	65.317	39.471
4	89.15	201.28	33.139	64.906	72.596	30.642
5	88.67	200.86	37.504	60.595	73.982	28.776
6	88.41	200.83	41.874	57.555	72.970	28.429
7	88.19	200.87	46.355	55.091	71.065	28.355
8	85.36	100.23	25.700	27.711	43.515	3.308
9	84.68	99.85	26.327	27.287	45.733	0.508
10	84.56	99.86	26.639	27.150	45.997	0.070
11	84.51	99.86	27.052	26.966	45.837	0.009
12	84.44	99.87	27.788	26.611	45.475	0.001
13	84.32	99.89	29.141	25.941	44.812	0.000
14	84.08	99.93	31.626	24.706	43.598	0.000
15	83.64	100.00	36.143	22.473	41.384	0.000

STAGE	TEMP.	TOTAL FLOW	COMPONENT FLOW (ALL FLOWS ARE VAPOR FLOWS)			
1	102.38	164.70	18.601	60.402	43.130	42.563
2	95.07	153.98	25.214	55.297	58.998	14.471
3	90.66	151.28	30.192	49.173	66.276	5.642
4	89.15	150.86	34.558	44.862	67.663	3.776
5	88.67	150.83	38.927	41.822	66.651	3.429
6	88.41	150.87	43.408	39.357	64.745	3.355
7	88.19	150.23	47.753	36.978	62.196	3.308
8	85.36	149.85	48.380	36.554	64.414	0.508
9	84.68	149.86	48.692	36.416	64.678	0.070
10	84.56	149.86	49.105	36.233	64.517	0.009
11	84.51	149.87	49.841	35.877	64.156	0.001
12	84.44	149.89	51.194	35.207	63.493	0.000
13	84.32	149.93	53.679	33.973	62.279	0.000
14	84.08	150.00	58.196	31.739	60.064	0.000
15	83.64	50.00	22.053	9.267	18.680	0.000

RESIDUE = 0.177E-03

PRODUCT FLOW, LIQIUD PHASE

1	2.947	15.733	6.320	25.000
2	0.0	0.0	0.0	0.0
3	0.0	0.0	0.0	0.0
4	0.0	0.0	0.0	0.0
5	0.0	0.0	0.0	0.0
6	0.0	0.0	0.0	0.0
7	0.0	0.0	0.0	0.0
8	0.0	0.0	0.0	0.0
9	0.0	0.0	0.0	0.0
10	0.0	0.0	0.0	0.0
11	0.0	0.0	0.0	0.0
12	0.0	0.0	0.0	0.0
13	0.0	0.0	0.0	0.0
14	0.0	0.0	0.0	0.0
15	0.0	0.0	0.0	0.0

PRODUCT FLOW, VAPOR PHASE

1	0.0	0.0	0.0	0.0
2	0.0	0.0	0.0	0.0
3	0.0	0.0	0.0	0.0
4	0.0	0.0	0.0	0.0
5	0.0	0.0	0.0	0.0
6	0.0	0.0	0.0	0.0
7	0.0	0.0	0.0	0.0
8	0.0	0.0	0.0	0.0
9	0.0	0.0	0.0	0.0
10	0.0	0.0	0.0	0.0
11	0.0	0.0	0.0	0.0
12	0.0	0.0	0.0	0.0
13	0.0	0.0	0.0	0.0
14	0.0	0.0	0.0	0.0
15	22.053	9.267	18.680	0.000

K-FACTOR PROFILE IN COLUMN

1	1.916	1.166	2.072	0.517
2	1.632	1.013	1.664	0.299
3	1.446	0.933	1.368	0.193
4	1.391	0.922	1.244	0.164
5	1.382	0.919	1.200	0.159
6	1.380	0.910	1.181	0.157
7	1.377	0.897	1.170	0.156
8	1.259	0.882	0.990	0.103
9	1.232	0.889	0.942	0.092
10	1.228	0.889	0.935	0.090
11	1.228	0.886	0.933	0.090
12	1.228	0.882	0.930	0.090
13	1.227	0.873	0.926	0.090
14	1.226	0.856	0.918	0.090
15	1.220	0.825	0.903	0.090

MOLEFRACTION PROFILE IN COLUMN (LIQUID PHASE)

1	0.059	0.315	0.126	0.500
2	0.100	0.355	0.230	0.315
3	0.138	0.348	0.320	0.194
4	0.165	0.322	0.361	0.152
5	0.187	0.302	0.368	0.143
6	0.209	0.287	0.363	0.142
7	0.231	0.274	0.354	0.141
8	0.256	0.276	0.434	0.033
9	0.264	0.273	0.458	0.005
10	0.267	0.272	0.461	0.001
11	0.271	0.270	0.459	0.000
12	0.278	0.266	0.455	0.000
13	0.292	0.260	0.449	0.000
14	0.316	0.247	0.436	0.000
15	0.361	0.225	0.414	0.000

MOLEFRACTION PROFILE IN COLUMN (VAPOR PHASE)

1	0.113	0.367	0.262	0.258
2	0.164	0.359	0.383	0.094
3	0.200	0.325	0.438	0.037
4	0.229	0.297	0.449	0.025
5	0.258	0.277	0.442	0.023
6	0.288	0.261	0.429	0.022
7	0.318	0.246	0.414	0.022
8	0.323	0.244	0.430	0.003
9	0.325	0.243	0.432	0.000
10	0.328	0.242	0.431	0.000
11	0.333	0.239	0.428	0.000
12	0.342	0.235	0.424	0.000
13	0.358	0.227	0.415	0.000
14	0.388	0.212	0.400	0.000
15	0.441	0.185	0.374	0.000

THE HEAT TO BE SUPPLIED IN THE REBOILER IS   0.1454E 07   (CAL/HR)

THE HEAT TO BE REMOWED BY THE CONDENSER IS   0.9686E 06   (CAL/HR)

```
C     ******************************************************************
C     *
C     *   MAIN DISTILLATION PROGRAM, SIMPLIFIED VERSION
C     *
C     ******************************************************************
C
C
      DIMENSION FLMAX(5)
      DIMENSION PROD(2,30,5)
      DIMENSION FEED(6),FL(30),FV(30),XKF(30,5),FLL(30,5),T(30)
      DIMENSION AMAT(5,6),BMAT(30,6,6),D(30,6)
      REAL*8 CM(6,11)
      DIMENSION SL(30),SV(30),FKV(30),FSTR(30,6),DFAC(30,5,6)
      DIMENSION ANT(3,5),NAVN(40)
      COMMON NK,Q(5),R(5),XL(5),PARAM(5,5,2)
      COMMON NOAC
C
C     MAXIMUM CHANGE IN TEMPERATURE BETWEEN ITERATIONS
C
      TMAX=10.
C
C     NUMBER OF COMPONENTS
C
      READ(5,8) NK
C
C     IF NOAC IS GREATER THAN ZERO, ACTIVITY COEFFICIENTS ARE CALCULATED
C     OTHERWISE THEY ARE EQUATED TO ONE
C
      READ(5,900) NOAC
  900 FORMAT(I5)
    8 FORMAT(16I5)
      NKO=NK
C
C     READ UNIQUAC PARAMETERS
C
      CALL PARIN(NAVN)
      WRITE(6,12) NAVN
   12 FORMAT(1H1,40A2)
      IF (NOAC.GT.0) WRITE(6,901)
  901 FORMAT(//,' INCLUDING ACTIVITY COEFFICIENTS',/)
      DO 17 J=1,NK
C
C     READ ANTOINE CONSTANTS
C
   17 READ(5,13)(ANT(K,J),K=1,3)
      DO 19 J=1,NK
      ANT(1,J)=2.302585*(ANT(1,J)-2.880814)
   19 ANT(2,J)=2.302585*ANT(2,J)
C
C     READ COLUMN CONFIGURATION
C
  333 READ(5,8) NST,NFEED,NSL,NSV,NCON
      IF (NCON .EQ. 0) GO TO 555
```

```
C
C      TT IS THE GUESSED TOP TEMPERATURE, TB THE BOTTOM TEMPERATURE
C
       READ(5,13) DEST,RFLX,P,TT,TB
   13  FORMAT(8F10.5)
       NK1=NK+1
       DO 200 I=1,NST
       SL(I)=0.
       SV(I)=0.
       FKV(I)=0.
       FSTR(I,NK1)=0.
       DO 200 J=1,NK
  200  FSTR(I,J)=0.
       DO 201 I=1,NFEED
       READ(5,8) NF
C
C      READ FEED SPECIFICATIONS
C
       READ(5,13)FKV(NF),(FSTR(NF,J),J=1,NK)
       DO 201 J=1,NK
  201  FSTR(NF,NK1)=FSTR(NF,NK1)+FSTR(NF,J)
       IF (NSL.EQ.0) GO TO 203
       DO 202 I=1,NSL
       READ(5,8) NF
  202  READ(5,13) SL(NF)
  203  IF (NSV.EQ.0) GO TO 204
       DO 205 I=1,NSV
       READ(5,8) NF
  205  READ(5,13) SV(NF)
C
C      ESTABLISH TOTAL VAPOR AND LIQUID FLOWS.
C
  204  FV(NST)=DEST
       FL(NST)=DEST*RFLX
       FV(NST-1)=FL(NST)-FSTR(NST,NK1)+SV(NST)+SL(NST)+DEST
       DO 210 II=3,NST
       IF (NST .LE. 2) GO TO 210
       I=NST+2-II
       FL(I)=FL(I+1)-SL(I)+(1.-FKV(I))*FSTR(I,NK1)
       FV(I-1)=FV(I)+SV(I)-FKV(I)*FSTR(I,NK1)
  210  CONTINUE
       FL(1)=FL(2)-SL(1)+(1.-FKV(1))*FSTR(1,NK1)
       FL(1)=FL(1)-FV(1)
       DO 211 J=1,NK1
       FEED(J)=0.
       DO 211 I=1,NST
  211  FEED(J)=FEED(J)+FSTR(I,J)
       WRITE(6,18)
   18  FORMAT(/,' FEED COMPOSITION AND ANTOINE PARAMETERS ',//)
       WRITE(6,23)(J,FEED(J),ANT(1,J),ANT(2,J),ANT(3,J),J=1,NK)
   23  FORMAT(I3,F12.3,5X,3F12.3)
C
C      FLMAX IS THE MAXIMAL ALLOWED CHANGE IN COMPONENT FLOW RATES
C      BETWEEN ITERATIONS
C
       DO 24 J=1,NK
   24  FLMAX(J)=FEED(J)*RFLX
       WRITE(6,16)NST,DEST,RFLX,P
```

```
   16 FORMAT(//,' NUMBER OF STAGES         ',I9/,' DISTILLATE
     1',F12.3/,' REFLUX RATIO             ',F15.6/,' TOTAL PRESSURE
     1 ',F12.3//)
      WRITE(6,215)
  215 FORMAT(//,' FLOW CONFIGURATION ',//,'  I        FL       FV       SL
     1       SV       FKV      FEED STREAMS',/)
      DO 216 I=1,NST
  216 WRITE(6,217) I,FL(I),FV(I),SL(I),SV(I),FKV(I),(FSTR(I,J),J=1,NK)
  217 FORMAT(I3,3X,10F8.1)
C
C     ESTABLISH INITIAL GUESS OF TEMPERATURE AND COMPONENT FLOW PROFILES
C
      DO 26 I=1,NST
   26 T(I)=TB+(I-1)*(TT-TB)/NST
      DO 3 I=1,NST
      DO 3 J=1,NK
    3 FLL(I,J)=FEED(J)/FEED(NK1)*FL(I)
  500 CALL KFAC(NST,NK,T,ANT,P,FLL,XKF,DFAC)
C
C     SET UP MATERIAL BALANCES AND CALCULATE DISCREPANCY FUNCTIONS
C
      DO 30 I=1,NST
      D(I,NK1)=FL(I)
      DO 31 J=1,NKO
      D(I,NK1)=D(I,NK1)-FLL(I,J)
      D(I,J)=FSTR(I,J)-(1.+(SL(I)+XKF(I,J)*(SV(I)+FV(I)))/FL(I))*FLL(I,J
     1)
      IF (I.GT.1) D(I,J)=D(I,J)+XKF(I-1,J)*FV(I-1)/FL(I-1)*FLL(I-1,J)
      IF (I.LT.NST) D(I,J)=D(I,J)+FLL(I+1,J)
   31 CONTINUE
C
C     DEFINITION OF THE JACOBIAN MATRIX
C
      DO 41 J=1,NK
      DO 42 K=1,NK1
   42 BMAT(I,J,K)=FLL(I,J)*DFAC(I,J,K)*(SV(I)+FV(I))/FL(I)
      BMAT(I,J,J)=BMAT(I,J,J)+1.+(SL(I)+XKF(I,J)*(SV(I)+FV(I)))/FL(I)
   41 BMAT(I,NK1,J)=1.
   30 BMAT(I,NK1,NK1)=0.
C
C     BLOCK ELIMINATION OF THE TRIDIAGONAL SYSTEM OF EQUATIONS
C
      DO 43 J=1,NK
      DO 44 K=1,NK1
   44 AMAT(J,K)=FLL(1,J)*DFAC(1,J,K)*FV(1)/FL(1)
   43 AMAT(J,J)=AMAT(J,J)+XKF(1,J  )*FV(1)/FL(1)
      DO 50 I=2,NST
      DO 51 J=1,NK1
      DO 51 K=1,NK1
   51 CM(J,K)=BMAT(I-1,K,J)
      DO 52 K=1,NK
      KK=K+NK1
      DO 59 J=1,NK1
      CM(J,KK)=AMAT(K,J)
   59 AMAT(K,J)=FLL(I,K)*DFAC(I,K,J)*FV(I)/FL(I)
   52 AMAT(K,K)=AMAT(K,K)+XKF(I,K  )*FV(I)/FL(I)
      CALL GAUSL(6,11,NK1,NK,CM)
      DO 53 J=1,NK1
```

```
          DO 53 K=1,NK
          KK=NK1+K
          CM(K,J)=CM(J,KK)
       53 CONTINUE
          DO 54 J=1,NKO
          DO 54 K=1,NKO
       54 BMAT(I,J,K)=BMAT(I,J,K)-CM(J,K)
          DO 55 J=1,NK
          DO 55 K=1,NK1
       55 D(I,J)=D(I,J)+CM(J,K)*D(I-1,K)
       50 CONTINUE
C
C     CORRECTION OF THE VARIABLES ACCORDING TO THE NEWTON-RAPHSON METHOD
C
          DO 60 KK=1,NST
          I=NST+1-KK
          DO 61 J=1,NK1
          CM(J,NK1+1)=D(I,J)
          DO 61 K=1,NK1
       61 CM(J,K)=BMAT(I,J,K)
          IF (I-NST) 62,63,63
       62 DO 64 J=1,NK
       64 CM(J,NK1+1)=CM(J,NK1+1)+D(I+1,J)
       63 CALL GAUSL(6,11,NK1,1,CM)
          DO 66 J=1,NK1
       66 D(I,J)=CM(J,NK1+1)
          DO 67 J=1,NK
C
C     EVALUATION OF THE DEVIATIONS. IF THEY ARE TOO GREAT, THEY ARE
C     EQUATED TO THEIR MAXIMUM ALLOWED VALUES. DETERMINATION OF NEW
C     VALUES FOR THE INDEPENDENT VARIABLES.
C
          QQ=1.
          IF (ABS(D(I,J))-FLMAX(J)) 222,222,223
      223 QQ=FLMAX(J)/ABS(D(I,J))
      222 CONTINUE
       67 FLL(I,J)=FLL(I,J)+QQ*D(I,J)
          QQ=1.
          IF (ABS(D(I,NK1))-TMAX) 233,233,234
      234 QQ=TMAX/ABS(D(I,NK1))
      233 CONTINUE
       60 T(I)=T(I)+D(I,NK1)*QQ
          DO 120 I=1,NST
C
C     CHECK SIGN
C
          DO 123 J=1,NKO
          IF (FLL(I,J)) 121,121,122
      121 FLL(I,J)=0.
      122 IF (FLL(I,J)-FL(I)) 123,123,124
      124 FLL(I,J)=FL(I)
      123 CONTINUE
      120 CONTINUE
C
C     INTERMEDIATE OUTPUT
C
          WRITE(6,21)
       21 FORMAT(//,' STAGE    TEMP.    TOTAL FLOW    COMPONENT FLOW'/)
```

```
      DO 6 I=1,NST
    6 WRITE(6,22) I,T(I),FL(I),(FLL(I,J),J=1,NK)
   22 FORMAT(I4,F9.2,F12.2,5X,5F12.3)
      RES=0.
C
C     SUM OF SQUARES OF CORRECTIONS TO INDEPENDENT VARIABLES
C
      DO 501 I=1,NST
      DO 501 J=1,NK1
  501 RES=RES+D(I,J)**2
      WRITE(6,502) RES
  502 FORMAT(/,'   RESIDUE = ',E12.3)
      IF (ABS(RES)-1.0) 600,600,500
  600 DO 612 I=1,NST
      DO 612 J=1,NK
      PROD(1,I,J)=FLL(I,J)*SL(I)/FL(I)
  612 PROD(2,I,J)=FLL(I,J)*XKF(I,J)/FL(I)*SV(I)
      DO 611 J=1,NK
      PROD(1,1,J)=FLL(1,J)
  611 PROD(2,NST,J)=FLL(NST,J)*XKF(NST,J)*FV(NST)/FL(NST)
C
C     FINAL OUTPUT
C
      DO 613 K=1,2
      IF (K.EQ. 1) WRITE(6,614)
      IF (K.EQ.2) WRITE(6,615)
  614 FORMAT(//,' PRODUCT FLOW, LIQUID PHASE ',/)
  615 FORMAT(//,' PRODUCT FLOW, VAPOR PHASE ',/)
      DO 613 I=1,NST
  613 WRITE(6,605)I,(PROD(K,I,J),J=1,NK)
      WRITE(6,603)
  603 FORMAT(////,'   K-FACTOR PROFILE IN COLUMN ',//)
      DO 604 I=1,NST
  604 WRITE(6,605)I,(XKF(I,J),J=1,NK)
  605 FORMAT(I4,7X,8F10.3)
      GO TO 333
  555 STOP
      END
```

```
C     ****************************************************************
C     *
C     *  SUBROUTINES FOR DISTILLATION PROGRAM, SIMPLIFIED VERSION
C     *
C     ****************************************************************
C
C
C     ****************************************************************
C     *
C     *  SUBROUTINE PARIN(NAME)
C     *
C     ****************************************************************
C
C
C     READ INPUT FOR UNIQUAC MODEL
C
      SUBROUTINE PARIN(NAME)
C
      DIMENSION NAME(40)
      COMMON NK,Q(5),R(5),XL(5),PARAM(5,5,2)
      COMMON NOAC
      READ(5,1) NAME
    1 FORMAT(40A2)
      IF (NOAC.LT.1) GO TO 200
      DO 5 I=1,NK
    5 READ(5,2) Q(I),R(I),XL(I)
    2 FORMAT(5X,3F15.5)
      DO 10 K=1,2
      DO 10 I=1,NK
   10 READ(5,3)(PARAM(I,J,K),J=1,NK)
    3 FORMAT(20X,5E12.5)
  200 CONTINUE
      RETURN
      END
```

```fortran
C     ****************************************************************
C     *
C     *   SUBROUTINE MQUAC(TEMP,X,GAM)
C     *
C     ****************************************************************
C
C
C     CALCULATION OF ACTIVITY COEFFICIENTS FROM THE UNIQUAC MODEL
C     TEMPERATURE IN DEG. K
C
      SUBROUTINE MQUAC(TEMP,X,GAM)
C
      DIMENSION X(5),GAM(5),THETA(5),PHI(5),THS(5),PAR(5,5)
      COMMON NK,Q(5),R(5),XL(5),PARAM(5,5,2)
      COMMON NOAC
      IF (NOAC.LT.1) GO TO 50
      THETS=0.
      PHS=0.
      XLS=0.
      DO 6 I=1,NK
      THETS=THETS+Q(I)*X(I)
      PHS=PHS+R(I)*X(I)
    6 XLS=XLS+XL(I)*X(I)
      DO 7 I=1,NK
      THETA(I)=Q(I)*X(I)/THETS
    7 PHI(I)=X(I)*R(I)/PHS
      DO 8 I=1,NK
      THS(I)=0.
      DO 8 J=1,NK
      PAR(J,I)=PARAM(J,I,1)+TEMP*PARAM(J,I,2)
    8 THS(I)=THS(I)+THETA(J)*PAR(J,I)
      DO 10 I=1,NK
      GA=ALOG(R(I)/PHS)+5.*Q(I)*ALOG(Q(I)/R(I)*PHS/THETS)
      GA=GA+XL(I)-R(I)/PHS*XLS
      GB=1.-ALOG(THS(I))
      DO 11 J=1,NK
   11 GB=GB-THETA(J)*PAR(I,J)/THS(J)
   10 GAM(I)=EXP(GA+Q(I)*GB)
      GO TO 51
   50 DO 52 J=1,NK
   52 GAM(J)=1.
   51 CONTINUE
      RETURN
      END
```

```
C     ****************************************************************
C     *
C     *   SUBROUTINE KFAC(NST,NKO,T,ANT,P,FLL,XFK,DFK)
C     *
C     ****************************************************************
C
C
C     CALCULATION OF K-VALUES AND ITS DERIVATIVES WITH RESPECT TO
C     TEMPERATURE AND COMPOSITION.  TEMPERATURE IN DEG. C
C
      SUBROUTINE KFAC(NST,NKO,T,ANT,P,FLL,XFK,DFK)
C
      DIMENSION XFK(30,5),DFK(30,5,6)
      DIMENSION T(30),ANT(3,5),FLL(30,5),PAR(5),XX(5),GAM(5),GAMX(5)
      DO 10 I=1,NST
      XT=0.
      DO 11 J=1,NKO
      XT=XT+FLL(I,J)
   11 PAR(J)=EXP(ANT(1,J)-ANT(2,J)/(T(I)+ANT(3,J)))/P
      DO 12 J=1,NKO
   12 XX(J)=FLL(I,J)/XT
      CALL MQUAC(T(I)+273.16,XX,GAM)
      DO 13 J=1,NKO
   13 XFK(I,J)=PAR(J)*GAM(J)
      DO 14 K=1,NKO
      DO 15 J=1,NKO
   15 XX(J)=FLL(I,J)/(XT+1.)
      XX(K)=(FLL(I,K)+1.)/(XT+1.)
      CALL MQUAC(T(I)+273.16,XX,GAMX)
      DO 14 J=1,NKO
   14 DFK(I,J,K)=PAR(J)*GAMX(J)-XFK(I,J)
      DO 10 J=1,NKO
   10 DFK(I,J,NKO+1)=XFK(I,J)*ANT(2,J)/(T(I)+ANT(3,J))**2
      RETURN
      END
```

```
***************************************************************************
*
*    SAMPLE INPUT (SIMPLIFIED VERSION)
*
***************************************************************************

    5
    4
    1
N-OCTANE(1)-ETHYLCYCLOHEXANE(2)-ETHYLBENZENE(3) EXTR. DIST. WITH PHENOL(4)
    1  4.936           5.8486          -0.2856
    2  4.316           5.3944           0.9976
    3  3.508           4.5972           1.84881
 2  4 4  2.68          3.5517           1.80681
 1  1  400.    450.    0.10000E 01 0.10000E 01 0.65737E 00-0.76135E-01
 2  1  400.    450.    0.10000E 01 0.10000E 01 0.65733E 00-0.76135E-01
 3  1  400.    450.    0.12762E 01 0.12762E 01 0.10000E 01 0.42845E-01
 4  1  400.    450.    0.18637E 01 0.18637E 01 0.17925E 01 0.10000E 01
 1  2  400.    450.    0.00000E 00 0.00000E 00 0.38239E-03 0.42876E-03
 2  2  400.    450.    0.00000E 00 0.00000E 00 0.38248E-03 0.42876E-03
 3  2  400.    450.   -0.31464E-03-0.31475E-03 0.00000E 00 0.90209E-03
 4  2  400.    450.   -0.75283E-04 0.75320E-04-0.86317E-03 0.00000E 00
6.92374    1355.126   209.516
6.87041    1384.036   215.128
6.95719    1424.255   213.206
7.13617    1518.1     175.
      30       2                  1
70.   5   10  8.      1.   25      115.        160.
    15
 0.         20.       40.          40.
    20
                                             300.
```

```
***********************************************************************
*
*    SAMPLE OUTPUT (SIMPLIFIED VERSION)
*
***********************************************************************
```

1 N-OCTANE(1)-ETHYLCYCLOHEXANE(2)-ETHYLBENZENE(3) EXTR. DIST. WITH PHENOL(4)

INCLUDING ACTIVITY COEFFICIENTS

FEED COMPOSITION AND ANTOINE PARAMETERS

1	20.000	9.309	3120.292	209.516
2	40.000	9.186	3186.860	215.128
3	40.000	9.386	3279.467	213.206
4	300.000	9.798	3495.553	175.000

```
NUMBER OF STAGES           30
DISTILLATE                 70.000
REFLUX RATIO               8.000000
TOTAL PRESSURE             1.000
```

FLOW CONFIGURATION

I	FL	FV	SL	SV	FKV	FEED STREAMS			
1	330.0	630.0	0.0	0.0	0.0	0.0	0.0	0.0	0.0
2	960.0	630.0	0.0	0.0	0.0	0.0	0.0	0.0	0.0
3	960.0	630.0	0.0	0.0	0.0	0.0	0.0	0.0	0.0
4	960.0	630.0	0.0	0.0	0.0	0.0	0.0	0.0	0.0
5	960.0	630.0	0.0	0.0	0.0	0.0	0.0	0.0	0.0
6	960.0	630.0	0.0	0.0	0.0	0.0	0.0	0.0	0.0
7	960.0	630.0	0.0	0.0	0.0	0.0	0.0	0.0	0.0
8	960.0	630.0	0.0	0.0	0.0	0.0	0.0	0.0	0.0
9	960.0	630.0	0.0	0.0	0.0	0.0	0.0	0.0	0.0
10	960.0	630.0	0.0	0.0	0.0	0.0	0.0	0.0	0.0
11	960.0	630.0	0.0	0.0	0.0	0.0	0.0	0.0	0.0
12	960.0	630.0	0.0	0.0	0.0	0.0	0.0	0.0	0.0
13	960.0	630.0	0.0	0.0	0.0	0.0	0.0	0.0	0.0
14	960.0	630.0	0.0	0.0	0.0	0.0	0.0	0.0	0.0
15	960.0	630.0	0.0	0.0	0.0	20.0	40.0	40.0	0.0
16	860.0	630.0	0.0	0.0	0.0	0.0	0.0	0.0	0.0
17	860.0	630.0	0.0	0.0	0.0	0.0	0.0	0.0	0.0
18	860.0	630.0	0.0	0.0	0.0	0.0	0.0	0.0	0.0
19	860.0	630.0	0.0	0.0	0.0	0.0	0.0	0.0	0.0
20	860.0	630.0	0.0	0.0	0.0	0.0	0.0	0.0	300.0
21	560.0	630.0	0.0	0.0	0.0	0.0	0.0	0.0	0.0
22	560.0	630.0	0.0	0.0	0.0	0.0	0.0	0.0	0.0
23	560.0	630.0	0.0	0.0	0.0	0.0	0.0	0.0	0.0
24	560.0	630.0	0.0	0.0	0.0	0.0	0.0	0.0	0.0
25	560.0	630.0	0.0	0.0	0.0	0.0	0.0	0.0	0.0
26	560.0	630.0	0.0	0.0	0.0	0.0	0.0	0.0	0.0
27	560.0	630.0	0.0	0.0	0.0	0.0	0.0	0.0	0.0
28	560.0	630.0	0.0	0.0	0.0	0.0	0.0	0.0	0.0
29	560.0	630.0	0.0	0.0	0.0	0.0	0.0	0.0	0.0
30	560.0	70.0	0.0	0.0	0.0	0.0	0.0	0.0	0.0

STAGE	TEMP.	TOTAL FLOW	COMPONENT FLOW			
1	162.49	330.00	0.0	4.000	22.485	305.054
2	148.50	960.00	37.864	97.792	195.208	629.136
3	147.00	960.00	74.130	171.382	307.267	407.222
4	145.50	960.00	102.469	228.734	396.233	232.564
5	144.00	960.00	122.043	268.308	416.000	109.269
6	142.50	960.00	133.235	291.077	416.000	34.569
7	141.00	960.00	137.283	299.134	416.000	1.598
8	139.50	960.00	135.698	295.494	416.000	1.425
9	138.00	960.00	129.927	283.726	416.000	23.550
10	136.50	960.00	121.708	266.894	416.000	58.858
11	135.00	960.00	112.326	247.910	416.000	100.011
12	133.50	960.00	102.719	228.982	416.000	141.925
13	132.00	960.00	93.860	212.010	416.000	181.180
14	130.50	960.00	86.395	198.899	416.000	215.906
15	129.00	960.00	81.631	191.391	416.000	244.485

Stage	Temp.	Total Flow	Component Flow			
16	127.50	860.00	62.004	150.428	381.014	266.554
17	126.00	860.00	53.565	135.424	370.084	300.927
18	126.91	860.00	47.595	126.292	360.111	326.001
19	128.36	860.00	44.848	123.802	347.730	343.620
20	128.51	860.00	47.284	128.676	329.415	354.625
21	120.00	560.00	59.300	140.193	301.435	59.073
22	118.50	560.00	74.460	170.341	353.203	0.0
23	117.00	560.00	80.223	182.236	376.000	0.0
24	115.50	560.00	80.079	183.194	376.000	0.0
25	114.00	560.00	76.955	179.431	376.000	0.0
26	112.50	560.00	73.176	176.139	376.000	0.0
27	111.00	560.00	70.874	178.101	363.537	0.0
28	109.50	560.00	72.649	189.444	336.008	0.0
29	108.00	560.00	82.808	212.995	293.700	0.0
30	106.50	560.00	109.798	248.607	230.881	0.0

RESIDUE = 0.111E 08

Stage	Temp.	Total Flow	Component Flow			
1	165.32	330.00	0.0	16.207	15.336	299.512
2	143.73	960.00	0.0	190.489	133.537	656.350
3	137.00	960.00	0.0	425.837	176.169	439.074
4	136.62	960.00	0.0	548.733	127.215	382.459
5	136.41	960.00	0.0	588.308	96.000	388.528
6	132.50	960.00	0.0	611.077	96.000	439.820
7	131.00	960.00	0.0	619.133	96.000	634.975
8	129.50	960.00	0.0	615.494	96.000	960.000
9	128.00	960.00	0.0	603.725	96.000	960.000
10	126.50	960.00	0.0	586.894	96.000	960.000
11	125.00	960.00	11.691	567.910	96.000	960.000
12	123.50	960.00	114.338	548.982	96.000	601.675
13	122.00	960.00	133.484	532.009	96.000	445.744
14	120.50	960.00	133.158	518.899	96.000	393.539
15	120.01	960.00	134.700	511.390	96.000	372.588
16	119.31	860.00	122.281	470.428	61.014	363.941
17	117.86	860.00	135.902	455.423	50.084	357.029
18	116.91	860.00	158.493	446.292	40.112	349.565
19	118.36	860.00	200.803	443.801	27.730	342.928
20	118.51	860.00	207.284	448.676	9.415	332.408
21	113.07	560.00	219.300	460.192	0.0	19.425
22	120.83	560.00	234.460	470.996	33.203	60.122
23	122.30	560.00	240.223	350.523	56.000	52.606
24	124.02	560.00	240.079	259.698	56.000	43.763
25	124.00	560.00	236.955	196.472	56.000	34.469
26	122.50	560.00	233.176	155.221	65.302	25.232
27	121.00	560.00	230.874	132.793	133.158	17.071
28	119.50	560.00	232.649	126.800	182.880	10.928
29	118.00	560.00	204.639	136.146	212.297	6.917
30	116.50	560.00	178.036	158.665	218.625	4.674

RESIDUE = 0.392E 08

STAGE	TEMP.	TOTAL FLOW	COMPONENT FLOW			
1	160.95	330.00	0.0	6.653	41.297	282.080
2	144.95	960.00	0.0	55.282	331.780	573.531
3	138.45	960.00	0.0	105.837	445.987	418.191
4	136.05	960.00	0.0	228.734	411.686	372.373
5	134.82	960.00	0.0	268.308	349.943	359.654
6	133.14	960.00	0.0	318.720	296.521	356.503
7	131.97	960.00	0.0	381.665	242.373	354.979
8	129.81	960.00	0.0	430.231	210.535	353.753
9	129.81	960.00	0.0	490.908	194.763	345.520
10	130.52	960.00	0.0	573.991	179.965	348.149
11	132.30	960.00	0.0	712.784	169.455	352.984
12	133.50	960.00	0.0	868.981	164.868	362.958
13	132.00	960.00	0.0	852.009	150.853	364.566
14	130.50	960.00	0.0	838.899	136.473	357.431
15	130.01	960.00	0.0	831.390	125.739	353.583
16	129.31	860.00	0.0	790.427	78.662	351.228
17	127.86	860.00	0.0	775.423	56.214	351.716
18	126.91	860.00	0.0	766.292	41.214	349.598
19	128.36	860.00	40.803	763.801	30.272	347.443
20	128.51	860.00	47.284	768.676	21.138	344.424
21	123.07	560.00	59.300	560.000	12.803	42.606
22	127.03	560.00	74.460	560.000	9.478	88.338
23	127.79	560.00	80.223	560.000	9.501	63.653
24	128.58	560.00	80.079	560.000	11.718	64.607
25	128.61	560.00	76.955	516.472	11.790	65.923
26	128.30	560.00	73.176	475.221	10.258	69.759
27	127.52	560.00	70.874	452.792	7.488	80.700
28	126.08	560.00	72.649	446.800	4.314	97.129
29	125.20	560.00	45.041	399.654	0.0	116.132
30	124.53	560.00	108.816	326.208	0.0	130.572

RESIDUE =   0.126E 08

STAGE	TEMP.	TOTAL FLOW	COMPONENT FLOW			
1	165.53	330.00	0.008	0.0	40.023	291.670
2	149.63	960.00	0.135	0.0	331.326	639.369
3	142.94	960.00	0.426	0.0	505.309	458.605
4	140.01	960.00	0.922	18.038	540.294	400.745
5	138.40	960.00	1.683	71.378	503.519	383.420
6	136.51	960.00	2.947	132.998	446.933	377.122
7	134.64	960.00	4.884	203.501	379.798	371.817
8	132.84	960.00	7.715	272.540	312.428	367.316
9	131.21	960.00	11.313	334.874	250.565	363.248
10	130.22	960.00	15.902	382.146	202.115	359.836
11	130.01	960.00	21.854	414.002	166.678	357.466
12	129.94	960.00	29.510	548.982	141.646	354.719
13	129.37	960.00	38.903	532.009	123.631	350.950
14	128.53	960.00	49.910	518.899	111.382	352.048
15	128.16	960.00	61.663	511.390	103.338	352.146

16	127.57	860.00		55.337	470.428	57.939	351.541
17	126.82	860.00		56.237	455.423	36.903	350.554
18	126.08	860.00		55.716	446.292	23.362	350.701
19	126.40	860.00		53.682	443.801	14.737	349.822
20	126.32	860.00		57.519	448.676	9.137	348.481
21	125.85	560.00		62.580	444.089	5.597	47.734
22	126.64	560.00		58.044	418.905	4.729	78.322
23	126.75	560.00		62.080	427.085	3.648	67.187
24	126.92	560.00		64.604	421.163	2.588	71.645
25	126.73	560.00		69.000	418.511	1.374	71.114
26	126.47	560.00		74.704	414.217	0.363	70.716
27	126.22	560.00		82.316	407.464	0.0	70.474
28	125.93	560.00		92.713	397.356	0.0	70.356
29	125.72	560.00		107.037	383.297	0.0	69.999
30	125.55	560.00		135.265	357.829	0.0	67.139

RESIDUE =   0.184E 07

STAGE	TEMP.	TOTAL FLOW	COMPONENT FLOW				
1	165.62	330.00	0.0	0.0	39.942	291.184
2	150.87	960.00	0.0	0.0	330.215	643.214
3	145.10	960.00	0.0	0.0	522.327	469.456
4	144.35	960.00	0.0	0.0	602.724	411.879
5	144.45	960.00	0.0	0.0	639.248	399.708
6	144.20	960.00	0.0	0.0	655.887	396.890
7	143.12	960.00	0.0	0.0	645.814	394.125
8	140.96	960.00	0.0	0.0	600.729	388.976
9	138.58	960.00	0.0	59.190	524.364	382.155
10	135.93	960.00	0.0	156.408	432.307	375.321
11	133.69	960.00	0.0	247.492	344.017	369.122
12	131.19	960.00	4.388	319.051	272.017	364.543
13	130.12	960.00	12.147	378.404	211.069	358.379
14	129.17	960.00	22.267	411.653	169.253	356.827
15	128.44	960.00	35.689	427.709	141.009	355.593
16	127.72	860.00	32.841	390.194	82.507	354.459
17	127.01	860.00	36.916	414.149	54.498	354.436
18	126.48	860.00	41.424	430.071	35.211	353.294
19	126.11	860.00	46.734	438.576	22.308	352.381
20	125.78	860.00	53.269	441.023	14.018	351.690
21	126.49	560.00	62.321	437.905	8.683	51.091
22	126.48	560.00	61.217	418.158	6.938	73.687
23	126.41	560.00	65.532	421.226	5.376	67.866
24	126.33	560.00	69.487	416.538	4.172	69.802
25	126.22	560.00	75.207	412.135	3.216	69.442
26	126.09	560.00	82.459	405.302	2.460	69.779
27	125.93	560.00	91.821	396.422	1.859	69.897
28	125.74	560.00	103.741	384.812	1.383	70.065
29	125.51	560.00	118.763	370.015	1.005	70.217
30	125.24	560.00	137.406	351.518	0.710	70.366

RESIDUE =   0.922E 06

STAGE	TEMP.	TOTAL FLOW	COMPONENT FLOW			
1	165.59	330.00	0.0	0.0	39.699	291.227
2	150.79	960.00	0.0	0.0	328.047	642.968
3	144.98	960.00	0.0	0.0	517.595	469.092
4	144.20	960.00	0.0	0.0	596.143	411.628
5	144.89	960.00	0.0	0.0	638.361	400.172
6	146.24	960.00	0.0	0.0	685.533	400.829
7	148.37	960.00	0.0	0.0	754.602	405.614
8	150.96	960.00	0.0	0.0	856.037	413.478
9	148.58	960.00	0.0	0.0	844.363	426.101
10	145.93	960.00	0.0	0.0	752.307	435.024
11	143.69	960.00	0.0	0.0	664.016	431.204
12	141.19	960.00	0.0	0.0	592.017	419.241
13	140.12	960.00	0.0	58.404	531.069	403.181
14	139.17	960.00	0.0	91.653	489.252	388.334
15	136.62	960.00	14.902	107.709	461.009	377.791
16	134.50	860.00	18.005	140.948	331.802	369.243
17	132.03	860.00	23.928	228.047	239.219	368.805
18	129.85	860.00	31.028	300.473	164.549	363.950
19	128.32	860.00	38.793	353.047	108.784	359.375
20	127.19	860.00	47.579	386.387	69.825	356.208
21	127.23	560.00	58.440	403.870	43.755	53.935
22	127.02	560.00	60.053	401.050	35.412	63.485
23	126.82	560.00	63.582	402.743	27.708	65.967
24	126.63	560.00	68.354	403.867	21.586	66.193
25	126.46	560.00	74.166	401.758	16.648	67.428
26	126.27	560.00	81.665	397.612	12.711	68.011
27	126.07	560.00	91.180	390.607	9.581	68.633
28	125.84	560.00	103.248	380.535	7.106	69.110
29	125.58	560.00	118.403	366.917	5.160	69.520
30	125.29	560.00	137.157	349.320	3.637	69.887

RESIDUE = 0.653E 07

STAGE	TEMP.	TOTAL FLOW	COMPONENT FLOW			
1	165.44	330.00	0.000	0.007	38.565	291.428
2	150.16	960.00	0.000	0.082	318.093	641.825
3	143.86	960.00	0.000	0.198	494.531	465.271
4	142.55	960.00	0.001	0.354	552.932	406.713
5	142.44	960.00	0.003	0.581	565.773	393.643
6	142.75	960.00	0.007	0.944	567.938	391.110
7	143.62	960.00	0.017	1.556	567.399	391.028
8	144.95	960.00	0.042	2.638	564.910	392.410
9	144.06	960.00	0.108	4.634	560.423	394.835
10	142.77	960.00	0.267	7.781	558.428	393.524
11	141.87	960.00	0.645	12.648	556.032	390.675
12	141.16	960.00	1.526	20.016	549.802	388.655
13	140.63	960.00	3.493	30.456	538.969	387.081
14	139.81	960.00	7.627	45.559	521.437	385.376
15	138.48	960.00	16.050	67.072	494.194	382.684

16	139.04	860.00	12.899	56.224	412.366	378.511
17	138.02	860.00	13.299	68.844	394.118	383.740
18	135.99	860.00	16.336	111.235	352.029	380.400
19	133.52	860.00	22.979	175.768	286.147	374.106
20	131.20	860.00	32.946	246.041	213.252	367.760
21	129.07	560.00	46.728	302.819	147.955	62.498
22	128.73	560.00	54.102	338.057	126.370	41.472
23	128.28	560.00	57.330	345.243	106.072	51.354
24	127.86	560.00	62.601	355.944	85.944	54.511
25	127.46	560.00	69.190	363.288	69.940	57.582
26	127.07	560.00	77.437	366.934	55.166	60.462
27	126.69	560.00	87.763	366.878	42.649	62.710
28	126.32	560.00	100.653	362.446	32.253	64.648
29	125.93	560.00	116.634	353.383	23.761	66.221
30	125.53	560.00	136.288	339.267	16.928	67.518

RESIDUE = 0.587E 06

STAGE	TEMP.	TOTAL FLOW	COMPONENT FLOW			
1	165.44	330.00	0.000	0.022	38.542	291.436
2	150.11	960.00	0.000	0.260	317.939	641.800
3	143.77	960.00	0.001	0.625	494.294	465.080
4	142.35	960.00	0.002	1.098	552.389	406.511
5	142.04	960.00	0.005	1.757	564.926	393.312
6	141.96	960.00	0.013	2.752	566.735	390.499
7	141.93	960.00	0.030	4.305	565.849	389.815
8	141.93	960.00	0.069	6.798	563.648	389.484
9	141.71	960.00	0.164	10.903	559.898	389.035
10	141.41	960.00	0.381	17.204	553.956	388.458
11	141.02	960.00	0.872	26.546	544.887	387.695
12	140.45	960.00	1.962	40.203	531.348	386.487
13	139.61	960.00	4.325	59.751	511.277	384.646
14	138.39	960.00	9.283	86.940	481.847	381.929
15	136.65	960.00	19.129	122.539	440.242	378.090
16	136.50	860.00	17.184	124.574	345.129	373.113
17	135.21	860.00	19.483	156.625	307.598	376.294
18	133.48	860.00	23.665	201.507	261.824	373.003
19	131.67	860.00	29.998	251.685	209.907	368.409
20	129.96	860.00	38.764	298.234	158.981	364.021
21	128.43	560.00	50.746	335.243	113.863	60.147
22	128.14	560.00	55.698	353.890	98.884	51.527
23	127.87	560.00	59.422	360.918	85.164	54.496
24	127.58	560.00	64.328	366.987	72.403	56.282
25	127.28	560.00	70.438	370.351	60.499	58.712
26	126.97	560.00	78.274	371.230	49.516	60.980
27	126.64	560.00	88.280	369.157	39.597	62.966
28	126.29	560.00	100.948	363.515	30.837	64.700
29	125.92	560.00	116.798	353.732	23.282	66.188
30	125.53	560.00	136.335	339.283	16.918	67.464

RESIDUE = 0.751E 05

STAGE	TEMP.	TOTAL FLOW	COMPONENT FLOW			
1	165.44	330.00	0.000	0.034	38.528	291.438
2	150.10	960.00	0.000	0.401	317.813	641.786
3	143.75	960.00	0.001	0.963	494.011	465.024
4	142.32	960.00	0.003	1.688	551.867	406.441
5	142.00	960.00	0.007	2.692	564.080	393.221
6	141.89	960.00	0.016	4.189	565.424	390.371
7	141.79	960.00	0.038	6.465	563.875	389.622
8	141.66	960.00	0.087	9.933	560.765	389.215
9	141.46	960.00	0.198	15.196	555.860	388.745
10	141.16	960.00	0.450	23.089	548.400	388.061
11	140.71	960.00	1.012	34.754	537.196	387.038
12	140.05	960.00	2.244	51.623	520.609	385.523
13	139.08	960.00	4.866	75.214	496.568	383.352
14	137.74	960.00	10.217	106.510	462.943	380.329
15	135.98	960.00	20.487	144.841	418.306	376.365
16	135.74	860.00	18.630	146.461	323.371	371.537
17	134.57	860.00	21.059	178.072	286.289	374.579
18	133.08	860.00	25.034	217.620	245.513	371.833
19	131.45	860.00	30.882	260.761	200.488	367.868
20	129.83	860.00	39.247	302.346	154.649	363.758
21	128.39	560.00	50.965	336.867	112.204	59.963
22	128.13	560.00	55.689	354.006	97.981	52.324
23	127.86	560.00	59.571	361.575	85.080	53.774
24	127.58	560.00	64.311	366.763	72.681	56.245
25	127.28	560.00	70.397	370.016	60.888	58.699
26	126.97	560.00	78.230	370.924	49.916	60.929
27	126.64	560.00	88.240	368.891	39.956	62.912
28	126.29	560.00	100.915	363.300	31.137	64.648
29	125.92	560.00	116.775	353.560	23.518	66.147
30	125.53	560.00	136.320	339.152	17.095	67.432

RESIDUE = 0.532E 04

STAGE	TEMP.	TOTAL FLOW	COMPONENT FLOW			
1	165.44	330.00	0.000	0.035	38.527	291.438
2	150.10	960.00	0.000	0.415	317.803	641.781
3	143.75	960.00	0.001	0.995	493.986	465.018
4	142.32	960.00	0.003	1.745	551.818	406.434
5	142.00	960.00	0.007	2.782	563.999	393.212
6	141.88	960.00	0.017	4.326	565.301	390.355
7	141.78	960.00	0.039	6.672	563.688	389.601
8	141.65	960.00	0.088	10.237	560.488	389.186
9	141.44	950.00	0.202	15.625	555.469	388.703
10	141.14	960.00	0.457	23.691	547.848	388.003
11	140.68	960.00	1.027	35.583	536.426	386.963
12	140.00	960.00	2.272	52.716	519.579	385.433
13	139.03	960.00	4.914	76.539	495.310	383.238
14	137.69	960.00	10.287	107.935	461.572	380.205
15	135.93	960.00	20.571	146.131	417.052	376.246

16	135.70	860.00	18.691	147.412	322.435	371.461
17	134.55	860.00	21.108	178.666	285.718	374.507
18	133.06	860.00	25.059	217.858	245.290	371.793
19	131.44	860.00	30.892	260.811	200.459	367.838
20	129.83	860.00	39.247	302.360	154.656	363.737
21	128.39	560.00	50.965	336.871	112.215	59.948
22	128.13	560.00	55.685	353.990	97.993	52.331
23	127.86	560.00	59.572	361.574	85.091	53.763
24	127.58	560.00	64.309	366.746	72.693	56.252
25	127.28	560.00	70.399	370.003	60.898	58.699
26	126.97	560.00	78.232	370.916	49.925	60.927
27	126.64	560.00	88.241	368.884	39.964	62.911
28	126.29	560.00	100.916	363.293	31.143	64.647
29	125.92	560.00	116.776	353.555	23.523	66.145
30	125.53	560.00	136.323	339.146	17.099	67.431

RESIDUE =   0.180E 02

STAGE	TEMP.	TOTAL FLOW	COMPONENT FLOW			
1	165.44	330.00	0.000	0.035	38.527	291.438
2	150.10	960.00	0.000	0.415	317.801	641.784
3	143.75	960.00	0.001	0.995	493.985	465.019
4	142.32	960.00	0.003	1.744	551.819	406.433
5	142.00	960.00	0.007	2.780	564.000	393.213
6	141.88	960.00	0.017	4.323	565.304	390.356
7	141.78	960.00	0.038	6.668	563.692	389.602
8	141.65	960.00	0.088	10.230	560.496	389.186
9	141.44	960.00	0.202	15.614	555.480	388.703
10	141.14	960.00	0.457	23.675	547.864	388.003
11	140.68	960.00	1.027	35.560	536.448	386.964
12	140.00	960.00	2.271	52.684	519.612	385.434
13	139.03	960.00	4.912	76.496	495.354	383.238
14	137.69	960.00	10.284	107.881	461.627	380.207
15	135.93	960.00	20.567	146.072	417.113	376.247
16	135.70	860.00	18.687	147.349	322.501	371.463
17	134.55	860.00	21.103	178.594	285.792	374.510
18	133.06	860.00	25.052	217.778	245.372	371.797
19	131.44	860.00	30.884	260.727	200.544	367.844
20	129.83	860.00	39.239	302.282	154.736	363.742
21	128.39	560.00	50.957	336.806	112.283	59.954
22	128.13	560.00	55.680	353.942	98.056	52.322
23	127.86	560.00	59.568	361.528	85.151	53.752
24	127.58	560.00	64.306	366.705	72.749	56.240
25	127.28	560.00	70.394	369.970	60.948	58.688
26	126.97	560.00	78.227	370.885	49.969	60.918
27	126.64	550.00	88.237	368.859	40.001	62.903
28	126.30	560.00	100.913	363.273	31.174	64.640
29	125.93	560.00	116.773	353.538	23.547	66.142
30	125.53	560.00	136.320	339.135	17.117	67.428

RESIDUE =   0.118E 00

PRODUCT FLOW, LIQUID PHASE

1	0.000	0.035	38.527	291.438
2	0.0	0.0	0.0	0.0
3	0.0	0.0	0.0	0.0
4	0.0	0.0	0.0	0.0
5	0.0	0.0	0.0	0.0
6	0.0	0.0	0.0	0.0
7	0.0	0.0	0.0	0.0
8	0.0	0.0	0.0	0.0
9	0.0	0.0	0.0	0.0
10	0.0	0.0	0.0	0.0
11	0.0	0.0	0.0	0.0
12	0.0	0.0	0.0	0.0
13	0.0	0.0	0.0	0.0
14	0.0	0.0	0.0	0.0
15	0.0	0.0	0.0	0.0
16	0.0	0.0	0.0	0.0
17	0.0	0.0	0.0	0.0
18	0.0	0.0	0.0	0.0
19	0.0	0.0	0.0	0.0
20	0.0	0.0	0.0	0.0
21	0.0	0.0	0.0	0.0
22	0.0	0.0	0.0	0.0
23	0.0	0.0	0.0	0.0
24	0.0	0.0	0.0	0.0
25	0.0	0.0	0.0	0.0
26	0.0	0.0	0.0	0.0
27	0.0	0.0	0.0	0.0
28	0.0	0.0	0.0	0.0
29	0.0	0.0	0.0	0.0
30	0.0	0.0	0.0	0.0

PRODUCT FLOW, VAPOR PHASE

1	0.0	0.0	0.0	0.0
2	0.0	0.0	0.0	0.0
3	0.0	0.0	0.0	0.0
4	0.0	0.0	0.0	0.0
5	0.0	0.0	0.0	0.0
6	0.0	0.0	0.0	0.0
7	0.0	0.0	0.0	0.0
8	0.0	0.0	0.0	0.0
9	0.0	0.0	0.0	0.0
10	0.0	0.0	0.0	0.0
11	0.0	0.0	0.0	0.0
12	0.0	0.0	0.0	0.0
13	0.0	0.0	0.0	0.0
14	0.0	0.0	0.0	0.0
15	0.0	0.0	0.0	0.0

16	0.0	0.0	0.0	0.0
17	0.0	0.0	0.0	0.0
18	0.0	0.0	0.0	0.0
19	0.0	0.0	0.0	0.0
20	0.0	0.0	0.0	0.0
21	0.0	0.0	0.0	0.0
22	0.0	0.0	0.0	0.0
23	0.0	0.0	0.0	0.0
24	0.0	0.0	0.0	0.0
25	0.0	0.0	0.0	0.0
26	0.0	0.0	0.0	0.0
27	0.0	0.0	0.0	0.0
28	0.0	0.0	0.0	0.0
29	0.0	0.0	0.0	0.0
30	20.000	39.965	1.478	8.568

K-FACTOR PROFILE IN COLUMN

1	9.798	5.702	3.797	0.630
2	5.713	3.527	2.184	0.412
3	4.023	2.619	1.583	0.377
4	3.618	2.399	1.451	0.382
5	3.531	2.351	1.423	0.383
6	3.508	2.338	1.416	0.383
7	3.495	2.330	1.411	0.382
8	3.479	2.321	1.405	0.381
9	3.456	2.307	1.397	0.379
10	3.422	2.287	1.385	0.375
11	3.371	2.256	1.367	0.370
12	3.296	2.212	1.340	0.363
13	3.190	2.148	1.302	0.353
14	3.047	2.063	1.250	0.340
15	2.866	1.954	1.184	0.324
16	3.002	2.025	1.216	0.305
17	2.914	1.970	1.179	0.293
18	2.772	1.885	1.124	0.281
19	2.618	1.792	1.063	0.268
20	2.468	1.701	1.004	0.257
21	1.320	1.040	0.788	0.903
22	1.270	1.008	0.785	1.059
23	1.258	1.000	0.775	1.072
24	1.249	0.994	0.763	1.063
25	1.240	0.987	0.750	1.052
26	1.230	0.980	0.738	1.043
27	1.218	0.972	0.726	1.034
28	1.205	0.963	0.714	1.027
29	1.190	0.953	0.702	1.021
30	1.174	0.943	0.691	1.017

APPENDIX 5

LIST OF PHASE EQUILIBRIUM DATA USED IN THE DETERMINATION OF UNIFAC PARAMETERS

CH_2 - C=C   Temperature Range: 333 - 341 K

   Hexane - 1-Hexene (885,999)

CH_2 - ACH   Temperature Range: 255 - 388 K

   Pentane - Benzene (1062)
   Cyclohexane - Benzene (1361)
   Hexane - Benzene (191,878,1360)
   Heptane - Benzene (528)
   Octane - Benzene (528)

CH_2 - $ACCH_2$ Temperature Range: 293 - 410 K

   Pentane - Toluene (1360)
   Hexane - Toluene (1357)
   Heptane - Toluene (20,21)
   Octane - p-Xylene (528)

CH_2 - CCOH   Temperature Range: 298 - 373 K

   Hexane - Ethanol (878,1362)
   Hexane - 1-Propanol (1363)
   Hexane - 2-Butanol (885)
   Heptane - Ethanol (1052)
   Heptane - 1-Propanol (1106)
   Heptane - 2-Propanol (1106)
   Heptane - 1-Octanol (1364)
   Heptane - 2-Octanol (1364)
   Heptane - 3-Octanol (1364)
   Heptane - 4-Octanol (1364)
   Decane - 1-Propanol (766)
   Decane - 2-Propanol (766)
   Decane - 1-Butanol (1049)

CH_2 - CH_3OH Temperature Range: 318 - 353 K

   Cyclohexane - Methanol (676,906,909)
   Hexane - Methanol (554)

CH_2 - H_2O Temperature Range: 279 - 298 K

   Pentane - Water (1365,1366)
   Cyclohexane - Water (1365,1366)

Hexane - Water (1365,1366)
Heptane - Water (1365,1366)
Octane - Water (1365,1366)

CH_2 - ACOH   Temperature Range: 450 - 490 K

Dodecane - Phenol (994)

CH_2 - CH_2CO Temperature Range: 298 - 368 K

Pentane - Acetone (182)
Cyclohexane - Acetone (620)
Hexane - Acetone (122)
Hexane - 5-Nonanone (1367)
Heptane - 2-Butanone (21)
Heptane - 3-Pentanone (468)
Octane - 2-Butanone (781)

CH_2 - CHO   Temperature Range: 318 K

Cyclohexane - Propionaldehyde (1055)

CH_2 - COOC   Temperature Range: 308 - 373 K

Cyclohexane - Methyl acetate (273,464)
Heptane - Butyl acetate (864)

CH_2 - CH_2O Temperature Range: 333 - 393 K

Cyclohexane - Tetrahydrofuran (1370)
Cyclohexane - 1,4-Dioxane (1368)
Hexane - 1,4-Dioxane (1046)
Heptane - 1,2-Dimethoxyethane (1369)
Heptane - 1,4-Dioxane (1046)
Heptane - Dipropyl ether (1369)
Octane - 1,4-Dioxane (1046)
Decane - Bis(2-Methoxyethyl)ether (1369)

CH_2 - CNH_2 Temperature Range: 293 - 333 K

Hexane - Methylamine (1372)
Hexane - Ethylamine (1372)
Hexane - Propylamine (1372)
Hexane - Butylamine (476)
Hexane - Hexylamine (476)
Nonane - Methylamine (1372)

CH_2 - CNH   Temperature Range: 273 - 333 K

Pentane - Diethylamine (1363)
Hexane - Dimethylamine (1372)
Hexane - Diethylamine (476,1375)
Hexane - Dipropylamine (476)
Heptane - Diethylamine (973)

```
CH2 - ACNH2                          Temperature Range:  325 - 373 K
-------------------------------------------------------------------
   Cyclohexane - Aniline (1089)
   Methylcyclohexane - Aniline (1376)
   Heptane - Aniline (314)

CH2 - CCN                            Temperature Range:  293 - 318 K
-------------------------------------------------------------------
   Heptane - Acetonitrile (616,828)

CH2 - COOH                           Temperature Range:  293 - 404 K
-------------------------------------------------------------------
   Heptane - Acetic acid (1377)
   Octane - Propanoic acid (237)

CH2 - CCl                            Temperature Range:  323 - 357 K
-------------------------------------------------------------------
   Cyclohexane - 1,2-Dichloroethane (59,884)
   Heptane - 1-Chlorobutane (526)

CH2 - CCl2                           Temperature Range:        333 K
-------------------------------------------------------------------
   Hexane - 1,1,2,2-Tetrachloroethane (885)

CH2 - CCl3                           Temperature Range:  308 - 343 K
-------------------------------------------------------------------
   Hexane - Chloroform (370)

CH2 - CCl4                           Temperature Range:  313 - 371 K
-------------------------------------------------------------------
   Cyclohexane - Tetrachloromethane (202,211,213,665,872)
   Methylcyclopentane - Tetrachloromethane (872)
   Hexane - Tetrachloromethane (872)
   3-Methylpentane - Tetrachloromethane (872)
   Heptane - Tetrachloromethane (872)

CH2 - ACCl                           Temperature Range:  338 - 373 K
-------------------------------------------------------------------
   Hexane - Chlorobenzene (305)
   Heptane - Chlorobenzene (973)

CH2 - CNO2                           Temperature Range:  298 - 400 K
-------------------------------------------------------------------
   2-Methylbutane - Nitromethane (385)
   Hexane - 1-Nitropropane (348)
   Heptane - Nitroethane (928)
   Heptane - 1-Nitropropane (928)

CH2 - ACNO2                          Temperature Range:        353 K
-------------------------------------------------------------------
   Cyclohexane - Nitrobenzene (305)
```

CH$_2$ - CS$_2$ Temperature Range: 290 K

  2-Methylbutane - Carbon disulfide (285)

C=C - ACH Temperature Range: 273 - 348 K

  Cyclohexene - Benzene (764)
  1-Hexene - Benzene (1378)
  1-Octene - Benzene (1378)

C=C - ACCH$_2$ Temperature Range: 283 - 408 K

  1-Hexene - Toluene (1378)
  1-Octene - Ethylbenzene (35)

C=C - CCOH Temperature Range: 333 K

  1-Hexene - Ethanol (1362)
  1-Hexene - 2-Butanol (885)

C=C - CH$_3$OH Temperature Range: 303 - 318 K

  Isoprene - Methanol (394)
  2-Methyl-2-Butene - Methanol (394)

C=C - H$_2$O Temperature Range: 283 - 298 K

  1,4-Pentadiene - Water (1365)
  1-Pentene - Water (1365)
  2-Pentene - Water (1365)
  1,5-Hexadiene - Water (1365,1379)
  1-Hexene - Water (1365)
  1,6-Heptadiene - Water (1365)
  1-Heptene - Water (1379)
  1-Octene - Water (1365)

C=C - CH$_2$CO Temperature Range: 273 - 353 K

  1-Pentene - Acetone (1381)
  1-Hexene - Acetone (1381)
  1-Hexene - 2-Butanone (885)
  1-Hexene - 5-Nonanone (1380)

C=C - CH$_2$O Temperature Range: 353 K

  1-Pentene - 1,4-Dioxane (1046)
  1-Hexene - 1,4-Dioxane (1046)
  1-Octene - 1,4 Dioxane (1046)

C=C - CNH$_2$ Temperature Range: 333 K

  1-Hexene - Butylamine (476)

C=C – CNH Temperature Range: 333 K

   1-Hexene – Diethylamine (476)
   1-Hexene – Dipropylamine (476)

C=C – CCN Temperature Range: 303 – 355 K

   1-Pentene – Acetonitrile (407)

C=C – COOH Temperature Range: 333 K

   Acetic acid – Styrene (434)

C=C – CCl Temperature Range: 352 – 357 K

   Cyclohexene – 1,2-Dichloroethane (884)

C=C – CCl$_2$ Temperature Range: 333 K

   1-Hexene – 1,1,2,2-Tetrachloroethane (885)

C=C – CCl$_3$ Temperature Range: 333 K

   1-Hexene – 1,1,1-Trichloroethane (885)

C=C – CCl$_4$ Temperature Range: 298 – 356 K

   Cyclohexene – Tetrachloromethane (872)
   1,3-Butadiene – Tetrachloromethane (874)

C=C – CNO$_2$ Temperature Range: 298 – 341 K

   1,3-Butadiene – Nitromethane (874)
   Isoprene – Nitromethane (385)
   2-Methyl-1-butene – Nitromethane (385)
   2-Methyl-2-butene – Nitromethane (385)

ACH – ACCH$_2$ Temperature Range: 354 – 379 K

   Benzene – Toluene (323)

ACH – CCOH Temperature Range: 298 – 390 K

   Benzene – Ethanol (680,878,1381)
   Benzene – 1-Propanol (501,1381)
   Benzene – 2-Propanol (84,688,1382,1383)
   Benzene – 1-Butanol (295,296)
   Benzene – 2-Butanol (1384)

ACH – CH$_3$OH Temperature Range: 293 – 363 K

   Benzene – Methanol (60,358,398,400,509)

ACH – H$_2$O   Temperature Range: 283 – 313 K

   Benzene – Water (1379)

ACH – ACOH   Temperature Range: 343 K

   Benzene – Phenol (553)

ACH – CH$_2$CO   Temperature Range: 318 – 353 K

   Benzene – Acetone (134,761,943)
   Benzene – 2-Butanone (21,943)

ACH – COOC   Temperature Range: 303 – 372 K

   Benzene – Methyl acetate (109,1146)
   Benzene – Ethyl acetate (281)
   Benzene – Propyl acetate (1385)

ACH – CH$_2$O   Temperature Range: 303 – 343 K

   Benzene – 1,2-Dimethoxyethane (1369)
   Benzene – 1,4-Dioxane (1386)
   Benzene – Dipropyl ether (941,1369)

ACH – CNH$_2$ Temperature Range: 323 – 343 K

   Benzene – Butylamine (973)

ACH – CNH   Temperature Range: 308 – 328 K

   Benzene – Diethylamine (973)

ACH – ACNH$_2$ Temperature Range: 323 – 393 K

   Benzene – Aniline (290,1089)

ACH – CCN   Temperature Range: 293 – 355 K

   Benzene – Acetonitrile (154,231,616)

ACH – COOH   Temperature Range: 293 – 384 K

   Benzene – Acetic acid (760,1377)

ACH – CCl   Temperature Range: 293 – 382 K

   Benzene – 1,2-Dichloroethane (127,300,301,438,475)

ACH – CCl$_3$ Temperature Range: 323 – 353 K

   Benzene – Chloroform (267,272,464)

ACH - CCl$_4$ Temperature Range: 283 - 368 K

   Benzene - Tetrachloromethane (127,159,205,207,351,353,354,355)

ACH - ACCl   Temperature Range: 299 - 405 K

   Benzene - Chlorobenzene (851,950)

ACH - CNO$_2$ Temperature Range: 298 - 318 K

   Benzene - Nitromethane (220,348)

ACH - ACNO$_2$ Temperature Range: 298 - 353 K

   Benzene - Nitrobenzene (348,553)

ACH - CS$_2$ Temperature Range: 293 - 303 K

   Benzene - Carbon disulfide (285,520,575)

ACCH$_2$ - CCOH   Temperature Range: 303 - 405 K

   Toluene - Ethanol (257,1106)
   Toluene - 1-Butanol (8)
   Toluene - 1-Pentanol (896)
   Ethylbenzene - 1-Propanol (536)
   Ethylbenzene - 1-Butanol (1111)

ACCH$_2$ - CH$_3$OH   Temperature Range: 336 - 377 K

   Toluene - Methanol (48,401)

ACCH$_2$ - H$_2$O   Temperature Range: 263 - 363 K

   Toluene - Water (1366)

ACCH$_2$ - ACOH   Temperature Range: 383 - 446 K

   Toluene - Phenol (12)

ACCH$_2$ - CH$_2$CO   Temperature Range: 308 - 348 K

   Toluene - Acetone (943)
   Toluene - 2-Butanone (943)

ACCH$_2$ - COOC   Temperature Range: 350 - 410 K

   Toluene - Ethyl acetate (281)
   p-Xylene - Ethyl acetate (281)

ACCH$_2$ – CH$_2$O   Temperature Range: 323 – 343 K

   Toluene – Dipropyl ether (941)
   Ethylbenzene – Dipropyl ether (941)

ACCH$_2$ – CNH   Temperature Range: 308 K

   Toluene – Diethylamine (816)
   Ethylbenzene – Diethylamine (816)

ACCH$_2$ – ACNH$_2$ Temperature Range: 353 – 373 K

   Toluene – Aniline (1376)

ACCH$_2$ – CCN   Temperature Range: 318 – 382 K

   Toluene – Acetonitrile (154,1013)

ACCH$_2$ – COOH   Temperature Range: 333 – 405 K

   Toluene – Propanoic acid (771)
   p-Xylene – Acetic acid (760,881)
   Ethylbenzene – Acetic acid (434,881)
   m-Xylene – Propanoic acid (1145)

ACCH$_2$ – CCl   Temperature Range: 360 – 382 K

   Toluene – 1,2-Dichloroethane (236)

ACCH$_2$ – CCl$_3$ Temperature Range: 338 – 372 K

   Toluene – Chloroform (532)

ACCH$_2$ – CCl$_4$ Temperature Range: 308 – 328 K

   Toluene – Tetrachloromethane (891)

ACCH$_2$ – CNO$_2$ Temperature Range: 318 K

   Toluene – Nitroethane (1013)

CCOH – CH$_3$OH   Temperature Range: 323 – 353 K

   Ethanol – Methanol (858)
   2-Propanol – Methanol (42,472)
   3-Methyl-1-butanol – Methanol (56)

CCOH – H$_2$O   Temperature Range: 303 – 363 K

   Ethanol – Water (830)
   1-Propanol – Water (766)
   2-Propanol – Water (1126)

1-Butanol - Water (1075)

CCOH - ACOH   Temperature Range: 375 - 394 K

Cyclohexanol - Phenol (703)

CCOH - CH_2CO Temperature Range: 298 - 372 K

Ethanol - Acetone (5)
Ethanol - 2-Butanone (14,720)
Ethanol - 2-Pentanone (391)
2-Propanol - Acetone (472,620)
2-Propanol - 2-Pentanone (591)

CCOH - COOC   Temperature Range: 298 - 395 K

Ethanol - Methyl acetate (1388)
Ethanol - Ethyl acetate (51,830)
Ethanol - Methyl propanoate (1041)
1-Propanol - Ethyl acetate (834)
2-Propanol - Ethyl acetate (51,834)
1-Butanol - Ethyl acetate (810)
1-Butanol - Butyl acetate (495)

CCOH - CH_2O Temperature Range: 273 - 381 K

Ethanol - Diethyl ether (171,173)
Ethanol - Dipropyl ether (838)
2-Propanol - Diisopropyl ether (1389)
1-Butanol - Tetrahydrofuran (1756)
2-Propanol - Tetrahydrofuran (1112)

CCOH - CNH_2 Temperature Range: 313 - 365 K

1-Propanol - Propylamine (921)
1-Butanol - Butylamine (725)

CCOH - CNH   Temperature Range: 293 - 428 K

Ethanol - Dimethylamine (1390)
Ethanol - Diethylamine (814)
1-Propanol - Dimethylamine (1390)
1-Propanol - Dipropylamine (921)
2-Propanol - Diisopropylamine (1391)
1-Butanol - Ethylbutylamine (1121)
1-Butanol - Dibutylamine (763)

CCOH - $ACNH_2$ Temperature Range: 370 - 456 K

Ethanol - Aniline (1392)
Cyclohexanol - Aniline (704)

CCOH - COOH   Temperature Range: 351 - 393 K

  Ethanol - Acetic acid (1393)
  1-Propanol - Acetic acid (1393)
  1-Butanol - Acetic acid (1393)

CCOH - CCl   Temperature Range: 313 - 382 K

  Ethanol - 1,2-Dichloroethane (56,242)
  1-Propanol - 1,2-Dichloroethane (106)
  1-Butanol - 1,2-Dichloroethane (643)
  3-Methyl-1-butanol - 1,2-Dichloroethane (106)

CCOH - CCl_2 Temperature Range: 313 - 354 K

  Ethanol - Dichloromethane (457)
  2-Methyl-1-Propanol - 1,1,2,2-Tetrachloroethane (1016)

CCOH - CCl_3 Temperature Range: 308 - 389 K

  Ethanol - Chloroform (75,363,364,559,838)
  1-Butanol - Chloroform (373)

CCOH - CCl_4 Temperature Range: 318 - 402 K

  Ethanol - Tetrachloromethane (101,664)
  1-Butanol - Tetrachloromethane (347)
  3-Methyl-1-Butanol - Tetrachloromethane (639)

CCOH - ACCl   Temperature Range: 347 - 368 K

  1-Propanol - Chlorobenzene (1048)

CCOH - CS_2 Temperature Range: 293 K

  Ethanol - Carbon disulfide (509)

CH_3OH - H_2O Temperature Range: 309 - 369 K

  Methanol - Water (27,28,29,42,104,174,193,194,196,387,504,
                                                   762,766)

CH_3OH - CH_2CO Temperature Range: 328 - 360 K

  Methanol - Acetone (4,104,118,120,387,472)
  Methanol - 4-Methyl-2-Pentanone (572)

CH_3OH - CHO   Temperature Range: 318 K

  Methanol - Propionaldehyde (1055)

CH$_3$OH - COOC Temperature Range: 323 - 386 K

  Methanol - Methyl acetate (388,557,837,873)
  Methanol - Ethyl acetate (46,51,55,834)

CH$_3$OH - CH$_2$O Temperature Range: 303 K

  Methanol - Diethyl ether (733)

CH$_3$OH - CNH$_2$ Temperature Range: 341 - 352 K

  Methanol - Butylamine (927)

CH$_3$OH - CNH Temperature Range: 331 - 341 K

  Methanol - Diethylamine (927)

CH$_3$OH - CCN Temperature Range: 303 K

  Methanol - Acetonitrile (338)

CH$_3$OH - COOH Temperature Range: 298 - 318 K

  Methanol - Propanoic acid (1402)
  Methanol - Butanoic acid (1402)

CH$_3$OH - CCl Temperature Range: 313 - 333 K

  Methanol - 1,2-Dichloroethane (56)

CH$_3$OH - CCl$_2$ Temperature Range: 310 - 336 K

  Methanol - Dichloromethane (382)

CH$_3$OH - CCl$_3$ Temperature Range: 308 - 336 K

  Methanol - Chloroform (46,70)

CH$_3$OH - CCl$_4$ Temperature Range: 293 - 350 K

  Methanol - Tetrachloromethane (19,60,509)

CH$_3$OH - CNO$_2$ Temperature Range: 337 - 370 K

  Methanol - Nitromethane (860)

CH$_3$OH - CS$_2$ Temperature Range: 293 - 332 K

  Methanol - Carbon disulfide (509,879)

H_2O - ACOH   Temperature Range: 329 - 385 K

    Water - Phenol (66,313)

H_2O - CH_2CO Temperature Range: 298 - 347 K

    Water - Acetone (142,890)
    Water - 2-Butanone (258)

H_2O - CHO   Temperature Range: 293 - 303 K

    Water - Acetaldehyde (1029)

H_2O - COOC   Temperature Range: 298 - 343 K

    Water - Methyl acetate (234)
    Water - Ethyl acetate (830)
    Water - Butyl acetate (327)

H_2O - CH_2O Temperature Range: 298 - 370 K

    Water - Tetrahydrofuran (471,1027,1394)
    Water - 1,4-Dioxane (264,429,510)
    Water - Diethyl ether (1394)

H_2O - CNH_2 Temperature Range: 325 - 360 K

    Water - Propylamine (921)
    Water - Butylamine (611)

H_2O - CNH   Temperature Range: 293 - 330 K

    Water - Diethylamine (179)
    Water - Dipropylamine (957)
    Water - Ethylbutylamine (1034)

H_2O - $ACNH_2$ Temperature Range: 371 - 425 K

    Water - Aniline (314,315)

H_2O - CCN   Temperature Range: 303 - 360 K

    Water - Acetonitrile (22,26)

H_2O - COOH   Temperature Range: 318 - 405 K

    Water - Formic acid (190)
    Water - Acetic acid (108,190,760)
    Water - Propanoic acid (190)
    Water - Butanoic acid (1395)

$H_2O - CCl$ Temperature Range: 303 - 353 K

    Water - 1,2-Dichloroethane (1366)

$H_2O - CCl_2$ Temperature Range: 273 - 303 K

    Water - Dichloromethane (1366)

$H_2O - CCl_3$ Temperature Range: 283 - 313 K

    Water - Chloroform (1366,1396)

$H_2O - CCl_4$ Temperature Range: 298 - 303 K

    Water - Tetrachloromethane (1366,1396)

$H_2O - ACCl$ Temperature Range: 298 K

    Water - Chlorobenzene (1366,1397)

$H_2O - CNO_2$ Temperature Range: 298 K

    Water - Nitromethane (1397)
    Water - Nitroethane (1397)
    Water - 1-Nitropropane (1397)
    Water - 2-Nitropropane (1397)

$H_2O - ACNO_2$ Temperature Range: 293 - 298 K

    Water - Nitrobenzene (1396,1397)

$H_2O - CS_2$ Temperature Range: 298 K

    Water - Carbon disulfide (1396)

ACOH - COOC   Temperature Range: 318 - 408 K

    Phenol - Butyl acetate (313,1398)

ACOH - CCl_4 Temperature Range: 297 - 323 K

    Phenol - Tetrachloromethane (356,842)

CH_2CO - CHO   Temperature Range: 318 - 329 K

    Acetone - Propionaldehyde (1399)
    2-Butanone - Propionaldehyde (1400)

CH_2CO - COOC   Temperature Range: 293 - 352 K

    Acetone - Methyl acetate (949)
    Acetone - Ethyl acetate (1401)

Acetone - Propyl acetate (1401)
2-Butanone - Methyl acetate (717)
2-Butanone - Ethyl acetate (720)

$CH_2CO - CH_2O$ Temperature Range: 303 K

Acetone - Diethyl ether (575)

$CH_2CO - CCN$ Temperature Range: 318 - 353 K

Acetone - Acetonitrile (92,648)

$CH_2CO - COOH$ Temperature Range: 323 - 429 K

Acetone - Acetic acid (11,1358)
3-Pentanone - Acetic acid (1359)
3-Pentanone - Propanoic acid (607)
3-Pentanone - Butanoic acid (607)
3-Hexanone - Propanoic acid (607)
3-Hexanone - Butanoic acid (607)
4-Methyl-2-Pentanone - Propanoic acid (931)

$CH_2CO - CCl$ Temperature Range: 329 - 357 K

Acetone - 1,2-Dichloroethane (475)

$CH_2CO - CCl_2$ Temperature Range: 313 - 329 K

Acetone - Dichloromethane (457)

$CH_2CO - CCl_3$ Temperature Range: 288 - 353 K

Acetone - Chloroform (67,127,131,267,269,365,366,367,368, 369,370,886)
2-Butanone - Chloroform (371)

$CH_2CO - CCl_4$ Temperature Range: 304 - 353 K

Acetone - Tetrachloromethane (134,135)
2-Butanone - Tetrachloromethane (342)

$CH_2CO - CNO_2$ Temperature Range: 318 K

Acetone - Nitromethane (92)

$CH_2CO - CS_2$ Temperature Range: 298 - 327 K

Acetone - Carbon disulfide (127,131,659)

$COOC - CH_2O$ Temperature Range: 273 - 374 K

Ethyl acetate - 1,4-Dioxane (478)
Ethyl acetate - Diethyl ether (1103)

COOC - CNH   Temperature Range: 329 - 350 K

    Ethyl acetate - Diethylamine (478)

COOC - CCN   Temperature Range: 332 - 396 K

    Methyl acetate - Acetonitrile (613)
    Propyl acetate - Acetonitrile (613)
    Butyl acetate - Acetonitrile (613)

COOC - COOH   Temperature Range: 315 - 398 K

    Ethyl acetate - Acetic acid (270,426)
    Butyl acetate - Acetic acid (815)

COOC - CCl_2 Temperature Range: 314 - 348 K

    Ethyl acetate - Dichloromethane (833)

COOC - CCl_3 Temperature Range: 323 - 351 K

    Methyl acetate - Chloroform (272,464)
    Ethyl acetate - Chloroform (46)

COOC - CCl_4 Temperature Range: 323 - 347 K

    Ethyl acetate - Tetrachloromethane (127,587)

COOC - CS_2 Temperature Range: 313 - 327 K

    Methyl acetate - Carbon disulfide (1065)

CH_2O - CNH   Temperature Range: 328 - 374 K

    1,4-Dioxane - Diethylamine (478)

CH_2O - COOH   Temperature Range: 293 - 393 K

    1,2-Dimethoxyethane - Acetic acid (595)
    1,4-Dioxane - Acetic acid (427)
    Dipropyl ether - Acetic acid (595)
    Ethylbutyl ether - Acetic acid (595)

CH_2O - CCl   Temperature Range: 277 - 307 K

    Diethyl ether - Chloroethane (442)

CH_2O - CCl_2 Temperature Range: 307 - 314 K

    Diethyl ether - Dichloromethane (383)

$CH_2O - CCl_3$ Temperature Range: 317 - 361 K

  Ethylpropyl ether - Chloroform (767)
  Dipropyl ether - Chloroform (767,838)

$CH_2O - CNO_2$ Temperature Range: 293 - 371 K

  1,4-Dioxane - Nitromethane (788)
  Diethyl ether - Nitromethane (338)

$CH_2O - CS_2$ Temperature Range: 293 - 308 K

  Dimethoxymethane - Carbon disulfide (127,660)
  Diethyl ether - Carbon disulfide (340)

$CNH_2 - CNH$ Temperature Range: 305 - 354 K

  Isopropylamine - Diisopropylamine (920)

$CNH_2 - ACCl$ Temperature Range: 333 - 353 K

  Butylamine - Chlorobenzene (973)

$CNH - CCl_4$ Temperature Range: 293 - 313 K

  Diethylamine - Tetrachloromethane (1374)

$CNH - ACCl$ Temperature Range: 308 - 333 K

  Diethylamine - Chlorobenzene (816,973)

$ACNH_2 - CCl_4$ Temperature Range: 298 - 323 K

  Aniline - Tetrachloromethane (951)

$ACNH_2 - ACNO_2$ Temperature Range: 356 - 393 K

  Aniline - Nitrobenzene (924)

$CCN - CCl_3$ Temperature Range: 335 - 355 K

  Acetonitrile - Chloroform (154)

$CCN - CCl_4$ Temperature Range: 318 K

  Acetonitrile - Tetrachloromethane (138)

$COOH - CCl$ Temperature Range: 319 - 391 K

  Formic acid - 1,2-Dichloroethane (181)
  Formic acid - 1-Chloropropane (595)
  Formic acid - 1-Chlorobutane (595)

Acetic acid - 1-Chloropropane (595)
Acetic acid - 1-Chlorobutane (595)

COOH - CCl$_2$ Temperature Range: 314 - 364 K

Acetic acid - Dichloromethane (445)

COOH - CCl$_4$ Temperature Range: 331 - 385 K

Acetic acid - Tetrachloromethane (770)

CCl - CCl$_2$ Temperature Range: 303 - 323 K

1,2-Dichloroethane - Dichloromethane (543)

CCl - CCl$_3$ Temperature Range: 347 - 356 K

1,2-Dichloroethane - 1,1,1-Trichloroethane (469)

CCl - CCl$_4$ Temperature Range: 348 - 356 K

1,2-Dichloroethane - Tetrachloromethane (469,544)

CCl$_2$ - CCl$_3$ Temperature Range: 314 - 334 K

Dichloromethane - Chloroform (87)
1,1-Dichloroethane - Chloroform (87)

CCl$_2$ - CCl$_4$ Temperature Range: 312 K

Dichloromethane - Tetrachloromethane (343)

CCl$_3$ - CCl$_4$ Temperature Range: 298 - 350 K

Chloroform - Tetrachloromethane (78,827)

CCl$_3$ - CS$_2$ Temperature Range: 298 K

Chloroform - Carbon disulfide (659,660)

CCl$_4$ - ACCl   Temperature Range: 350 - 405 K

Tetrachloromethane - Chlorobenzene (950)

CCl$_4$ - CNO$_2$ Temperature Range: 298 - 318 K

Tetrachloromethane - Nitromethane (220)
Tetrachloromethane - Nitroethane (348)

CCl$_4$ - ACNO$_2$ Temperature Range: 298 K

Tetrachloromethane - Nitrobenzene (348)

$CCl_4 - CS_2$ Temperature Range: 298 - 348 K

   Tetrachloromethane - Carbon disulfide (131,435)

$ACCl - CNO_2$ Temperature Range: 348 - 393 K

   Chlorobenzene - 1-Nitropropane (288)

REFERENCES

4   H.H. Amer, R.R. Paxton and M. Van Winkle, Ind.Eng.Chem., 48(1956)142.
5   A.R. Gordon and W.G. Hines, Can.J.Res., 24B(1946)254.
8   R.S. Mann and L.W. Shemilt, J.Chem.Eng.Data, 8(1963)189.
11 R. York and R.C. Holmes, Ind.Eng.Chem., 34(1942)345.
12 H.G. Drickamer, G.G. Brown and R.R. White, Trans.Am.Inst.Chem.Eng., 41(1945)555.
14 L.R. Hellwig and M. Van Winkle, Ind.Eng.Chem., 45(1953)624.
19 H. Hipkin and H.S. Myers, Ind.Eng.Chem., 46(1954)2524.
20 A. Rose and E.T. Williams, Ind.Eng.Chem., 47(1955)1528.
21 H.H. Steinhauser and R.H. White, Ind.Eng.Chem., 41(1949)2912.
22 A.L. Vierk, Z.Anorg.Allg.Chem., 261(1950)283.
26 D.S. Blackford and R. York, J.Chem.Eng.Data, 10(1965)313.
27 W. Schroeder, Chem.Ing.Tech., 30(1959)523.
28 R.S. Ramalho, F.M. Tiller, W.J. James and D.W. Bunch, Ind.Eng.Chem., 53(1961)895.
29 H.E. Hughes and J.O. Maloney, Chem.Eng.Progr., 48(1952)192.
35 J.H. Weber, Ind.Eng.Chem., 48(1956)134.
42 J.G. Dunlop, Ph.D. Thesis, Brooklyn 1948.
46 I. Nagata, J.Chem.Eng.Data, 7(1962)367.
48 M. Benedict, C.A. Johnson, E. Solomon and L.C. Rubin, Trans.Am.Inst.Chem.Eng., 41(1945)371.
51 P.S. Murti and M. Van Winkle, Chem.Eng.Data Ser., 3(1958)72.
55 K. Akita and F. Yoshida, J.Chem.Eng.Data, 8(1963)484.
56 V.V. Udovenko and Ts.B. Frid, Zh.Fiz.Khim., 22(1948)1263.
59 C.R. Fordyce and D.R. Simonsen, Ind.Eng.Chem., 41(1949)104.

60 G. Scatchard, S.E. Wood and J.M. Mochel, J.Amer.Chem.Soc. 68(1946)1960.
66 F.A.H. Schreinemakers, Z.Phys.Chem., 35(1900)459.
67 D. Tyrer, J.Chem.Soc. (London), 101(1912)1104.
70 V.A. Kireev and I.P. Sitnikov, Zh.Fiz.Khim., 15(1941)492.
75 H. Roeck and W. Schroeder, Z.Phys.Chem. (Frankfurt), 9 (1956)277.
78 M.L. McGlashan, J.E. Prue and J.E.J. Sainsbury, Trans. Faraday Soc., 50(1954)1284.
84 I. Nagata, J.Chem.Eng.Data, 10(1965)106.
87 S.I. Kaplan and Z.D. Monakhova, Zh.Obshch.Khim., 7(1937) 2499.
92 I. Brown and F. Smith, Austr.J.Chem., 13(1960)30.
101 J.A. Barker, J. Brown and F. Smith, Disc.Faraday Soc., 15 (1953)142.
104 J.C. Chu, R.J. Getty, L.F. Brennecke and R. Paul, Distillation Equilibrium Data, New York 1950.
106 V.V. Udovenko and Ts.B. Frid, Zh.Fiz.Khim., 22(1948)1135.
108 E. Sebastiani and L. Lacquaniti, Chem.Eng.Sci., 22(1967) 1155.
109 J.W. Hudson and M. Van Winkle, J.Chem.Eng. Data, 14(1969) 310.
118 S. Uchida, S. Ogawa and M. Yamagushi, Japan Sci.Rev.Eng. Sci., 1(1950)41.
120 B.G. Harper and J.C. Moore, Ind.Eng.Chem., 49(1957)411.
122 K. Schaefer and W. Rall, Z.Elektrochem., 62(1958)1090.
127 V.J. Zawidski, Z.Phys.Chem., 35(1900)129.
131 M.A. Rosanoff and C.W. Easeley, J.Amer.Chem.Soc., 31(1909) 953.
134 I. Brown and F. Smith, Austr.J.Chem., 10(1957)423.
135 K.C. Bachman and E.L. Simons, Ind.Eng.Chem., 44(1952)202.
138 I. Brown and F. Smith, Austr.J.Chem., 7(1954)269.
142 W.G. Beare, G.A. McVicar and J.B. Fergusson, J.Phys.Chem. 34(1930)1310.
154 F. Mato and M. Sanchez, An.Real.Soc.Espan.Fis.Y Quim.B, 63(1967)1.
159 International Critical Tables, McGraw Hill 1928.
171 A.R. Gordon and E.J. Hornibrook, Can.J.Res., 24B(1946) 263.

173 J. Nagai and N. Isii, J.Soc.Chem.Ind.Jap., 38(1935)8.
174 S.I. Green and R.E. Vener, Ind.Eng.Chem., 47(1955)103.
179 J.L. Copp and D.H. Everett, Disc.Faraday Soc., 15(1953)174.
181 V.V. Udovenko and L.P. Aleksandrova, Zh.Fiz.Khim.,34(1960) 1366.
182 T.Ch. Lo, H.H. Bieber and A.E. Karr, J.Chem.Eng. Data, 7 (1962)327.
190 T. Ito and F. Yoshida, J.Chem.Eng. Data, 8(1963)315.
191 P.S. Prabhu and M. Van Winkle, J.Chem.Eng. Data,8(1963)210.
193 V.M. Olevsky and I.F. Golubev, Tr.Giap.Vyp., 6(1956)45.
194 J. Ocon and F. Rebolleda, An.Real.Soc.Espan.Fis.Y Quim. 54B(7-8)(1958)525.
196 J. Ocon and C. Taboada, An.Real.Soc.Espan.Fis.Y Quim. 55B (3)(1959)255.
202 I. Brown and A.H. Ewald, Austr.J.Sci.Res., A3(1950)306.
205 R.T. Fowler and S.C. Lim, J.Appl.Chem., 6(1956)74.
207 G. Scatchard, S.E. Wood and J.M. Mochel, J.Amer.Chem.Soc., 62(1940)712.
211 K. Dvorak and T. Boublik, Collect.Czech.Chem.Commun., 28 (1963)1249.
213 G. Scatchard, S.E. Wood and J.M. Mochel, J.Amer.Chem.Soc., 61(1939)3206.
220 I. Brown and F. Smith, Austr.J.Chem., 8(1955)501.
231 I. Brown and F. Smith, Austr.J.Chem., 8(1955)62.
234 A. McKeown and F.P. Stowell, J.Chem.Soc.(London)(1927)97.
236 C.A. Jones, E.M. Schoenborn and A.P. Colburn, Ind.Eng. Chem., 35(1943)666.
237 A.I. Johnson, W.F. Furter and T.W. Barry, Can.J.Technol., 32(1954)179.
242 V.V. Udovenko and L.G. Fatkulina, Zh.Fiz.Khim., 26(1952) 719.
257 C.B. Kretschmer and R. Wiebe, J.Amer.Chem.Soc., 71(1949) 1793.
258 S.R.M. Ellis and R.D. Garbett, Ind.Eng.Chem., 52(1960)385.
264 E.R. Smith and M.J. Wojciechowski, J.Res.Nat.Bur.Stand. Sect. A, 18(1937)461.
267 W. Reinders and C.H. De Minjer, Rec.Trav.Chim.Pays-Bas,59 (1940)369.

269 J.A. Tallmadge and L.N. Canjar, Ind.Eng.Chem., 46(1954) 1279.
270 F.H. Garner, S.R.M. Ellis and C.J. Pearce, Chem.Eng.Sci., 3(1954)48.
272 I. Nagata, J.Chem.Eng. Data, 7(1962)360.
273 I. Nagata, J.Chem.Eng. Data, 7(1962)461.
281 A.D. Carr and H.W. Kropholler, J.Chem.Eng.Data, 7(1962)26.
285 J. Hirschberg, Bull.Soc.Chim.Belg., 41(1932)163.
288 J.R. Lacher, W.B. Buck and W.H. Parry, J.Amer.Chem.Soc., 63(1941)2422.
290 G. Kortuem and H.J. Freier, Chem.Ing.Tech., 26(1954)670.
295 R.S. Mann, L.W. Shemilt and M. Waldichuck, J.Chem.Eng. Data, 8(1963)502.
296 S. Yerazunis, J.D. Plowright and F.M. Smola, AIChE J., 10 (1964)660.
300 V.A. Kireev and A.A. Skvortsova, Zh.Fiz.Khim., 6(1936)63.
301 M.A. Rosanoff and C.W. Easeley, J.Amer.Chem.Soc., 36(1914) 979.
305 I. Brown, Austr.J.Sci.Res., 5(1952)530.
313 V. Kliment, V. Fried and J. Pick, Collect.Czech.Chem. Commun., 29(1964)2008.
314 H. Roeck and L. Sieg, Z.Phys.Chem. (Frankfurt), 3(1955) 355.
315 I. Horyna, Collect.Czech.Chem.Commun., 24(1959)3253.
323 M.A. Rosanoff, C.W. Bacon and J.F.W. Schulze, J.Amer.Chem. Soc., 36(1914)1999.
327 H. Schuberth, J.Prakt.Chem., 6(1958)129.
338 N.I. Joukowsky, Bull.Soc.Chim.Belg., 43(1934)397.
340 J.W. Knowlton, N.C. Schielt and D. McMillan, J.Amer.Chem. Soc., 68(1946)208.
342 R.T. Fowler and G.S. Norris, J.Appl.Chem., 5(1955)266.
343 C.R. Mueller and A.J. Ignatowski, J.Chem.Phys., 32(1960) 1430.
347 B.V.S. Rao and C.V. Rao, J.Chem.Eng. Data, 8(1963)368.
348 D.F. Saunders and A.I.B. Spaull, Z.Phys.Chem. (Frankfurt) 28(1961)332.
351 A. Schulze, Z.Phys.Chem., 86(1914)309.
353 A.N. Campbell and W.J. Dulmage, J.Amer.Chem.Soc., 70 (1948)1723.

354 J. Ocon and J. Espantoso, An.Real.Soc.Espan.Fis. Y Quim., 54B(6)(1958)401.

355 T. Boublik, Collect.Czech.Chem.Commun., 28(1963)1771.

356 J. Chevalley, Bull.Soc.Chim.Fr., (1961)510.

358 J. Ocon and J. Espantoso, An.Real.Soc.Espan.Fis. Y Quim., 54B(6)(1958)421.

363 E. Hala, V. Fried, J. Pick and O. Vilim, Chem.Listy, 47 (1953)1423.

364 A.G. Morachevsky and R.Sh. Rabinovich, Zh.Prikl.Khim.,32 (1959)458.

365 H. Roeck and W. Schroeder, Z.Phys.Chem. (Frankfurt), 11 (1957)41.

366 C.R. Mueller and E.R. Kearns, J.Phys.Chem., 62(1958)1441.

367 A.A. Kotsyuba, Tr.Dnepropetr,Khim.-Tekhn.Inst.Vyp., 12 (1959)195.

368 I.B. Rabinovich and P.N. Nikolaev, Zh.Fiz.Khim., 34(1960) 2289.

369 K.B. Schnelle and L.N. Canjar, Chem.Eng.Sci., 17(1962)189.

370 L.S. Kudryavtseva and M.P. Susarev, Zh.Prikl.Khim., 36 (1963)1231.

371 V.V.G. Krishnamurty and C.V. Rao, J.Sci.Ind.Research,14B (1955)55.

373 B.V.S. Rao and C.V. Rao, Chem.Eng.Sci., 17(1962)574.

382 F.G. Tenn and R.W. Missen, Can.J.Chem.Eng., 41(1963)12.

383 O. Vilim and J. Szlaur, Collect.Czech.Chem.Commun., 29 (1964)1878.

385 S.K. Ogorodnikov, V.B. Kogan and M.S. Nemtsov, Zh.Prikl. Khim., 34(1961)841.

387 S. Uchida, S. Ogawa, M. Hirata, Y. Shimada and S. Schimokawa., Kagaku Kikai Chem.Eng., 17(1953)191.

388 A.G. Crawford, G. Edwards and D.S. Lindsay, J.Chem.Soc. (London), 5(1949)1054.

391 E.C. Britton, H.S. Nutting and L.H. Horsley, Anal.Chem., 19(1947)601.

394 S.K. Ogorodnikov, V.B. Kogan and M.S. Nemtsov, Zh.Prikl. Khim., 33(1960)2685.

398 G. Scatchard, S.E. Wood and J.M. Mochel, J.Amer.Chem.Soc. 68(1946)1957.

400   W. Jost, H. Roeck, W. Schroeder, L. Sieg and H.G. Wagner, Z.Phys.Chem. (Frankfurt),10(1957)133.
401   B.C.Y. Lu, Can.J.Technol., 34(1957)468.
407   S.K. Ogorodnikov, V.B. Kogan, M.S. Nemtsov and G.V.Burova, Zh.Prikl.Khim., 34(1961)1096.
426   L.L. Shmidt, Zh.Russ.Fiz.Khim.Obshch., 62(1930)1847.
427   K.N. Kovalenko and N.I. Balandina, Uch.Zap.Rost. Na Donu Univ., 41(1958)39.
429   F. Hovorka and D. Dreisbach, J.Amer.Chem.Soc., 56(1934) 1664.
434   O. Vilim, E. Hala, J. Pick and V. Fried, Chem.Listy, 48 (1954)989.
435   K. Hlavaty, Collect.Czech.Chem.Commun., 35(1970)2881.
438   J.J. Kipling and D.A. Tester, J.Chem.Soc. (London)(1952) 4123.
442   M.B. Neimann and S.Z. Demikhovskaya, Zh.Obshch.Khim., 19 (1949)593.
445   L.L. Dobroserdov and V.D. Shakhanov, Zh.Prikl.Khim. (Leningrad), 44(1971)445.
457   M.V. Aleksandrova, L.A. Boldina, V.F. Komarova, N.V. Kistereva and T.S. Fedorova, Sb.Nauch.Tr.Vladimir Politekh.Inst., 12(1971)140.
464   I. Nagata and H. Hayashida, J.Chem.Eng.Jap., 3(1970)161.
468   G. Geiseler and H. Koehler, Ber.Bunsenges.Phys.Chem., 72 (1968)697.
469   M. Sagnes and V. Sanchez, J.Chem.Eng. Data, 16(1971)351.
471   J. Matous, J.P. Novak, J. Sobr and J. Pick, Collect.Czech. Chem.Commun., 37(1972)2653.
472   D.C. Freshwater and K.A. Pike, J.Chem.Eng. Data, 12(1967) 179.
475   L.N. Canjar, E.C. Horni and R.R. Rothfus, Ind.Eng.Chem., 48(1956)427.
476   J.L. Humphrey and M. Van Winkle, J.Chem.Eng. Data, 12 (1967)526.
478   K. Engelmann and H.-J. Bittrich, Wiss.Z.Tech.Hochsch.Chem. Leuna-Merseburg, 8(1966)148.
495   A.S. Brunjes and C.C. Furnas, Ind.Eng.Chem., 27(1935)396.
501   K. Strubl, V. Svoboda, R. Holub and J. Pick, Collect. Czech.Chem.Commun., 40(1975)1647.

504   D.F. Othmer and R.F. Benenati, Ind.Eng.Chem., 37(1945) 299.
509   A. Niini, Ann.Acad.Sci.Fenn., A55, 8(1940).
510   G. Kortuem, D. Moegling and F. Woerner, Chem.Ing.Tech., 22(1950)453.
520   J. Sameshima, J.Amer.Chem.Soc., 40(1918)1503.
526   C.P. Smith and E.W. Engel, J.Amer.Chem.Soc., 51(1929) 2646.
528   L. Sieg, Chem.Ing.Tech., 22(1950)322.
532   M.A. Rosanoff, C.W. Bacon and R.H. White, J.Amer.Chem. Soc., 36(1914)1803.
536   S.R.M. Ellis and B.A. Froome, Chem.Industry, (1957)237.
543   W. Davies, J.B. Jagger and H.K. Whalley, J.Soc.Chem.Ind. (London) 68(1949)26.
544   W.A. Kireev and S.D. Monachowa, Zh.Fiz.Khim., 7(1936)71.
553   A.R. Martin and C.M. George, J.Chem.Soc. (London)(1933) 1413.
554   J.B. Ferguson, J.Phys.Chem., 36(1932)1123.
557   G. Bredig and R. Bayer, Z.Phys.Chem., 130(1927)15.
559   G. Scatchard and C.L. Raymond, J.Amer.Chem.Soc., 60(1938) 1278.
572   W.D. Hill and M. Van Winkle, Ind.Eng.Chem., 44(1952)208.
575   J. Sameshima, J.Amer.Chem.Soc., 40(1918)1482.
587   P.W. Schutz, J.Amer.Chem.Soc., 61(1939)2691.
591   L.H. Ballard and M. Van Winkle, Ind.Eng.Chem., 45(1953) 1803.
595   W. Hunsmann and K.H. Simmrock, Chem.Ing.Tech., 38(1966) 1053.
607   O.V. Skvorosova, A.M. Chashchin and L.A. Serafimov, Gidroliz.Lesokhim.Prom., 5(1973)10.
611   V.M. Komarov and B.K. Krichevtsov, Zh.Prikl.Khim., 39 (1966)2834.
613   F. Mato, J.L. Cabezas and J. Coca, An.Real Soc.Espan. De Fis.Y Quim., 69(1973)1.
616   G. Werner and H. Schuberth, J.Prakt.Chem., 31(1966)225.
620   P.S. Puri, J. Polak and J.A. Ruether, J.Chem.Eng. Data, 19(1974)87.

639 A. Doniec, R. Krauze, St. Michalowski and M. Serwinski, Zesz.Nauk.Politech.Lodz.Chem., 139(1973)85.
643 D. Subramanian, G.D. Nageshwar and P.S. Mene, J.Chem.Eng. Data, 14(1969)421.
648 H.R.C. Pratt, Trans.Inst.Chem.Eng., 25(1947)43.
659 N.D. Litvinov, Zh.Fiz.Khim., 26(1952)1144.
660 N.D. Litvinov, Zh.Fiz.Khim., 26(1952)1152.
664 N.D. Litvinov, Zh.Fiz.Khim., 26(1952)1405.
665 K.S. Yuan, B.C-Y. Lu, J.C.K. Ho and A.K. Keshpande, J.Chem.Eng. Data, 8(1963)549.
676 A.G. Morachevsky and E.G. Komarova, Vestn.Leningrad.Univ. 12,Ser.Fiz.Khim., 1(1957)118.
680 R. Nielsen and J.H. Weber, J.Chem.Eng. Data, 4(1959)145.
688 A.V. Storonkin and A.G. Morachevsky, Zh.Fiz.Khim., 30(1956)1297.
703 D.R. Cova, J.Chem.Eng. Data, 5(1960)282.
704 J. Novak, J. Matous and J. Pick, Collect.Czech.Chem.Commun., 25(1960)2405.
717 S.V. Babich, R.A. Ivanchikova and L.A. Serafimov, Zh.Prikl.Khim., 42(1969)1354.
720 S.V. Babich, I.V. Borozdina, T.M. Kushner and L.A. Serafimov, Zh.Prikl.Khim., 41(1968)589.
725 F. Ratkovics, J. Liszi and M. Laszlo, Acta Chim.Acad.Sci. Hung., 79(1973)387.
733 H.W. Schulte, Diplomarbeit Dortmund 1974.
756 V.A. Shnitko, V.B. Kogan and N.V. Petrova, Zh.Prikl.Khim. (Leningrad),44(1971)2126.
760 J. Marek, Collect.Czech.Chem.Commun., 20(1955)1490.
761 A.N. Campbell, E.M. Kartzmark and S.C. Anand, Can.J.Chem. 49(1971)2183.
762 L. Verhoeye and H. De Schepper, J.Appl.Chem.Biotechn., 23(1973)607.
763 V.M. Komarov and B.K. Krichevtsov, Zh.Prikl.Khim., 39(1966)2838.
764 T. Morisue, K. Noda and K. Ishida, J.Chem.Eng.Jap., 6(1973)355.
766 G.A. Ratcliff and K.C. Chao, Can.J.Chem.Eng., 47(1969)148.

767 J.L. Chevalier, J.Chim.Phys.Physicochim.Biol., 66(1969) 1457.

770 J. Wisniak and A. Tamir, J.Chem.Eng. Data, 20(1975)168.

771 J.K. Donnelly, S.N. Malpani and R.G. Moore, J.Chem.Eng. Data 20(1975)170.

781 V.O. Maripuri and G.A. Ratcliff, J.Chem.Eng.Data, 17 (1972)366.

788 J.S. Stadnicki, Bull.Acad.Polon.Sci.,Ser.Sci.Chim., 11 (1963)293.

810 S.V. Mainkar and P.S. Mene, Indian J.Technol., 3(1965) 228.

814 F. Ratcovics, J. Liszi, M. Laszlo, B. Szeiler and J. Devay, Acta.Chim. (Budapest),77(1973)249.

815 M. Hirata and Y. Hirose, Kagaku Kogaku, 30(1966)121.

816 Ch. Dehmelt, M. Finke and H.J. Bittrich, Z.Phys.Chem. (Leipzig),255(1974)251.

827 R. Krauze and M. Serwinski, Zesz.Nauk.Politech.Lodz, 24 (1973)53.

828 D.A. Palmer and D.B. Smith, J.Chem.Eng. Data 17(1972)71.

830 I. Mertl, Collect.Czech.Chem.Commun., 37(1972)366.

833 I.V. Bagrov, L.L. Dobroserdov, V.D. Shakhanov and M.I. Mikhailova, Khim.Khim.Tekhnol., 13(1970)499.

834 I. Nagata, T. Yamada and S. Nakagawa, J.Chem.Eng. Data, 20(1975)271.

837 K. Nagahama and M. Hirata, J.Chem.Eng.Jap., 4(1971)205.

838 P. Ernst, Diplomarbeit Dortmund 1975.

842 H. Brusset and J. Chevalley, Compt.Rend., 250(1960)3016.

851 G. Schay, G. Varsanyi and F. Billes, Roczn.Chem., 32 (1958)375.

858 A.O. Delzenne, Ind.Eng.Chem., Chem.Eng. Data Ser., 3 (1958)224.

860 K. Nakanishi, H. Shirai and K. Nakasato, J.Chem.Eng.Data, 13(1968)188.

864 D.R. Laurance and G.W. Swift, J.Chem.Eng. Data, 19(1974) 61.

872 A.J. Rodger, C.C. Hsu and W.F. Furter, J.Chem.Eng. Data, 14(1969)362.

873 I. Nagata, J.Chem.Eng. Data, 14(1969)418.

874 K.F. Wong and C.A. Eckert, J.Chem.Eng.Data, 14(1969)432.
878 V.C. Smith and R.L. Robinson Jr., J.Chem.Eng. Data, 15(1970)391.
879 M. Iino, J. Sudo, M. Hirata, Y. Hirose and A. Nakae, J.Chem.Eng.Data, 15(1970)446.
881 O.P. Bagga and K.S.N. Raju, J.Chem.Eng. Data, 15(1970)531.
884 J. Mesnage and A.A. Marsan, J.Chem.Eng. Data, 16(1974)434.
885 D.O. Hanson and M. Van Winkle, J.Chem.Eng. Data, 12(1967)319.
886 K. Kojima, A. Kato, H. Sunaga and G. Hasimoto, Kagaku Kogaku, 32(1968)337.
890 K. Kojima, H. Sunaga and G. Hashimoto, Chem.Eng., 32(1968)480.
891 J.L.H. Wang, L. Boublikova and B.C.Y. Lu, J.Appl.Chem., 20(1970)172.
896 L.Y. Sadler, D.W. Luff and M.D. McKinley, J.Chem.Eng. Data, 16(1971)446.
906 A.N. Marinichev and M.P. Susarev, Zh.Prikl.Khim., 38(1965)1619.
909 S. Madhavan and P.S. Murti, Chem.Eng.Sci., 21(1966)465.
920 V.M. Komarov and B.K. Krichevtsov, Zh.Prikl.Khim., 42(1969)2772.
921 B.K. Krichevtsov and V.M. Komarov, Zh.Prikl.Khim.(Leningrad),43(1970)112.
924 S. Sakujama, J.F. Industrie-Chemie (in Japanese), 44(1941)522.
927 K. Nakanishi, H. Shirai and T. Minamiyama, J.Chem.Eng. Data, 12(1967)591.
928 T.S. Tolstova, V.B. Kogan, V.M. Fridmann and T.G. Romanova, Zh.Prikl.Khim. (Leningrad),46(1973)907.
931 J. Wisniak and A. Tamir, J.Chem.Eng. Data, 21(1976)88.
941 J. Linek, I. Wichterle and J. Polednova, Collect.Czech.Chem.Commun., 37(1972)2820.
943 J. Kraus and J. Linek, Collect.Czech.Chem.Commun., 36(1971)2547.
949 V. Bekarek, Collect.Czech.Chem.Commun., 33(1968)2608.

950 M. Serwinski and S. Michalowski, Zeszyty Nauk.Politech. Lodz.Chem., 16(1965)45.

951 G. Pannetier, L. Abello and M. Kern, Ann.Chim., 10(1965) 403.

957 R.R. Davison, J.Chem.Eng. Data, 13(1968)348.

973 T.M. Letcher and J.W. Bayles, J.Chem.Eng. Data, 16(1971)266.

994 A.Ja. Aarna and T.K. Kaps, Trudy Tallinskovo Politechn. Inst.Ser. A, 285(1970)3.

999 Y.S. Suryanarayana and M. Van Winkle, J.Chem.Eng. Data, 11(1966)7.

1013 R.V. Orye and J.M. Prausnitz, Trans.Faraday Soc., 61(1964)1338.

1016 M.S. Reddy, M.S. Krishna and V.K.C. Rao, Chem.Age India, 16(1965)576.

1027 K.L. Pinder, J.Chem.Eng. Data, 18(1973)275.

1029 S.G. D'Avila and R.S.F. Silva, J.Chem.Eng. Data, 15(1970)421.

1034 R.R. Davison and W.H. Smith, J.Chem.Eng. Data, 14(1969)296.

1041 J. Polak and B.C.Y. Lu, J.Chem.Eng. Data, 17(1972)456.

1046 D. Tassios and M. Van Winkle, J.Chem.Eng. Data, 12(1967)555.

1048 S.R.M. Ellis, C. McDermott and J.C.L. Williams, in P.A. Rothenburg (ed.) "Proceedings of the International Symposium on Distillation", (1960)89 (London: The Institution of Chemical Engineers).

1049 L.L. Lee and W.A. Scheller, J.Chem.Eng. Data, 12(1967)497.

1052 H.C. Van Ness, C.A. Soczek and N.K. Kochar, J.Chem.Eng. Data, 12(1967)346.

1055 I. Matsunaga and T. Katayama, J.Chem.Eng.Jap., 6(1973)397.

1062 W.W. Bowden, J.C. Staton and B.D. Smith, J.Chem.Eng.Data, 11(1966)296.

1065 A.F.M. Fahmy and N.I.M. El-Ghannam, Bull.Acad.Pol.Sci., Ser.Sci.Chim., 20(1972)493.

1075 E. Schreiber, E. Schuettau, D. Rant and H. Schuberth, Z. Phys.Chem. (Leipzig), 247(1976)23.

1089 S.M. Hosseini and G. Schneider, Z.Phys.Chem. (Frankfurt), 36(1963)137.

1103 G.C. Schmidt, Z.Phys.Chem., 121(1926)221.
1106 H.C. Van Ness, C.A. Soczek, G.L. Peloquin and R.L. Machado, J.Chem.Eng.Data, 12(1967)217.
1111 S.R.M. Ellis and M. Razavipour, Chem.Eng.Sci., 11(1959) 99.
1112 V.A. Shnitko, V.B. Kogan and T.V. Sheblom, Zh.Prikl. Khim., 42(1969)2389.
1121 B.K. Krichevtsov, and V.M. Komarov, Zh.Prikl.Khim. (Leningrad),43(1970)703.
1126 E. Sada and T. Morisue, J.Chem.Eng.Jap., 8(1975)191.
1145 S.O. Jain, O.P. Bagga and K.S.N. Raju, J.Chem.Eng.Jap., 9(1976)273.
1146 I. Nagata, T. Ohta, T. Takahashi and K. Gotoh, J.Chem. Eng.Jap., 6(1973)129.
1357 H.S. Myers, Ind.Eng.Chem., 47(1955)2215.
1358 G.F. Meehan and N.F. Murphy, Chem.Eng.Sci., 20(1965)757.
1359 P.O. Haddad and W.C. Edmister, J.Chem.Eng. Data, 17(1972) 275.
1360 I.P.C. Li, Y.W. Wong, S. Chang and B.C.Y. Lu, J.Chem.Eng. Data, 17(1972)492.
1361 H.J. Gumpert, H. Koehler, W. Schille and H.-J. Bittrich, Wiss.Z.Tech.Hochsch.Chem. Leuna-Merseburg, 15(1973)179.
1362 G.W. Lindberg and D. Tassios, J.Chem.Eng. Data, 16(1971) 52.
1363 K. Susokowska-Keheiaian, K. Orsel and H. Keheiaian, Bull.Acad.Poln.Sci.Ser.Chim., 14(1966)711.
1364 G. Geiseler, K. Quitzsch, H.-G. Vogel, D. Pilz and H. Sachse, Z.Phys.Chem. (Frankfurt),56(1967)288.
1365 C. McAuliffe, Nature, 200(1963)1092.
1366 H. Stephen and T. Stephen, Solubilities of Inorganic and Organic Compounds, Pergamon Press, Oxford 1963.
1367 H. Renon, Int. Data Ser.(A)(1973)64.
1368 H.W. Prengle and M.A. Pike, J.Chem.Eng. Data, 6(1961) 400.
1369 T. Treszczanowicz, Personal Communication 1975.
1370 H. Arm and D. Bankay, Helv.Chim. Acta, 52(1969)279.
1372 H. Wolff, A. Hoepfner and H.M. Hoepfner, Ber.Bunsenges, Phys.Chem., 68(1964)410.

1373 H. Wolff and H.E. Hoeppel, Ber.Bunsenges.Phys.Chem., 70 (1966)874.

1374 H. Kilian and H.J. Bittrich, Z.Phys.Chem. (Leipzig), 230 (1965)383.

1375 R. Siedler and H.J. Bittrich, J.Prakt.Chem., 311(1969)721.

1376 G. Schneider, Z.Phys.Chem. (Frankfurt), 24(1960)165.

1377 G. Werner, J.Prakt.Chem., 29(1965)26.

1378 J.H. Vera and J.M. Prausnitz, J.Chem.Eng. Data, 16(1971) 149.

1379 C. Black, G.G. Joris and H.S. Taylor, J.Chem.Phys., 16 (1948)537.

1380 H. Renon, Int. Data Ser.(A)(1973)67.

1381 A.H. Wehe and J. Coates, AIChE J., 1(1955)241.

1382 V.V. Udovenko and T.F. Mazanko, Zh.Fiz.Khim., 41(1967) 1615.

1383 I. Nagata, T. Ohta and Y. Uchiyama, J.Chem.Eng. Data, 18 (1973)54.

1384 I. Brown, W. Fock and F. Smith, J.Chem.Thermodyn., 1 (1969)273.

1385 A.F.M. Fahmy and N.I. El-Ghannam, Bull.Acad.Poln.Sci., 20(1972)157.

1386 D.D. Deshpande and S.L. Oswal, J.Chem.Soc. Faraday Trans. 1,68(1972)1059.

1388 I. Nagata, T. Ohta and T. Takahashi, J.Chem.Eng.Jap., 5 (1972)227.

1389 H.C. Miller and H. Bliss, Ind.Eng.Chem., 32(1940)123.

1390 W. Niepel, J.P. Novak, J. Matous and J. Sobr, Chem.Zvesti 26(1972)44.

1391 V.M. Komarov and B.K. Krichevtsov, Zh.Prikl.Khim., 42 (1969)2772.

1392 J. Hollo, G. Ember, T. Lenghel and A. Wieg, Acta Chim. Acad.Sci.Hung., 13(1958)307.

1393 A. Rius, J.L. Otero and A. Macarron, Chem.Eng.Sci., 10 (1959)105.

1394 R. Signer, H. Arm and H. Daenecker, Helv.Chim. Acta, 52 (1969)2347.

1395 R.S. Hansen, F.A. Miller and S.D. Christian, J.Phys.Chem. 59(1955)391.

1396 D. Donahue and F.E. Bartell, J.Phys.Chem., 56(1952)480.
1397 J.A. Riddick and W.B. Bunger, Organic Solvents, Wiley-Interscience, New York 1970.
1398 R. Weller, H. Schuberth and E. Leipnitz, J.Prakt.Chem., 21(1963)234.
1399 E. Danciu, Rev.Chim. (Bucharest), 21(1970)149.
1400 I. Matsunaga and T. Katayama, J.Chem.Eng.Jap., 6(1973)397.
1401 V. Subrahmanyan and P.D. Murty, J.Appl.Chem. (London), 14(1964)500.
1402 A. Apelblat and F. Kohler, J.Chem.Thermodynamics, 8(1976)749.

NOMENCLATURE

(Chapters 1-9 only)

a_{nm}	UNIFAC binary interaction parameter (related to Ψ_{nm})
A_k	van der Waals group surface area
A_{ji}	UNIQUAC parameter, see Equation (3.10)
B	second virial coefficient
B_{ij}	second virial coefficient for i-j interaction
f_i^o	pure-component reference fugacity of component i
$f_{n,i}$	molar feed rate of component i on stage n
$F_{k(n,i)}$	discrepancy function for distillation calculations
F_n	molar feed rate on stage n
g	reduced excess Gibbs free energy (G^E/RT)
G	Gibbs free energy
$\bar{G}_i$	partial molar Gibbs free energy of component i
G^E	excess Gibbs free energy
G_{ji}	NRTL parameter
H_n	enthalpy of vapor from stage n
h^E	liquid phase excess enthalpy (Chapter 9)
H^E	molar excess enthalpy
h_n	enthalpy of liquid from stage n
ΔH^{vap}	enthalpy of vaporization
K	equilibrium constant
$K_{n,i}$	equilibrium ratio of component i on stage n
$\ell_{n,i}$	liquid molar flow rate of component i from stage n
ℓ_i	pure-component constant defined in Equation (3.8)
L_k	Legendre polynomial of degree k
L_n	total liquid molar flow rate from stage n
n_T	total number of moles
n_i	number of moles of component i

P	pressure
P_i^s	pure-component vapor pressure of component i
P'	parachor
q_i	pure-component area parameter of component i
Q_k	group area parameter for group k
r_i	pure-component volume parameter of component i
RD	mean radius of gyration
R_k	group volume parameter of group k
R	the gas constant
S	entropy
S_n^V	vapor phase sidestream from stage n
S_n^L	liquid phase sidestream from stage n
T	temperature
u_{ji}	UNIQUAC binary interaction parameter (related to τ_{ji} and A_{ji})
V_k	van der Waals group volume
V_n	total vapor molar flow rate from stage n
$v_{n,i}$	vapor phase molar flow rate of component i from stage n
v	molar volume
v_i	molar liquid volume of component i
x_i	liquid phase mole fraction of component i
X_k	group fraction of group k
y_i	vapor phase mole fraction of component i
z	compressibility factor
z	lattice coordination number (here equal to ten)
Z_i	true mole fraction of species i
α_{ji}	NRTL parameter
γ_i	activity coefficient of component i

Γ_k	activity coefficient of group k
$\Gamma_k^{(i)}$	activity coefficient of group k in pure component i
$\varsigma_{n,i}$	stripping factor of component i on stage n
$\eta_{n,i}$	Murphree stage efficiency for component i on stage n
θ_i	area fraction of component i
θ_k	area fraction of group k
Λ_{ij}	Wilson parameter
$\nu_k^{(i)}$	number of groups of kind k in molecular species i
τ_{ji}	UNIQUAC parameter
τ_{ji}	NRTL parameter
φ_i	fugacity coefficient of component i
Φ_i	segment fraction of component i
Ψ_{nm}	UNIFAC parameter defined in Equation (4.9)